Resources and Sustainable Development in the Arctic

Over the past thirty years we have witnessed a demand for resources such as minerals, oil, and gas, which is only set to increase. This book examines the relationship between Arctic communities and extractive resource development.

With insights from leading thinkers in the field, the book examines this relationship to better understand what, if anything, can be done in order for the development of non-renewable resources to be of benefit to the long-term sustainability of these communities. The contributions synthesize circumpolar research on the topic of resource extraction in the Arctic, and highlight areas that need further investigation, such as the ability of northern communities to properly use current regulatory processes, fiscal arrangements, and benefit agreements to ensure the long-term sustainability of their culture communities and to avoid a new path dependency

This book provides an insightful summary of issues surrounding resource extraction in the Arctic, and will be essential reading for anyone interested in environmental impact assessments, globalization and Indigenous communities, and the future of the Arctic region.

Chris Southcott is the Principal Investigator for the Resources and Sustainable Development in the Arctic (ReSDA) research project as well as Theme Coordinator for the Sustainable Communities research work. He is a Professor of Sociology at Lakehead University in Thunder Bay, Ontario, Canada.

Frances Abele is the ReSDA Theme Coordinator for the Sustainable Regions research and is a Professor in the School of Public Policy and Administration and Academic Director of the Carleton Centre for Community Innovation, both at Carleton University in Ottawa, Canada.

David Natcher is the ReSDA Theme Coordinator for the Sustainable Cultures research and is a Professor in the Department of Agricultural and Resource Economics at the University of Saskatchewan, Canada.

Brenda Parlee is the ReSDA Theme Coordinator for the Sustainable Environments research and currently Associate Professor and a Canada Research Chair in the Department of Resource Economics and Environmental Sociology in the Faculty of Agricultural, Life & Environmental Sciences at the University of Alberta, Canada.

Routledge Research in Polar Regions
Series Editor: Timothy Heleniak, Nordregio International
Research Centre, Sweden

The Routledge series in *Polar Regions* seeks to include research and policy debates about trends and events taking place in two important world regions: the Arctic and Antarctic. Previously neglected on the periphery, with climate change, resource development, and shifting geopolitics, these regions are becoming increasingly crucial to happenings outside these regions. At the same time, the economies, societies, and natural environments of the Arctic are undergoing rapid change. This series seeks to draw upon fieldwork, satellite observations, archival studies, and other research methods that inform about crucial developments in the Polar regions. It is interdisciplinary, drawing on the work from the social sciences and humanities, bringing together cutting-edge research in the Polar regions with the policy implications.

For more information about this series, please visit: www.routledge.com/ Routledge-Research-in-Polar-Regions/book-series/RRPS

Resources and Sustainable Development in the Arctic

Edited by
Chris Southcott, Frances
Abele, David Natcher, and
Brenda Parlee

LONDON AND NEW YORK

First published 2019
by Routledge
2 Park Square, Milton Park, Abingdon, Oxon OX14 4RN

and by Routledge
52 Vanderbilt Avenue, New York, NY 10017, USA

First issued in paperback 2020

Routledge is an imprint of the Taylor & Francis Group, an informa business

British Library Cataloguing-in-Publication Data
A catalogue record for this book is available from the British Library

Library of Congress Cataloging-in-Publication Data
A catalog record has been requested for this book

ISBN 13: 978-0-367-58554-9 (pbk)
ISBN 13: 978-1-138-49729-0 (hbk)

Typeset in Times New Roman
by Florence Production Ltd, Stoodleigh Court, Stoodleigh, Devon, UK

Contents

Illustrations

Figures

Tables

Contributors

Matthew Berman, Professor of Economics, Institute of Social and Economic Research, University of Alaska-Anchorage, Anchorage, AL.

Jill Blakly, Associate Professor, Department of Geography & Planning, University of Saskatchewan, Canada.

Jean-Sébastien Boutet, PhD candidate, KtH (The Royal Institute of Technology), Stockholm, Sweden.

Ben Bradshaw, Associate Professor, Department of Geography, University of Guelph, Guelph, ON, Canada.

Philip Cavin, Project Manager, Federal Home Loan Bank of Chicago, Chicago, IL.

Ken Coates, Canada Research Chair in Regional Innovation, Johnson-Shoyama Graduate School of Public Policy at the University of Saskatchewan, Canada.

Matthew Cooney, Master's student, Department of Geography, University of Northern Iowa, Cedar Falls, IA.

David Cox, Master's student, McMaster University, Hamilton, ON, Canada.

Anne Dance, visiting researcher, University of Ottawa, ON, Canada.

Martha Dowsley, Associate Professor, Departments of Anthropology and Geography and the Environment, Lakehead University, Thunder Bay, ON, Canada.

Cecilie Ebsen, Department of Anthropology, University of Alaska, Fairbanks, AL.

Courtney Fidler, Honorary Research Fellow, Sustainable Minerals Institute, University of Queensland, Brisbane, Australia.

Jessica Graybill, Associate Professor, Department of Geography, Colgate University, Hamilton, New York.

Kevin Hanna, Associate Professor, I.K. Barber School of Arts and Sciences, University of British Columbia, Okanagan Campus, Canada.

Lee Huskey, Emeritus Professor of Economics, University of Alaska-Anchorage, Anchorage, AL.

Henry P. Huntington, Huntington Consulting, Eagle River, AL.

Aytalina Ivanova, Vice Dean, Faculty of Law, North Eastern Federal University, Yakutsk, Russian Federation.

Arn Keeling, Professor, Department of Geography, Memorial University of Newfoundland, Canada.

Vera Kuklina, Research Professor of Geography at the George Washington University, Senior Researcher at the Sochava Institute of Geography, Siberian Branch of Russian Academy of Sciences.

Francis Lévesque, Assistant Professor, School of Indigenous Studies, Université du Québec en Abitibi-Témiscamingue, Val-d'Or, Quebec, Canada.

Irina Litvina, University of Lapland, Rovaniemi, Finland.

Hereward Longley, PhD candidate, University of Alberta, Canada.

Suzanne Mills, Associate Professor, School of Labour Studies, McMaster University, Hamilton, ON, Canada.

Bram Noble, Professor, Department of Geography & Planning, University of Saskatchewan, Canada.

Andrey N. Petrov, Associate Professor of Geography and Geospatial Technology; Directory of the Arctic, Remote, and Cold Territories Interdiciplinary Center at the University of Northern Iowa, Cedar Falls, IA.

Rasmus Ole Rasmussen, Senior Research Fellow, Nordregio, Stockholm, Sweden.

Thierry Rodon, Associate Professor, Department of Political Science, Laval University, Quebec City, Quebec, Canada.

John Sandlos, Department of History, Professor, Memorial University of Newfoundland, Canada.

Peter Schweitzer, Professor of Social and Cultural Anthropology, University of Vienna, and Professor Emeritus, University of Alaska, Fairbanks, AL.

Florian Stammler, Research Professor, the Arctic Centre, University of Lapland, Rovaniemi, Finland.

Adam Wright, University of Guelph, Guelph, ON, Canada.

1 Introduction

Chris Southcott, Frances Abele,
David Natcher, and Brenda Parlee

As the world economy continues to expand, demand for energy and other natural resources is increasing. Reserves of some resources such as oil are becoming more difficult to replace. Natural resource industries are increasingly interested in new sources of supply in non-traditional yet politically stable regions such as the Arctic.[1] This is occurring at a time when climate change has the potential to make Arctic resources increasingly accessible. The Arctic has the potential to become a major source of commodity wealth in future years.

While northern resources have the potential to produce great wealth for some people in the future, past experience has showed that most Arctic communities have benefited little from resource exploitation. In the case of Canada, northern communities have experienced enormous social and economic challenges over the past half century and these challenges can be closely linked to impacts of past resource exploitation. Resource dependence is seen as one of the most important challenges facing the region. In the past this dependence has failed to produce sustained benefits for northern communities (Berardi, 1998; Berger, 1977; Bradbury, 1984; Duhaime, 2004; M. M. R. Freeman, 2000; Green, 1972; Hall, 1991; Honnigman, 1965; House, 1981; Jorgensen, 1990; Loney, 1987; Lotz, 1970; Niezen, 1993; Osherenko and Young, 1989; Riabova, 2001; Robinson, 1962; Simard, 1996; Waldram, 1993; Watkins, 1977; Weller, 1977). The "resource curse" identified in other regions of the world has been very much present in Northern Canada although in a different version (R. M. Auty, 1994; Banta, 2006; Ross, 2007; Sachs and Warner, 2001; Tynkkynen, 2007). Communities are disrupted to serve the interests of a type of resource development where few jobs go to local peoples and the arrival and departure of migrant workers creates great social problems (Huskey and Morehouse, 1992; Stabler, 1989; Storey, 2010). Added to these issues are attempts by the region's Indigenous population to ensure that their traditional activities and cultures are maintained in the face of multiple stressors (Duhaime, 1991; Hobart, 1984; Kleinfeld, Kruse, and Travis, 1983; Robbins and Little, 1988).

Resource development has often been linked to an increase in the disruption of these communities, leading to a variety of social and health challenges. Resource production often represents a threat to the northern environment upon which the traditional economy of the region's Indigenous population still depend. In the past, the region was forced to deal with the negative impacts both during the exploitation phase and the closure phase. Despite attempts to diversify the economy of Northern Canada, communities in the region remain heavily dependent on the exploitation of natural resources. Most projections for the future point to a continued dependence on these resources (GNWT, 2015; GY, 2017; NEF, 2014).

There is some indication however that the worst aspects of resource dependence can be countered through the introduction of new policies and models of development that increase local control of development and ensure a higher share of resource rents and other benefits are passed on to northern communities (Alcantara, Cameron, and Kennedy, 2012; Baena, Sévi, and Warrack, 2012; Banta, 2006; CGMRBS, 2014; Dana, 2008; Humphreys, Sachs, and Stiglitz, 2007; O'Faircheallaigh and Ali, 2008). In Canada, new land claims agreements, impact-benefit agreements, and co-management boards offer the potential for the development of natural resources in Northern Canada in a manner that increases the benefits of these developments for local communities and helps ensure that development is done in an environmentally sound manner (Coates and Crowley, 2013; Cournoyea, 2009; Dahl and Hicks, 2000; Fidler, 2010; Fitzpatrick, 2008; Fugmann, 2009; Hitch, 2006; McPherson, 2003; Natcher, Davis, and Hickey, 2005; Saku, 2002).

This book is an examination of the possibilities for resource development benefiting the long-term sustainability of northern communities. It is an initial discussion of the past and current relationship between resource development and Arctic communities with a view to better understand what, if anything, can be done in order for the development of non-renewable resources to be of benefit to the long-term sustainability of these communities. The chapters contained in this book are a result of the initial findings of the Resources and Sustainable Development in the Arctic (ReSDA) research project. The main objective of this project is to work closely with communities and stakeholders to conduct research on the best ways of developing northern natural resources in a manner that maximizes benefits to communities and minimizes dangers to the environment.[2] The initial work of this project was centred around a series of gap analyses of key questions concerning how resource development could potentially contribute positively to the long-term future of these communities and regions and help the people living in the Arctic deal with the many challenges that they face. Following discussions between researchers and community partners a series of 14 reports were prepared which serve as the base for the chapters in this volume. Initially the work has focused on extractive non-renewable types of resource development largely because these are the types of projects that are increasingly present in the region and which are causing the most concern among Arctic communities. Most discussion in this book centres on the situation in Northern Canada but also attempts to include other regions of the Arctic in a comparative perspective.

Extractive industries and sustainable development

Despite earlier suggestions that a post-industrial society would lead to a decrease in the importance of natural resource extraction, over the past 30 years we have seen a continued desire for commodities such as minerals and oil and gas. Increasing need from a number of sources means that many analysts are suggesting demand in commodities will increase further over the next 20 to 30 years (Ali et al., 2017). Production in existing areas will intensify and expand into new areas such as the Arctic (Rudra and Jensen, 2011). Along with the increased demand in commodity production comes an increasing realization among socio-economic researchers that extractive industries are often problematic for producing countries, regions, and communities. Building on earlier Canadian staples theory, researchers have shown that, despite an intuitive belief that natural resource development will increase the wealth, and therefore the well-being of producing regions, a resource curse exists. The paradox of the resource curse is that extractive resource development often leads to a decrease in development possibilities in these regions and produces other problems (Auty, 2002; Collier, 2007; Davis and Tilton, 2005; Sachs and Warner, 2001).

The resource curse findings may have emerged in the 1990s but communities that have had to deal with extractive industries have known about it for much longer. Much of the resource curse work has dealt with cross-national econometric regression analysis but another important, and older, literature exists that has shown the negative aspects of extractive resource development on sub-national regions and communities – most notably Indigenous communities. While the academic impact of this "localist" perspective has not been nearly as great as that of the resource curse, there is no shortage of research, much of it "grey literature", that has shown the damages inflicted upon communities in Latin America, Africa, Australia, North America, and other regions, by mining and oil and gas development (Bebbington et al., 2008; Berger, 1977; Langton and Longbottom, 2012).

Despite past problems with resource development, there is a great hope that the paradox can be resolved and that extractive industries can actually assist communities and regions to become sustainable. There exists cases where the resource curse has not affected regions (Acemoglu et al., 2003; Davis and Tilton, 2005). A consensus has emerged among the resource curse researchers that the curse can be avoided with the right institutions (R. M. Auty, 2007; Boschini, 2013; Pegg, 2006). "Localist" researchers are detecting new sub-national, national, and international power structures emerging that enable communities to better control aspects of resource development (Bebbington et al., 2008; Caulfield, 2004; O'Faircheallaigh and Ali, 2008). Even industry is realizing that efficiency of production now requires externalities such as environmental damage and socio-economic problems be controlled (ICMM, 2006).

While there is an increased belief that, properly managed, extractive industries can actually contribute to sustainability, there is still no coordinated effort by researchers on how exactly to do this. How can resource extraction be best used to create successful societies? What are the best ways to avoid the staples trap/resource curse? How can resource extraction best contribute to the well-being of producing regions and communities? What institutions best ensure that producing regions and communities benefit from resource extraction? Because

Canada is a country that has been largely built on, and continues to be dependent on, extractive industries, these questions are important for this country. At the same time, these are complex questions that represent a challenge of global importance.

Finding answers to these questions is the reason behind the formation of the ReSDA project. Funded from 2011 to 2018 by the Social Sciences and Humanities Research Council of Canada as a Major Collaborative Research Initiative, the project is a partnership of Arctic communities and over 50 researchers from 29 institutions and all eight circumpolar nations. As the Arctic is increasingly the site of extractive industry interest, the ReSDA network is looking for ways to increase the benefits of extractive industry developments to communities and to mitigate negative impacts. This is an opportune time to undertake this endeavour. New commodity pressures, the current state of the resource curse debate, the search for new global standards, a seemingly more concerned industry, increasing Indigenous control of decision-making in the region, and the fact that various levels of government in the Arctic are looking for new answers means that there is currently a great opportunity for progress on this question.

Researchers are starting to find answers to many aspects of the central questions. What is surprising is that this is being done largely in isolation from one another. There is little comparative work. While the resource curse literature has had a significant impact on international research discourses, other findings are limited to much more localized discussions. Many researchers are more focused on responding to the immediate needs of communities they are working with and, as a result, are isolated from the larger research discussions. This book is a first attempt to try and tie together the research that is being done around extractive resource development and the sustainability of Arctic communities and regions.

The Arctic and resource development

While most of the attention given to resource development in the Arctic has occurred starting in the twentieth century, the history of this relationship starts much before this. Archeological research suggests that about 2,000 years ago, the ancestors of the modern day Inuit were heavily involved in the iron trade between China's Han Empire and other northern peoples (McGhee, 2005: 119). McGhee suggests that the arrival of the early Inuit into North America was tied to the discovery that iron was available in the Eastern Arctic and that this iron would be valuable for trading. Indigenous peoples in the region periodically gathered minerals to use in fashioning tools used in subsistence hunting and gathering and for trade.

While the search for renewable resources such as fish, furs, and whales was responsible for the initial interest of Europeans in the Arctic, the possibility of discovering precious metals was a factor present in some incursions into the region (Vaughn, 2007: 247). It was not until the industrial era however that the Arctic started to be viewed as a potential source of extractive resource wealth. By the nineteenth century, the entire circumpolar region was divided by colonialism into northern peripheries of national states. Under colonialism, these northern peripheries were used to increase wealth in the national centres. Industrialism served to intensify this role. Under industrialism, the circumpolar world became increasingly important for those nations that had a piece of it. This region became the storehouse of those natural resources that were required by industrialism. For most of these nations, during the nineteenth and twentieth centuries the future of the entire nation was increasingly linked to the ability of the north to supply those resources required by industrialism.

Due to the transportation costs involved in accessing mineral resources in the Arctic, initially interest was shown in only those resources that had a high value, such as precious metals, or were easily accessible. Gold mining existed in the Yenisey and Lena river basins of Siberia in the 1840s but most of the exploitation was done by hand using primitive mining techniques and an unskilled labour force (Armstrong, 1965: 95). By the 1860s artisanal gold mining extended itself to Finland and other areas of the Arctic. Iron mining had existed in Northern Sweden since the 1600s but this activity became much more intense when railroads were built into the region in the 1880s (Eilu, 2012). Coastal deposits that were easily accessible by ship were some of the earliest non-precious mineral mines to develop in the Arctic such as the cryolite mine in Greenland and the coal mines in Svalbard.

A more detailed discussion of the mining history of Northern Canada is found in Chapter 2 of this volume but undoubtedly the Klondike Gold Rush at the end of the nineteenth century did much to bring the potential mineral wealth of the Arctic to the attention of the world. Spreading quickly into Alaska, this rush immediately demonstrated that precious metals such as gold could be found in the Arctic and, despite problems with accessibility, could produce great wealth for some.

The main factors underlying the exploitation of other types of mineral resources in these northern regions was the relative availability of these resources closer to home and a satisfactory transportation system to exploit these resources. Industrialism brought with it new forms of transportation to the circumpolar world. The construction of railways at the end of the nineteenth and beginning of twentieth centuries in Canada, Siberia, and Fenno-Scandia helped to integrate this region with the more central regions and allow for the more intensive utilization of the resources of the north.

As the twentieth century progressed, increased access to the Arctic led to increased resource exploitation. The development of the iron mines of Northern Sweden are very much linked to the construction of a railway but also an increasingly important vision of the north as the source of future national wealth and power (Sörlin, 1988). In Alaska, the building of a railway to the sea allowed the development of the Kennecott copper mines in 1911. In Canada, exploitation of silver and lead mines in the Yukon during the 1920s and 1930s was made possible by a river-based shipping system combined with a railway from Whitehorse to the Pacific port of Skagway. In a similar fashion, in the Makenzie Valley region, oil developments at Fort Norman in the 1920s and radium and gold mining in the 1930s were all facilitated by the existence of a seasonal river-based shipping system based in Fort Smith (Armstrong, Rogers, and Rowley, 1978: 84).

The greatest resource development expansion occurred in the Russian Arctic. The southern area of Siberia experienced some development following the construction of the Trans-Siberian Railway but its impact on the more northern areas was limited until the Soviet period. The extension of the railway system to Murmansk also provided opportunities for industrial expansion in the north but these too were primarily developed during the Soviet period.

Almost all of the large-scale industrial exploitation in the Russian north was developed initially by the Soviet Union. The development of natural resources in Northern Russia under the Soviet Union was done to provide the materials necessary for an isolated Soviet Union to industrialize (Armstrong, Rogers, and Rowley, 1978). Mining was the most important new activity to occur in the Soviet north. The need for reliable national sources of minerals meant that the new Soviet Union had to create new mining centres relatively quickly. The railway extension to the Kola Peninsula allowed the easiest access to new northern deposits. Communities were created to mine apatite starting in 1929. Monchegorsk was created in 1935 to mine nickel as well as copper and cobalt.

Elsewhere in the Soviet north, Noril'sk was created in the late 1930s to supply nickel and copper. Coal mining was expanded in Vorkuta to meet war needs in the early 1940s. The Soviet need for foreign currency led to the expansion of gold mining activities in several locations in the Sahka Republic starting in the 1920s. Gold mining expanded into the upper Kolyma basin by the 1930s.

The period following World War II represents the continuation and intensification of pre-war trends in mining along with a new interest in oil and gas development. In terms of mining, development continues as previously in the Soviet Union with the addition of diamond mining in Yakutia starting in the 1950s. The building of the Alaska Highway by the US Army during World War II gave birth to a road system in the Yukon which enabled more mines to open such as the large Faro lead-zinc mine which opened during the late 1960s. During the 1950s the federal government initiated a "Roads to Resources" program which made some deposits more accessible. A railway was built during the 1950s to Pine Point Northwest Territories in order to allow access to a zinc deposit. Other smaller mines are created where ship access was possible such as the Rankin Inlet copper and nickel mine which opened in 1957.

By the 1960s however, oil and gas development starts to become more interesting to outside investors than mining. While oil deposits were developed in Norman Wells in Canada starting in the 1920s, it is again in the Soviet Union where large-scale northern oil and gas development starts to occur. Gas was found in 1953 in Western Siberia and oil was found in 1960 (Armstrong, Rogers, and Rowley, 1978: 32). Development started in the 1960s. Starting in the southern Siberian slopes of the Ural Mountains, discoveries and new developments moved progressively north so that by the 1990s the Yamal-Nenets region became a major producing region. In North America, during the 1960s, oil and gas deposits were discovered in the North Slope region of Alaska and the Mackenzie delta regions of Canada. In Alaska this quickly lead to a settlement of a long-lasting Native land claim and the subsequent building of an oil pipeline from Prudhoe Bay on the Arctic Ocean to Valdez, Alaska on the Pacific. In Canada, attempts were made to build a similar gas pipeline from the Mackenzie delta to the south but a special government inquiry led by Justice Thomas Berger highlighted opposition to the pipeline among the local Indigenous population until outstanding land claims were settled. The inquiry concluded that the pipeline should not go ahead until Indigenous concerns were dealt with. Even outside of the north, the Berger Inquiry had an important impact on the way that society looked at the development of natural resources. Social Impact Assessment as a valid and important form of investigation received an increase in importance following the Berger Inquiry (Freudenberg, 1986).

The impacts of resource development on northern communities

The Berger Inquiry represents the start of an era where cultural, social, and economic impacts became important considerations for resource development. Prior to the Berger Inquiry, whether or not resource development was beneficial to northern communities was rarely mentioned in academic research. Certain regional political figures would occasionally protest at the way resources were being developed but they rarely wanted these developments to stop or even put on hold.

Prior to the Berger Inquiry, the main idea behind most discussions dealing with resource development in the north was on whether or not development was profitable. As Huskey has noted, this is exemplified in the classic article by well-known writer Jack London, "The Economics of the Klondike Gold Rush" written in 1900 (London, 1900). It the article he notes that much more money and effort was spent on the Klondike Gold Rush than gold produced. He calculates that $220 million dollars in time and expenditures was spent on the gold rush but only $22 million dollars in gold was extracted. While this may appear to be a waste of resources to many people, London notes that "it has been of inestimable benefit to the Yukon country, to those who will remain in it, and to those yet to come" (London, 1990: n.p.). According to London, because of the gold rush, famine, a common occurrence prior to 1896, will no longer exist in the Yukon. The transportation infrastructure built as a result of the mining operations will allow food to be accessed throughout the territory. In addition, this same infrastructure will permit the continued existence of mining at a more reasonable level and that "the resources of the Yukon district will be opened up and developed in a steady, business-like way" (London, 1990: n.p.)

Prior to the Berger Inquiry there was no major work on the relationship of the local Indigenous population to resource development or discussion of negative impacts of resource development on the region. Still, doubts about resource development for regions such as the territories started to emerge prior to the 1970s. By the 1960s a discourse had developed around resource development communities in Canada, most of which were in the provincial or territorial north (Robinson,

1962). These communities were seen to have a number of problems related to their "boomtown" nature but also related to a dependence on outside actors and limited possibilities for future development (Himelfarb, 1982). In addition, the notion that resource development led to community and regional prosperity was increasingly criticized by the emergence of a more critical version of staples theory in the 1960s – one that highlighted the problem of a "staples trap" (Watkins, 1963). Finally, starting in the 1960s a series of reports on the conditions of Indigenous peoples in Canada started to bring a certain awareness to the fact that past resource development in certain areas of Canada had not benefited the local Indigenous population and, in fact, may have contributed to the poor conditions in these communities (Hawthorn, Cairns, and Tremblay, 1967; Honnigman, 1965; Lotz, 1970).

Concern about the impacts of resource development on the physical environment started to surface in the 1960s and led the United States government to pass the National Environmental Policy Act in 1969. While it was concentrated on the physical environment, the Act also took into consideration the impact that the environmental changes would have on communities. The term Social Impact Assessment was first introduced in 1973 when it became apparent that the building of the Alyeska Pipeline in Alaska would have important impacts on the Native people of Alaska. While some research concerning the social and economic impacts of the Alaska pipeline took place (Dixon, 1978), it was the Berger Inquiry which was "the first case where social impacts were considered in project decision making" (Burdge and Vanclay, 1996: 63).

The Berger Inquiry was the first major report to critically examine the impacts of resource development on the north. Based on evidence presented to the inquiry, an industrial project such as the pipeline would continue to add to the problems of the region. From this point on, social science research concentrated on the critical themes elaborated in volume 1 of the Final Recommendations and increasingly stressed the negative impacts of resource development on northern communities.

The potential impacts were clearly stated in the first volume of the report and were based largely on what was happening in Alaska with its pipeline and on community and academic observations of what impacts previous projects and industrial development in general were having on communities in the Mackenzie Valley. Environmental impacts were discussed in the report but what was highlighted was the negative implications of these environmental impacts on the traditional activities of the Indigenous people of the region. This theme continued through the report where Berger devotes chapters to cultural, economic, and social impacts. The primary focus of these impacts are what it would mean for the cultural survival of the Indigenous peoples of the region. Berger discounted evidence that Indigenous peoples were abandoning traditional subsistence activities and actively seeking out industrial jobs (Berger, 1977: 109). Based on evidence from community members and others, he noted that what he referred to as the "Native economy" continued to exist and remained essential for the region's Indigenous communities. While he was not opposed to industrial development, he noted the negative impacts of a situation where projects such as the proposed pipeline were creating the image that there was no other choice for these communities other than industrial development. His report suggested that it would be much better for the region if industrial development could be put on hold until ways were found to protect the "renewable resource" sector and to use industrial employment as a supplement to traditional subsistence activities (Berger, 1977: 115).

He considered the potential positive economic impacts of the pipeline. It had been proposed by industry and government officials that the pipeline would have positive impacts on the region in the form of jobs. Based on the evidence presented to the inquiry he discounted these benefits and stated: "we have always over-estimated the extent to which native people are unemployed or underemployed" and "we have never fully recognized that industrial development has, in itself, contributed to social, economic and geographic dislocations among native peoples" (Berger, 1977: 123). He looked at the impacts of the Alaskan pipeline and other industrial developments in the Canadian north to show that few jobs went to local Indigenous people and when they did, the jobs were unskilled and short-lived. In addition, for the few that did get jobs, they had to leave their home communities and were introduced to alcoholism and drug abuse. Most importantly, the traditional economy would be increasingly seen as less relevant and this would have extremely negative impacts on communities.

He noted that as far as the regional economy was concerned, very little of the money from the project would stay in the Northwest Territories. According to his estimates, the biggest source of new funds for the Government of the Northwest Territories would be from the sale of liquor since the territorial government had no other way to source revenues from the project. While the GNWT would be responsible for dealing with many of the negative consequences of the project, it would not have additional resources to do so adequately. As far as local northern businesses were concerned, they would perhaps benefit somewhat but most services and goods would be provided by outside contractors. In addition, these local businesses would be the ones who would have to deal with the negative impacts of the post-construction bust period.

While some proponents suggested that social impacts from the pipeline would be positive in that they would "reduce the unemployment, welfare dependence, crime, violence and alcoholism" that exist in these communities (Berger, 1977: 148), Berger disagreed. While this could occur in some communities, the unique cultural aspects of northern communities would bring about an opposite effect. According to Berger the evidence presented to the inquiry showed that industrial development had, in the past, brought on a large number of negative social impacts and this would likely continue with the pipeline project. It would destroy existing social relations, increase welfare dependency as families were torn apart, increase crime and violence as traditional values and beliefs were damaged, it would expose the region to increased sexually transmitted and other diseases, cause important changes to diets, and drastically increase alcohol and drug abuse. Berger noted the pipeline would be particularly hard on women as the housing crisis would worsen and sexual exploitation would increase. Social inequality would become worse as Indigenous workers would likely be employed in an ad hoc manner as unskilled workers and "find themselves on the bottom rungs of the ladder and most of them are likely to remain there" (Berger, 1977: 159). All this meant that the pipeline would likely result in an increase in the loss of Indigenous identity and of self-respect.

The Berger Inquiry contributed to the decision of the government to place the Mackenzie pipeline on hold. While other aspects were involved, the federal government largely accepted Berger's ideas that large-scale resource developments should only occur once the local population were better able to control and benefit from these developments. While the report of the Berger Inquiry, and publications associated with it (Watkins, 1977), was largely considered the first widely known criticism of extractive resource development in the north, this theme has dominated social science research on the topic ever since.

This is seen in publications related to another large-scale industrial project in the Canadian north: The James Bay Hydro Electric project. At the same time that Berger was listening to what Northerners thought about the pipeline, the Cree and Inuit of Northern Quebec were dealing with a hydro-development project in their lands. While the government of Quebec did have a social impact report done before the project started, unlike the Berger Inquiry, there was little or no extensive consultation with local communities. Action on the part of the Cree and Inuit of the region resulted in the first modern treaty in Canada when the James Bay and Northern Quebec Agreement was signed in 1975, after work on the project had already started.

The period following the Berger Inquiry is one where social scientists concentrated on highlighting the negative findings of the inquiry. A recent literature review of resource development and Indigenous culture by Angell and Parkins refers to this period as "the community impacts phase" (Angell and Parkins, 2011) where resource developments where viewed as negatively impacting communities through social disruption which caused social pathologies. A number academic studies elaborated on the findings of the Berger Inquiry by enumerating the negative impacts of resource development on northern communities either due to hydro electric developments (Loney, 1987; Niezen, 1993; Simard, 1996; Waldram, 1993), oil and gas (Berardi, 1998; D. M. Freeman, 1981; House, 1981), or mining (Bradbury, 1984; Stabler, 1989). Though these works we gained a better understanding of many of the costs of resource development to northern communities.

A new perspective on extractive resource development

Things started to change in the mid-1990s (Angell and Parkins, 2011). Most notably the change can be seen in the changing political climate. The signing of modern land claim treaties in the years following the Berger Inquiry were extremely important and this, combined with the processes of decolonization and self-government, resulted in a new consideration of resource development in many northern communities (Abele, 2009).

While there would undoubtedly be many potential negative impacts from these developments, the legal, political, and cultural situation had changed since the 1970s. New land claim settlements gave many northern communities a perceived increased ability to control extractive resource projects. Increasingly, northern actors saw the potential of using the wage economy associated with resource development to enhance and re-invigorate their communities. A new vision of a mixed economy emerged which saw sustainable development of these communities dependent on a balance of traditional, wage, and transfer economies (Elias, 1991; Kruse, 1986). Added to this, new political powers for territorial governments and new mechanisms such as impact-benefit agreements meant that there was a greater chance of capturing and controlling the benefits from extractive resource projects.

While very much aware of the dangers of extractive resource development, northern communities saw few realistic opportunities for sustainable economic development for their regions. They increasingly noted the need to examine how mining and oil and gas projects could be used to improve community well-being (SERNNoCa, 2010). It was this need that led to the formation of the ReSDA project and its key research question: What is the potential for extractive resource development to contribute to the well-being and long-term sustainability of Arctic communities? Starting in 2011 ReSDA brought a team of Canadian and international researchers, representatives of northern governments, treaty organizations, educational institutions, and community organizations together around the need to find ways that more benefits from resource development could flow to northern communities and that any negative impacts could be mitigated. As one of its first activities, ReSDA sponsored a series of gap analyses which would look at what had already been done and what needed to be done.

The point of departure for these gap analyses was the underlying belief that things have improved in terms of the impacts of resource development in northern communities. These improvements indicate that the potential exists for resource development to help make communities sustainable. The gap analyses are built around the key question: What are some of the key points communities need information on in order for them to decide whether resource development proposals will assist in their efforts at sustainability (or will continue to work against this)? In order to do this each gap analysis examined what exists now in terms of how to increase benefits and eliminate negative impacts of resource development going to communities and what research needs to be done to find more ways of increasing benefits and eliminating negative impacts.

Initial discussions between researchers and partners resulted in list of 13 areas of research: the history of resource development in the Arctic, main social impacts, the measurement of social impacts, differing revenue regimes, socio-economic impact assessment processes, regional economic development strategies, community well-being, community-industry relations, impact benefits agreements, subsistence activities and resource development, traditional knowledge and resource development, environmental issues and resource development, and gender and resource development. Researchers later added an additional area, climate change and resource development, as it was felt the amount of work being done by academics on climate change required ReSDA to investigate the relationship. The results of these gap analyses are presented in this book.

Chapter 2 deals with the issue from a historical perspective. Coates looks at contributions of existing historical literature to our understanding of the impact of resource developments on northern communities. He tries to understand how historians have contributed to "the collective understanding of the problems for northern communities associated with resource development" as well as, more recently, constructive impacts that resource developments have had on the region. After a brief historical introduction to resource development in the Canadian Arctic he identifies a series of themes that emerge from the literature on resource development in the north. One tendency that he notes is the change from an uncritical perspective on resource development prior to the 1970s, to a predominantly critical one for the decades that follow, to a more recent nuanced perspective. Linked to this trend is the shift from a southern-based perspective of the region to a more northern "insider" perspective. Related to this shift to a more northern perspective is a recent attempt by historians to look at impacts of resource development on the environment and on changes in the management of these impacts.

Coates also notes however that historians have been relatively silent on evaluations of the resource development process. There has been a reluctance to engage in discussions of whether social, cultural, or economic impacts have been positive or negative for northern communities. He outlines a list of areas of research where historians and others could assist communities looking at resource development. This list includes such issues as a better idea of the earlier impacts of resource development on these communities; changes in Indigenous relationships to resource development; and the development of new institutions in the region and their impacts on resource development. While historians are, by their nature, biased toward the important task of informing the public about past developments, Coates notes that historians and historical research can be important in "defining and clarifying public policy issues". By examining the past impacts of resource development in terms of evaluating their influence on northern communities, historians can provide communities with a knowledge base that they can use to determine how to shape their relations to resource development in the future. In this way communities would have a better idea of how to maximize benefits and minimize costs.

The next section of the book examines the social impacts, their regulation, and their measurement. We have a much better idea than ever before about the social impacts of extractive resource development. Indeed, prior to the 1970s social impacts were rarely even considered. The Berger Inquiry played a major role in changing this situation. Berger's insistence on looking at potential negative social, cultural, and economic impacts, especially those impacting the local subsistence activities, fundamentally changed the way we look at resource developments in the north. This new understanding of social impacts is described in the chapter by Peter Schweitzer et al. Looking at the situation in Alaska, Greenland, and Russia,

Schweitzer et al. point out that while social impacts are considered, how they are considered varies from region to region. In most instances cultural impacts are not properly understood.

Chapter 4 looks at the formal regulatory situation concerning extractive resource development in the Arctic. Noble et al. point out that we now have laws to ensure that environmental, social, and other impacts are examined prior to a project going ahead. This is an obvious improvement from the situation prior to the 1970s. Despite this integration, problems remain and the chapter lists a number of research gaps that need to be filled in order to enhance the effectiveness of the processes and enhance communities' involvement and ability to use the EA process to ensure that resource development occurs in a way that maximizes their benefits. According to Noble et al. there needs to be a better understanding between the actors and participants of what EA is designed to do so that expectations are closely aligned between actors. In particular we need to know what communities want and need out of the EA process. The current process is seen by almost all actors as cumbersome with limited sharing of information. Communities often do not have the capacity for meaningful participation in the EA process. Other issues are the limitations placed on our understanding of cumulative impacts due to the project-based nature of EA, inadequate indicators, and a lack of understanding of the relationship between EA and new negotiated agreements with communities such as Impact Benefit Agreements (IBAs).

Rodon and Levesque look in greater detail at the situation concerning social impacts and the EA process in Canada. They list the various impacts that are often considered in resource development projects in Canada and note that there are still some problems associated with properly understanding and measuring these impacts which limits our ability to mitigate some negative impacts. Building on the findings in Chapter 4 they note in particular that socio-economic indicators are problematic – especially as concerns culturally related indicators. They also note a need to develop instruments to better monitor socio-economic impacts.

These issues concerning socio-economic indicators and the monitoring of impacts is the central focus of Chapter 6. While current indicators are useful for understanding some basic impacts, Petrov et al. note that socio-economic impact indicators need to be improved in a number of ways. This improvement must involve more community-relevant indicators. In addition, indicators must be readily available and understandable to communities. To ensure that communities have greater control of the monitoring process indicators must be designed so that they can be collected and analysed by communities themselves.

Chapter 7 deals with the issue of community well-being. Increasingly we are aware that community well-being needs to be enhanced in the north and that extractive resource development can sometimes be destructive of this goal. Parlee notes the importance of a better understanding of well-being in order to properly understand how northern communities can benefit from extractive resource development. Planning has to be based on a more profound appreciation for those aspects of communities that most contribute to a sense of well-being in these communities. In addition to the traditional notions of material well-being, Parlee notes the importance of knowledge and education, culture, the land, social relations, self-determination, and a general notion of human security. A better understanding of well-being in northern communities can help us find ways that extractive resource development can improve rather than harm life in the region.

In their chapter looking at resource revenue regimes, Huskey and Southcott examine the notion of financial benefits from resource development. They start out by examining the notion of the resource curse. While the resource curse discourse is of interest to Arctic communities looking at resource extraction as a potential means of development, the authors point out that staples theory is a more appropriate model to use when trying to increase benefits to northern communities.

This analytical framework grew out of Canada's experience with resource development. The use of four types of linkages as a conceptual framework allows researchers to understand how the development of a resource commodity can assist a community or region to diversify into a sustainable economy, or how it can provoke economic decline and produce a "staples trap" (Watkins, 2007). If production-related inputs are sought and developed locally, this produces additional economic growth through backward linkages. If the commodities are used to develop new products then the benefits of value-added would produce economic development through forward linkages. If the local wages and earnings are spent locally on demand-related products then economic growth is enhanced through final demand linkages. Finally, if the resource rents are captured and reinvested locally, new economic growth would result from these fiscal linkages. If these linkage opportunities did not exist then linkage benefits would become leakages. Under this situation the community or region would remain in a staples trap.

Against the background of this framework, Huskey and Southcott focus on comparing resource revenue regimes across the Arctic and show how fiscal linkages vary due to differing taxation and royalty regimes. They point out that certain regions, notably Norway and Alaska, have been able to capture and control resource rents through use of wealth funds that are designed to provide for future sustainable benefits and avoid many of the pitfalls associated with the resource curse.

Frances Abele's chapter on regional economic development notes that while resource development has long been a part of regional policies, these policies have failed to provide sustainability to northern communities. She argues that a more balanced form of development can only occur with a proper understanding of regional dynamics. In particular she notes the importance of developing policy from the "ground up". At the same time, she freely admits that, given the colonial history of the region, this is not easy to do. She notes the need for innovation in attempt to develop a sustainable future for northern communities and that these innovations much be both technical and social. More needs to be done to develop effective and better coordinated policy instruments and programs need to be re-designed to be more appropriate for the regional settings.

In Chapter 10, Keeling et al. look at the environmental legacies of extractive resource development but from the perspective of what these mean for northern communities. They note that since the Berger Inquiry there have been improve-ments in the way the environmental impacts have been taken into consideration in resource development projects in the region. Part of these improvements has been the integration of local communities into the planning and assessment processes.

They point out however that there are still areas where we need more research in order for us to understand how to deal with environmental legacies. They note a need to better understand the "intersection" of past environmental legacies with present impacts by examining cumulative impacts. There is also a need to under-stand community involvement in the management of environmental legacies and, finally, a need to better understand the relationship between communities and the remediation of these legacies. They note the need to develop new interpretive frameworks to better understand the long-term impacts of these legacies on the sustainability of communities.

Central to their argument is the idea that research needs to focus on how communities can become more appropriately involved in the processes surrounding environmental legacies. A more central place for traditional knowledge in the management and remediation of environmental legacies, along with a better dialogue with communities on how remediation activities can better benefit the region is key to ensuring more adequate long-term sustainability.

Tangible benefits to communities are often institutionalized in IBAs. These IBAs are signed agreements between companies and communities to provide the community with a range of benefits in order to for these communities to better mitigate potential negative impacts. As noted in Chapter 11, IBAs have been portrayed as a relative improvement from what existed previously. They represent the potential for many new benefits to be passed on to communities and relative empowerment of the community in decision-making. At the same time there are problems that have emerged related to these agreements.

These problems have to be examined and solutions suggested if IBAs are to be used to maximize benefits from extractive resource development being passed on to these communities. One of these areas is the debate over the effectiveness of these agreements. Many benefits are promised and some are certainly delivered, but the overall effectiveness of the agreements is sometimes challenged. Since these agreements are so new, communities often need more time to evaluate their impacts. We need to better understand whether these agreements are bene-fiting these communities or whether, as some suggest, they are perpetuating existing problems. An issue that was also echoed in the chapter dealing with Environmental Assessment (EA) is the relationship of these agreements to the regulation process. Are these agreements supplements to the EA process or replacements to the EA process? Can we harmonize these two processes. Another important question is how can communities negotiate the best possible agreement for their needs?

In 1977 the Berger Inquiry radically altered the view of extractive industries and the north. One of the main, if not the main, points of the inquiry's recom-mendations was that extractive resource development would be "devastating" for the subsistence economy which is so important for Northern Indigenous communities. Forty years have passed since Berger heard that traditional hunting and gathering economies were in danger of disappearing due to the changes that were occurring in the region.

Where are we now? The relationship between extractive resource development and the subsistence economy is less clear than it was in the 1970s. Studies show the subsistence economy remains strong. New comprehensive land claims have provided new supports and protections for these activities. New arrangements now mean that, in some instances, extractive resource development can be used as way to re-invigorate the traditional economy. What we have seen recently though has been a series of seemingly conflicting studies that show these projects as being both positive and negative for subsistence activities. In his chapter on the subsistence economy Natcher points to the need to "normalize" our understanding of the subsistence economy – we need to consider these activities as a normal part of a healthy northern economy. Once we do this we need to find out more about how projects help or hinder these normal economic processes.

In Chapter 13, Huntington examines the issue of traditional knowledge and resource development. He points out that, largely due to the political activities of Northern Indigenous peoples, traditional knowledge is increasingly accepted as a legitimate source of knowledge in resource development projects. While this is especially the case in the planning and management of renewable resources, it has also been applied to non-renewable projects. A number of benefits can be expected as a result. The use of traditional knowledge in resource development decision-making ensures that communities have more of a voice in these decisions. Their knowledge is the basis for community concerns about development projects. To know that their knowledge is given at least some legitimacy means that there is a greater chance of these concerns being heeded.

Huntington points out, however, that we have little information on the extent to which these expectations are justified by outcomes. We need to know more about the extent that traditional knowledge has influenced these decisions and how existing power relations affect this influence. We would also benefit from a better idea of how Indigenous participants feel about their inclusion in the process: whether they feel their voice has been heard and whether existing infrastructures are sufficient for their participation. Finally, we need to know more about the relationship between traditional knowledge and resource development generally in order to know how the two can or should fit together.

This issue of gender in northern resource development is dealt in the chapter by Mills et al. They note that much of the research looking at gender and resource development narrowly focuses on how women receive few benefits and bear much of the cost. This chapter suggests that research in the north is more complex. Although women are underrepresented in resource employment they often benefit from resource development through employment in associated sectors due to their higher levels of education. Other research is founded on the idea that employment in extractive industries will have a negative impact on subsistence activities and this will have a negative impact on women. When taken as a whole, the literature focuses on how resource development negatively impacts women. This focus adopts a lens that reduces gender to an individual attribute and homogenizes differences among communities, regions and forms of development, with the result that little attention is paid to broader shifts in gender relations or to the resilience and successes of Indigenous women. The chapter also reviews a promising area of literature examining gender and resource governance. This research proposes ways that researchers and policy makers can ensure that gender is considered in efforts to increase community benefits from extractive resource development.

Finally, most of the research in the Arctic over the past 20 years has been related to the issue of climate change. A belief often noted in this research is that the increased interest in resource development in the Arctic is the result of climate change. The melting ice is uncovering resources in the Arctic and making them more accessible. Chapter 15 looks at the relationship between climate change and resource development. A review of the literature shows there is little evidence-based research indicating that climate change is currently having any impact on resource development. Most of the commentary in the "scientific" discourse is hypothetical in nature. Based on their assessments of current climate change models, scientists believe climate change will impact resource development in the future. They are not sure however whether these impacts will be positive or negative for resource development. In the long term, climate change may lead to more Arctic resource development but in the short term it is more likely to be a barrier. While some may believe that climate change is the main reason for increased interest in resource development, there is evidence that the main determining factors in Arctic resource development are the needs of the global economy. So far, there is little proof that climate change has created a situation where increased ease of access through increased shipping will lead to more resource development in the Arctic, however this may change in the future.

The chapters in this book are a start to a more in-depth discussion of resource development in the Arctic. Generally they point to a situation where in the past, resource development has had serious negative impacts on the long-term sustainability of northern communities. This started to change following the Berger Inquiry of the 1970s. The situation has improved to the point where all main actors agree that resource development has to occur in a manner that does not negatively impact these communities. Increasingly, the actors are agreeing that for resource development to occur, communities must be assured that they benefit from these developments.

At the same time, as the chapters point out, we need to better understand many of the aspects of these developments to be assured that communities will indeed benefit to the maximum degree possible. We need to increase our understanding of past and current social impacts. We need to better understand the involvement of northern communities in the regulatory processes that have been developed around resource development as well as their role in the collection and analysis of impact indicators and monitoring. Before we can understand how resource development can enhance well-being, we need to better understand what well-being means to these communities. It is possible that the financial benefits from resource development can be captured by these communities to a greater extent than in the past but first we need to have a better idea of where the money earned from these projects goes now. How can subsistence activities be enhanced through resource development? What are the best economic policies for these communities? How can these communities become more closely involved in the assessment and monitoring of environmental impacts? Are IBAs the best instruments to use for northern communities? These and other questions need to be at least partially answered for us to ensure that extractive resource development can indeed help Arctic communities become more sustainable.

Notes

1   The definition of the Arctic used in this book roughly follows those definitions used in standard Arctic Council research documents such as the Arctic Climate Impact Assessment (2004) and the Arctic Human Development Report (2004). This includes the northern parts of Fenno-scandia, European Russia and Siberia, Alaska, Greenland, and Iceland. In Canada the region consists primarily of the three territories, Nunavik, and Labrador. Due to the similarity of conditions in some areas of the provincial norths, communities in these regions are also sometimes discussed.
2   For more information on the ReSDA project see www.resda.ca.

References

Abele, F. (2009). Northern Development Past, Present and Future. In Abele, F., Courchene, T., Seidle, L., and St-Hilaire, F. (Eds.), *Northern Exposure: Peoples, Powers and Prospects in Canada's North.* Montreal: IRPP.

Acemoglu, D., Johnson, S., Robinson, J., and Thaicharoen, Y. (2003). Institutional Causes, Macroeconomic Symptoms: Volatility, Crises and Growth. *Journal of Monetary Economics, 50*(1), 49–123.

Alcantara, C., Cameron, K., and Kennedy, S. (2012). Assessing Devolution in the Canadian North: A Case Study of the Yukon Territory. *Arctic, 65*(3), 328–338.

Ali, S. H., Giurco, D., Arndt, N., Nickless, E., Brown, G., Demetriades, A., . . . Yakovleva, N. (2017). Mineral Supply for Sustainable Development Requires Resource Governance. *Nature, 543*(7645), 367–372.

Angell, A. C. and Parkins, J. R. (2011). Resource Development and Aboriginal Culture in the Canadian North. *Polar Record, 47*(1), 67–79.

Armstrong, T. (1965). *Russian Settlement in the North.* Cambridge, UK: Cambridge University Press.

Armstrong, T., Rogers, G., and Rowley, G. (1978). *The Circumpolar North: A Political and Economic Geography of the Arctic and Sub-Arctic.* London: Methuen & Company.

Auty, R. M. (1994). The Resource Curse Thesis – Minerals in Bolivian Development, 1970–90. *Singapore Journal of Tropical Geography, 15*(2), 95–111.

Auty, R. M. (2003). *Sustaining Development in Mineral Economies: The Resource Curse Thesis.* London: Routledge.

Auty, R. M. (2007). Natural Resources, Capital Accumulation and the Resource Curse. *Ecological Economics, 61*(4), 627–634.

Baena, C., Sévi, B., and Warrack, A. (2012). Funds from Non-Renewable Energy Resources: Policy Lessons from Alaska and Alberta. *Energy Policy*, *51*, 569–577.

Banta, R. (2006). The Resource Curse and the Mackenzie Gas Project. *Policy Options* (December 2006–January 2007), 81–86.

Bebbington, A., Hinojosa, L., Bebbington, D. H., Burneo, M. L., and Warnaars, X. (2008). Contention and Ambiguity: Mining and the Possibilities of Development. *Development and Change*, *39*(6), 887–914.

Berardi, G. (1998). Natural Resource Policy, Unforgiving Geographies, and Persistent Poverty in Alaska Native Villages. *Natural Resources Journal*, *38*(1), 85–108.

Berger, T. R. (1977). *Northern Frontier, Northern Homeland* (Vol. 1). Ottawa: Mackenzie Valley Pipeline Inquiry.

Boschini, A. (2013). The Resource Curse and Its Potential Reversal. *World Development*, *43*, 19–41.

Bradbury, J. H. (1984). Declining Single-Industry Communities in Quebec-Labrador, 1979–1983. *Journal of Canadian Studies*, *19*(3), 125–139.

Burdge, R. and Vanclay, F. (1996). Social Impact Assessment: A Contribution to the State of the Art Series. *Impact Assessment*, *14*(1), 59–86.

Caulfield, R. A. (2004). *Resource Governance. Arctic Human Development Report*. Winnipeg: University of Manitoba Press, 121–138.

CGMRBS. (2014). *To the Benefit of Greenland*. Copenhagen: Committee for Greenlandic Mineral Resources to the Benefit of Society.

Coates, K. and Crowley, B. (2013). *New Beginnings: How Canada's Natural Resource Wealth Could Re-shape Relations with Aboriginal People*. Ottawa: MacDonald-Laurier Institute.

Collier, P. (2007). *The Bottom Billion: Why the Poorest Countries Are Failing and What Can Be Done About It*. Oxford: Oxford University Press.

Cournoyea, N. (2009). Navigating and Managing Economic, Environmental and Social Change in the Inuvialuit Settlement Region. In Abele, F., Courchene, T., Seidle, L., and St-Hilaire, F. (Eds.), *Northern Exposure: Peoples Powers and Prospects in Canada's North* (pp. 389–393). Montreal: IRPP.

Dahl, J. and Hicks, J. (2000). *Nunavut: Inuit Regain Control of Their Lands and Their Lives*. Copenhagen: IWGIA, International Work Group for Indigenous Affairs.

Dana, L.-P. (2008). Oil and Gas and the Inuvialuit People of the Western Arctic. *Journal of Enterprising Communities: People and Places in the Global Economy*, *2*(2), 151–167.

Davis, G. A. and Tilton, J. E. (2005). The Resource Curse. *Natural Resources Forum*, *29*(3), 233–242.

Dixon, M. (1978). *What Happened to Fairbanks: The Effects of the Trans-Alaska Oil Pipeline on the Community of Fairbanks, Alaska*. Boulder, CO: Westview Press.

Duhaime, G. (1991). Le pluriel de l'Arctique. Travail salarié et rapports sociaux en zone périphérique. *Sociologie et sociétés*, *23*(2), 113–128.

Duhaime, G. (2004). Economic Systems. In Einarsson, N., Nymand Larsen, J., Nilsson, A., and Young, O. R. (Eds.), *The Arctic Human Development Report*. Akureyri: Stefansson Institute.

Eilu, P. E. A. (2012). Mining History of Fennoscandia. In Eilu, P. (Ed.), *Mineral Deposits and Metallogeny of Fennoscandia* (pp. 19–32). Helsinki: Geological Survey of Finland.

Elias, P. D. (1991). *Development of Aboriginal People's Communities*. Concord: Captus Press.

Fidler, C. (2010). Increasing the Sustainability of a Resource Development: Aboriginal Engagement and Negotiated Agreements. *Environment, Development and Sustainability, 12*, 233–244.

Fitzpatrick, P., Sinclair, A. J., and Mitchell, B. (2008). Environmental Impact Assessment Under the Mackenzie Valley Resource Management Act: Deliberative Democracy in Canada's North? *Environmental Management, 42*(1), 1–18.

Freeman, M. M. R. (2000). *Endangered Peoples of the Arctic: Struggles to Survive and Thrive*. Westport, CT: Greenwood Press.

Freudenburg, W. R. (1986). Social Impact Assessment. *American Review of Sociology, 12*, 451–478.

Fugmann, G. (2009). Development Corporations in the Canadian North – Examples for Economic Grassroots Initiatives among the Inuit. *Erdkunde, 63*(1), 69–79.

GNWT. (2015). *Economic Outlook 2014–15 Northwest Territories*. Yellowknife: Government of the Northwest Territories.

Green, L. (1972). *The Gold Hustlers*. Vancouver: J. J. Douglas.

GY. (2017). *Yukon Economic Outlook April 2017*. Whitehorse: Government of Yukon.

Hall, T. (1991). Blockades and Bannock: Aboriginal Protests and Politics in Northern Ontario, 1980–1990. *Wicazo Sa Review, 7*(2), 58–77.

Hawthorn, H. B., Cairns, H., and Tremblay, M. A. (1967). *A Survey of the Contemporary Indians of Canada: A Report on Economic, Political, Educational Needs and Policies*. Indian Affairs Branch.

Himelfarb, A. (1982). The Social Characteristics of One-industry Towns in Canada. In Bowles, R. (Ed.), *Little Communities and Big Industries*. Toronto: Butterworths.

Hitch, M. (2006). *Impact and Benefit Agreements and the Political Ecology of Mineral Development in Nunavut* (Ph.D.), University of Waterloo, Waterloo.

Hobart, C. W. (1984). Impact of Resource Development Projects on Indigenous People. In Detomasi, D. D. and Gartrell, J. W. (Eds.), *Resource Communities: A Decade of Disruption* (pp. 111–124). Boulder, CO: Westview Press.

Honnigman, J. (1965). Social Disintegration in Five Northern Canadian Communities. *Canadian Review of Sociology and Anthropology, 2*(4), 199–214.

House, J. D. (1981). Big Oil and Small Communities in Coastal Labrador – The Local Dynamics of Dependency. *Canadian Review of Sociology and Anthropology, 18*(4), 433–452.

Humphreys, M., Sachs, J., and Stiglitz, J. E. (Eds.). (2007). *Escaping the Resource Curse*. New York: Columbia University Press.

Huskey, L. and Morehouse, T. A. (1992). Development in Remote Regions: What Do We Know? *Arctic, 45*(2), 128–137.

ICMM. (2006). *Community Development Toolkit*, London: International Council on Mining and Metals.

Jorgensen, J. G. (1990). *Oil Age Eskimos*. Berkeley, CA: University of California Press.

Kleinfeld, J., Kruse, J., and Travis, R. (1983). Inupiat Participation in the Wage Economy: Effects of Culturally Adapted Jobs. *Arctic Anthropology, 20*(1), 1–21.

Kruse, J. (1986). Subsistence and the North Slope Inupiat: The Effects of Energy Development. In Langdon, S. (Ed.), *Contemporary Alaskan Native Economies* (pp. 121–152). Langham, MD: University Press of America.

Langton, M. and Longbottom, J. (2012). *Community Futures, Legal Architecture: Foundations for Indigenous Peoples in the Global Mining Boom*. London: Routledge.

London, J. (1900). The Economics of the Klondike Gold Rush. *The American Monthly Review of Reviews*.

Loney, M. (1987). The Construction of Dependency: The Case of the Grand Rapids Hydro Project. *Canadian Journal of Native Studies, 6*(1), 57–78.

Lotz, J. (1970). *Northern Realities: The Future of Northern Development in Canada.* Toronto: New Press.

McGhee, R. (2005). *The Last Imaginary Place: A Human History of the Arctic World.* Oxford and New York: Oxford University Press.

McPherson, R. (2003). *New Owners in their Own Lands: Minerals and Inuit Land Claims.* Calgary: University of Calgary Press.

Natcher, D., Davis, S., and Hickey, C. G. (2005). Co-management: Managing Relationships, Not Resources. *Human Organization, 64*(3), 240–250.

NEF. (2014). *2013 Nunavut Economic Outlook.* Iqaluit: Nunavut Economic Forum.

Niezen, R. (1993). Power and Dignity – The Social Consequences of Hydroelectric Development for the James Bay Cree. *Canadian Review of Sociology and Anthropology, 30*(4), 510–529.

O'Faircheallaigh, C. and Ali, S. H. (2008). *Earth Matters: Indigenous Peoples, The Extractive Industries and Corporate Social Responsibility.* Sheffield, UK: Greenleaf.

Osherenko, G. and Young, O. (1989). *The Age of the Arctic: Hot Conflicts and Cold Realities.* Cambridge, UK: Cambridge University Press.

Pegg, S. (2006). Mining and Poverty Reduction: Transforming Rhetoric into Reality. *Journal of Cleaner Production, 14*(3–4), 376–387.

Riabova, L. (2001). Coping with Extinction: The Last Fishing Village on the Murman Coast. In Aarsæther, N. and Bærenholdt, J. O. (Eds.), *The Reflexive North.* Copenhagen: Nordic Council of Ministers.

Robbins, L. A. and Little, R. L. (1988). Subsistence Hunting and Natural-Resource Extraction – St Lawrence Island, Alaska. *Society & Natural Resources, 1*(1), 17–29.

Robinson, I. (1962). *New Industrial Towns on Canada's Resource Frontier.* Chicago, IL: University of Chicago.

Ross, M. (2007). How Mineral-Rich States Can Reduce Inequality. In Humphreys, M., Sachs, J., and Stiglitz, J. E. (Eds), *Escaping the Resource Curse* (pp. 237–255). New York: Columbia University Press.

Rudra, N. and Jensen, N. (2011) Globalization and the Politics of Natural Resources. *Comparative Political Studies, 44*(6), 639–661.

Sachs, J. D. and Warner, A. M. (2001). The Curse of Natural Resources. *European Economic Review, 45*(4–6), 827–838.

Saku, J. C. (2002). Modern Land Claim Agreements and Northern Canadian Aboriginal Communities. *World Development, 30*(1), 141–151.

SERNNoCa. (2010). *Nunavut Summit on the Social Economy Proceedings.* Iqaluit: Social Economy Research Network for Northern Canada and the Nunavut Economic Forum.

Simard, J. J. (1996). *Tendances nordiques, les changements sociaux 1970–1990 chez les Cris et les Inuit du Québec.* Quebec: GETIC.

Sörlin, S. (1988). *Framtidslandet: debatten om Norrland och naturresurserna under det industriella genombrottet.* Stockholm: Carlssons.

Stabler, J. C. (1989). Dualism and Development in the Northwest Territories. *Economic Development and Cultural Change, 37*(4), 805–839.

Storey, K. (2010). Fly-in/Fly-out: Implications for Community Sustainability. *Sustainability, 2,* 1161–1181.

Tynkkynen, V.-P. (2007). Resource Curse Contested – Environmental Constructions in the Russian Periphery and Sustainable Development. *European Planning Studies, 15*(6), 853–870.

Vaughn, R. (2007). *The Arctic: A History*. Chalford, UK: Alan Sutton Publishing.
Waldram, J. B. (1993). *As Long as the Rivers Run: Hydroelectric Development and Native Communities in Western Canada*. New York: University of Manitoba Press.
Watkins, M. (1963). A Staple Theory of Economic Growth. *Canadian Journal of Economics and Political Science, 29*(2), 141–158.
Watkins, M. (Ed.). (1977). *Dene Nation, the Colony Within*. Toronto: University of Toronto Press.
Weller, G. R. (1977). Hinterland Politics – Case of Northwestern Ontario. *Canadian Journal of Political Science-Revue Canadienne De Science Politique, 10*(4), 727–754.

2 The history and historiography of natural resource development in the Arctic

The state of the literature

Ken Coates

Natural resource development is a prominent and controversial topic in the Far North. While media attention focuses largely on the potential for oil and natural gas discoveries – the extension of the Norway-Russia extraction frontier to North American Arctic waters – the active exploration for and development of mineral deposits has come to feature prominently in the region. For the Governments of Nunavut, Northwest Territories and Yukon, and for the communities across the territorial and provincial Arctic regions, the prospect for economic and social improvements based on cooperative exploitation of natural resources is viewed as an important part of the political puzzle in the North. The recent developments, however, build on a long and complicated history of resource development, albeit one marked more by hope and promise than practical results.

This chapter focuses on the history of the Canadian Arctic and does not examine developments in other parts of the Arctic. The Canadian situation, understandably, reflects the realities of the Canadian North, national and territorial policies, and the actions of corporations active in Canada. The analysis shows that historical research to date has focused on government policies and the macro-level study of the mining sector. Only recently have scholars turned their attentions to the central concern of the ReSDA project, namely the impact of resource developments on communities and methods and approaches that might improve community benefits from resource development.

One of the first major forays by Europeans into the Far North, the expedition led by Martin Frobisher that arrived in the Baffin Island region in 1576, returned with what some believed was gold bearing ground. Frobisher secured support for two more expeditions, which focused more on mining than mapping the desired but still unidentified Northwest Passage, and started mining operations around the bay that bears his name. Frobisher's gold turned out to be iron pyrite – Fool's Gold – thereby marking the first, but not the last, time that the promise of Arctic riches would fail to be realized (McGhee, 2001: 200).

This chapter examines the history of natural resource developments in the Arctic.[1] While the geographic scope is broadly defined to include portions of the provincial North in Canada and comparable areas in other Arctic countries, it is important to recognize that the vast majority of the mining activity in the

Canadian North has occurred in the lower reaches of the provincial Norths.[2] The Arctic regions, in comparison, attracted little resource development until after World War II, and even then the attention was sporadic and typically short term. Much of the scholarship on northern resources development, therefore, focuses on more southerly districts and not on the Arctic as that geographic term is generally understood. This holds, incidentally, for much of the Circumpolar World, with Russia being a major exception. This chapter will look, specifically, at the contributions of existing historical literature to our understanding of the impact of resource developments on northern communities. In this way, the chapter will describe how historians and historically minded scholars (geographers have long been vital to the historical understanding of the North)[3] have contributed to the collective understanding of the problems for northern communities associated with resource development and the more constructive impacts that resource developments have played in shaping the human history of the region.

For the purposes of this chapter, natural resources will be taken to include only non-renewable resources. There is, of course, a rich literature on such economic development sectors as fur trading, northern agriculture, fishing, whaling and forestry.[4] This is very important scholarship and has contributed significantly to our understanding of the evolution of northern societies. The issues raised by this literature are, however, quite different from those related to the extractive industries (mining and oil and gas development). For this reason, the paper will adopt a more narrow focus on mining and oil and gas activities in the Arctic regions.

Despite the long-term significance of natural resource development to the Canadian society and economy, historians have devoted a surprisingly limited amount of attention to the topic. There are some major exceptions, such as the excellent work done by Morris Zaslow (Zaslow, 1971: 339; Zaslow, 1988: 332) and Parks Canada historians on the mining aspects of the Klondike Gold Rush (Neufeld and Norris, 1996: 182; Guest, 1985; Stuart, 1980) and, more recently, top notch scholarship by Kerry Abel, John Sandlos, Arn Keeling and Liza Piper on aspects of post-World War II mining activity in the North (Sandlos and Keeling, 2012 and 2015; Boutet, Keeling and Sandlos, 2015; Cater and Keeling, 2013; Keeling, 2012; Piper, 2004 and 2009; Piper and Sandlos, 2007). Mining, however, has not attracted systematic scholarship, largely because of the southern and urban focus of the Canadian historical profession. Moreover, the presence of the remarkable documentary record left by the Hudson's Bay Company has encouraged an historiographical preoccupation with the fur trade era that, with archival material much more scattered and limited in scope and volume, has never occurred with the mining sector. Indeed, the historical documentation on the mining community is sparse indeed, particularly when mining communities closed down following the closure of the mines.[5]

Until the arrival of more north-centred scholars in the 1970s and 1980s, northern historiography also lacked the regional orientation that might have led scholars to examine the impact of mining on the people, communities, ecosystems and overall economy (Coates and Morrison, 1985; Grant and Hodgins, 1986). Northern mines were isolated, their populations very transient, and the mining communities themselves often short-lived. Without a regional imperative driving the historical agenda, the nuances of hard-rock mining camps, Indigenous displacement by mining activity and community formation and restructuring associated with the resource cycles seemed of little interest in the broader national narrative. This is starting to change, a function of more regionally engaged scholarship, growing national interest in northern resources and ongoing debates about the socio-economic benefits of resource activities on northern peoples and ecosystems.

Natural resource development in the Arctic

From Frobisher's humble and forgettable beginnings, Arctic natural resource development got off to a very slow start. Generations would pass before outsiders turned their attentions more systematically to the mineral potential of the Far North. There were, as Morris Zaslow has documented, forays into the middle and provincial Norths in the nineteenth century, most notable perhaps being the steady northward extension of the placer gold mining activities during the Cariboo Gold Rush in what became British Columbia. The development of Barkerville in the middle of British Columbia in the 1860s hardly constituted a northern development, but it directed prospectors' attentions further northward (Barman, 1991). In the 1870s, the first prospectors reached the upper Yukon River basin and, after more than 20 years of hard-scrabble work, touched off the Klondike Gold Rush with the discovery at Rabbit (Discovery) Creek in 1896 (Webb, 1993; Wright, 1976; Coates and Morrison, 2005; Blum, 2011; Grey, 2010). The boom that followed – based on real gold and not Frobisher-like ore – was as important for how it changed international assumptions about the Far North as for the millions of dollars of gold dust and nuggets wrestled from the permafrost.

Mining interest in the North escalated during and after the Klondike Gold Rush, with prospectors fanning out across vast territories in search of the next bonanza. They found pockets of gold and, over time, hard-rock mineral properties like the lead-zinc at Elsa-Keno in the Yukon, copper near Whitehorse, gold at Yellowknife and uranium at Great Bear Lake (Hunt, 1974). The discovery of oil at Norman Wells along the Mackenzie River added a new commodity into the mix and generated a surge of interest in northern petroleum. None of these properties matched the Klondike discovery in scale or impact, but the identification of commercial grade deposits kept dozens of prospecting teams in the field. Across the provincial North, particularly in northern Ontario and Manitoba, promising discoveries matched with a speculative zeal generated ongoing pre-World War II interest in sub-Arctic mining and saw several mines come into production. The war accelerated the development of the Eldorado Mine on Great Bear Lake and resulted in a massive but ill-conceived American plan to delivery Norman Wells oil to a poorly designed refinery in Whitehorse, Yukon, a wartime experiment that was abandoned shortly after the war (Bothwell, 2011; Coates and Morrison, 1992; Murphy, 1948; Nassichuk, 1987).

Mining activity accelerated after the Second World War, due to a fortuitous combination of surging continental and international demand, improved exploratory and extractive technologies, higher prices and Cold War military imperatives. Resource booms swept across the North, from Labrador to British Columbia, and through the three territories. Company towns emerged, like Schefferville, Quebec, Thompson, Manitoba, Uranium City, Saskatchewan, Granisle, British Columbia, Pine Point, NWT[6] and Faro in the Yukon. At Rankin Inlet, a nickel and copper property became the first to make extensive use of Inuit miners in their operations. The resource projects attracted a great deal of subsidiary investment, much of it by governments, in road, rail, airfields, telecommunications and power-generating capacity. The resource boom did not last. By the 1980s, many of the mines and company towns had been closed, the grand promise of long-term prosperity fuelled by resource development in tatters. Promising individual discoveries and developments followed, including the Voisey Bay in Labrador, the Mary River project on Baffin Island and most famously the Diavik Diamond Mine in the Northwest Territories, the first of the Canadian diamond mines (Gibson, 2006; Wismer, 1996; Hood and Blaikie, 1998; Nassichuk, 1987; Brigham, 2013; Hall, 2013; Galbraith et al., 2007).

The oil and gas industry followed a similar cycle, with a slow start in the 1950s and 1960s, vast Beaufort Sea expansion plans in the 1970s (including the Mackenzie Valley Pipeline project) and a sustained downturn in the 1980s and 1990s. Panarctic Oils and other companies ran test projects in the high Arctic Islands, but the grand vision of national prosperity based on northern resource development faded (Page, 1986; Lackenbauer, 2009; Isard, 2010). The most successful initiative – the oil sands around Fort McMurray, Alberta – also proved to be the most controversial, attracting considerable local opposition, particularly from Aboriginal communities, and from the international environmental movement. The oil sands also proved central to national economic development plans in the early twenty-first century (Ferguson, 1986; Cowie, James and Mayer, 2015; Atkins and MacFadyen, 2008).

A quick glance at the natural resource landscape across the Canadian North reveals as many closed and abandoned mines as active ones, grandiose visions of resource-driven prosperity and dashed dreams of regional expansion. Several of the mines – Thompson, Manitoba, Fort McMurray and a few others – have had substantial lives. But many, including the planned natural gas developments in the Beaufort Sea, absorbed billions of dollars in investment but did not produce lasting northern prosperity. The vast majority of the mines were located far from established settlements. In some instances, like Rankin Inlet, Aboriginal people coalesced around the mining developments (Keeling and Sandlos, 2009; Sukuk and Blakney, 2008; Rodon and Lévesque, 2015; Wonders, 2003: 449). The inability to sustain the company towns created from the 1950s through to the 1970s convinced mining firms to look for alternate models. The firms increasingly turned to the transient workforce approach, typically described as fly in/ fly out, offering high wages and excellent local conditions for workers who generally maintained their southern permanent addresses (Finnegan and Jacobs, 2015). These same mines have developed strong collaborative (or impact and benefit) agreements with regional Indigenous communities, offering skills training and jobs, business preferences and other economic benefits.

Major themes in the historiography of natural resource development, northern peoples and community development

There are some key themes that emerge from a review of the scholarly literature on northern resource development. The list is designed to illustrate some of the general patterns, leading to the last part of the chapter which focuses on major gaps requiring additional scholarly research.

Natural resources and community development have attracted comparatively little research. The first theme is the most disappointing. In contrast to other prominent themes in northern history, particularly the fur trade, Christian missionaries and the role of the Government of Canada, mining has not drawn sustained scholarly interest. The limited amount of scholarly writing, while often very good, has underscored the need for much more research on this subject. This problem is not unique to the North. Mining remains seriously understudied in the historical profession at large, with little sustained interest in community formation and social relations associated with mining activity. The problem, it seems, rests with the comparatively low status of mining in the broader scholarly community, a most unfortunate trend that needs to be addressed.

Favourable portrayals of developers. After the 1970s the advent of ecological awareness and greater concern for Indigenous peoples turned scholars' attentions toward the disruptive impact of natural resource development. Before that time, resource development was viewed in a more pro-development and celebratory mode. The prospectors and developers were viewed in generally positive terms, as leaders in the economic development of the region and country. Throughout this period, with the country encouraging resource development and with government policy oriented towards "opening" the North, the individuals behind mineral and oil and gas developments were singled out for attention, much of it favourable.[7]

Southern-focused scholarship. Led by the invaluable work of Morris Zaslow, the pre-1980s scholarship was dominated by a southern orientation.[8] The focus was squarely on, to use the title of one of Zaslow's major works, *The Northward Course of Canada*. In this important formulation, the scholarly attention was on what the country (by which the author's generally meant Ontario and Quebec) thought about the North and on the agents of government, commerce and culture that were dispatched to the region to extend national influence. The southern-focused scholarship was concerned primarily with relations between the North and the South and much less so on the nuances of regional or community life. Indeed, in much of this work, northerners appear as a largely undifferentiated group, largely passive in their reaction to the forces of development and rapid change.

Policy-oriented work. Much of northern scholarship, and not just on natural resource development, is actually Ottawa-centric, focused primarily on demonstrating and understanding the role of national government policy in the shaping of northern history. From studies on Aboriginal policy through to extensive coverage of Prime Minister John Diefenbaker's interest in northern development and to the lengthy political debates about northern oil and gas exploitation, scholars have retained their interest in federal policies toward the North (Isard, 2010; Lithwick, 1987; Dodds, 2015). This is valuable context-setting work, but it generally lacks two key elements: an examination of how the policies worked on in practice and the responses of northerners to the policies that influenced and shaped the evolution of the region.

Toward critical assessments of resource development. For the past 30+ years, scholarly analysis of northern resource development has shifted from the celebratory tone of the pre-1970 era to the more critical evaluation of mining and oil and gas development. The academy has, in recent decades, taken a more sceptical stance toward the development impulse and has offered more analytical evaluations of short- and long-term environmental impacts, social disruption and contradictory influences on the North itself. This work is part of a much larger analytical re-evaluation of resource and economic development generally, and differs sharply from the more supportive writing of the pre-1970s period.[9]

Toward North and community-centred scholarship. One of the greatest transitions in northern scholarship has been the shift from south-centred studies (concerned primarily with policy and the work of southern actors) to north-centred approaches (concerned principally with the socio-economic transitions in the North).[10] This work has the added benefit of prioritizing the experiences of Indigenous peoples and focusing on long-term transitions in the region. The scholarship has been driven by the growing number of scholars raised in the North, and reflects their "insider's" perspectives on northern realities. Importantly, the focus on the North also increased the attention given to community-level studies, which emphasize the unique nature of the various cities, towns and villages across the region, differences driven by a combination of Indigenous cultures, economic activity, Indigenous-newcomer relations and historic transitions.[11]

Growing awareness of the impact of resources on Aboriginal communities. From the vantage point of the 2010s, it is difficult to imagine the limited attention given to Indigenous peoples by scholars only half a century ago. Indeed, the pendulum has swung sharply in the opposite direction, to the point that non-Aboriginal people are given much less attention than in the past. Scholarship has now documented the fact that resource activities caused considerable, sometimes devastating, impacts on local Indigenous populations. As Aboriginal voices – as interview subjects, as writers and more regularly now as scholars – have entered the debate, a much greater sense has emerged of the impact of mining, oil and gas activities on Aboriginal communities. This has resulted in growing attention to Aboriginal protests against resource development, particularly related to the Mackenzie Valley Pipeline project.[12] Interestingly, there has been less attention to the engagement of Indigenous peoples as resource workers, a pattern that emerged at the outset of the resource expansion, and that has some important contemporary manifestations. Most importantly, recent scholarship has demonstrated that the transitions associated with major resource projects have been pivotal, along with social welfare programming, in accelerating the transformation of Aboriginal life in the Far North. While the specifics and nature of the transitions remain largely uncharted, there is a growing realization that resource development accelerated the rate and complexity of change in Aboriginal communities associated with the sector.

Newcomer populations, community formation and change, and natural resource development. One of the interesting patterns in northern scholarship is how research on the territories focuses on Aboriginal people while studies of the provincial North has much more comprehensive coverage of non-Indigenous peoples. There are major exceptions to this generalization; the Klondike Gold Rush work is largely focused on newcomers and there is good writing on aspects of northern Manitoba's development.[13] Across the North, however, there is surprisingly limited attention to the experience of newcomers, both in the pre- and post-World War II era. Even the resource boom of the 1950s and 1960s has generally been studied at the macro level, focusing largely on policy issues and industry trends, with much less attention given to community-level and social issues in the non-Aboriginal populations. In this period, newcomers made up the vast majority of the northern population, with a significant number making the transition from short-term resource workers to long-time northerners. The non-Aboriginal people, starting in the nineteenth century, both shaped the evolution of the North and were transformed by the North themselves. To a quite surprising degree, historians have devoted comparatively little effort to understanding the role of non-Aboriginal northerners in the resource sector. There has been some work on single industry and company towns, although little of this was done by historians. Indeed, the expansion of fly in/fly out operations has lessened interest in the company town phenomena, which is increasingly looking like a time-limited experiment in northern and resource social organization. At present and historically, the North has been shared space from the fur trade era on, with prominent non-Aboriginal populations across much of the North. Ultimately, the historical understanding of the North must draw on the experiences of Aboriginal people, newcomers and the interactions between Indigenous and non-Indigenous peoples. In this important respect, the history of the North remains substantially incomplete.

Infrastructure, natural resources and northern communities. The resource sector brought many changes to the Far North, one of the most important being the development of transportation, communications and community infrastructure. While some of the projects were undertaken for traditional government-centred reasons, such as connecting a distant territory to the rest of the nation, most were connected to resource projects. Even the largest set of infrastructure projects in northern history – the Northwest Staging Route – include a substantial resource component. While the major project, the Alaska Highway, was driven by para-military priorities, the substantial side-initiative, the Canol pipeline and Whitehorse refinery, had its routes in continental excitement about the potential of the Norman

Wells oil field (Coates and Morrison, 1992; Coates and Morrison, 1994). Scholars looking at the post-World War II period have focused on the Diefenbaker government's Road to Resources program, which brought roads and railways into the North. Other crucial initiatives attracted a great deal less attention. These undertakings included the development of port facilities and shipping capabilities generally, the slow and expensive expansion of the communications and electrical systems in the northern communities, and the enhancement of satellite television and, later, Internet services across the region. To this point, historians have written little on these vital developments and have not contributed a great deal to the understanding of how infrastructure investments responded to community needs and demands, how the national government prioritized and funded northern projects and how these infrastructure initiatives reshaped the Far North. This is one of the major lacunae in the historical understanding of the Canadian North, and requires much more attention than it has received to this point.

Arctic environments and the long-term effects of natural resource development on northern peoples. The combination of resource development, climate change/ global warning and greater scientific understanding of Arctic ecosystems has sparked global interest in the environment in the Far North. Much of the southern and international engagement in the Arctic is tied to global concern about Arctic sustainability and the unique vulnerability of the ecosystems of the region.[14] In the 1950s and 1960s, analysts paid more attention to the best means of developing the North than to the proper approaches to protecting the environment. A series of major initiatives, including the proposed Mackenzie Valley Pipeline and the massive James Bay hydroelectric developments (and comparable hydro projects across the Middle North) (Richardson, 1975; Manore, 1999; Berkes, 1988; Rynard, 2000) and the oil sands in northern Alberta, generated widespread interest in northern ecosystems and, in the process, spurred scholarly and general interest in Arctic ecology. Historians have responded impressively to this growing concern about the Far North, studying the long-term effects of the Klondike Gold Rush, the environmental and socio-environmental impacts of specific mining projects and the effect of the industrial economy generally. Historians have a particularly important opportunity to contribute at the micro-level to the greater understanding of the environmental manifestations of resource development. Given the intense global interest in Arctic adaptation, resilience and ecological change, it is likely that the historical work will continue to expand and contribute to the better understanding of the environmental effects of resource development.

Historians and the natural resource evaluation process: why have historians' been largely silent? Northern development is highly contentious. Many Aboriginal communities support resource development; others are either cautious or opposed to a specific project or to extraction activities in general. All across the North, debate continues about northern pipelines, mines and oil and gas development. The courts are involved in the process, as are politicians at multiple levels. To date, historians have been comparatively silent in this often controversial and very important regional conversation. Historians, it must be noted, have not been overly prominent in public policy debates generally, so the concern is not limited to the North and resource development. The lack of historical engagement works in the opposite direction, as government officials and industry proponents have demonstrated little interest in the history of northern resource development. The historical silence or, to be more fair, near silence, has implications for the North. Research on early infrastructure projects demonstrated the transformative effects of roads, railways and other initiatives on the region. Detailed studies of Aboriginal engagement or, conversely, marginalization demonstrate the manner in which Indigenous peoples and communities responded to major economic and environmental changes. Historians believe, with good reason, that historical understand is central to a society's abilities to confront the present and major constructive choices about the future. There is ample evidence that such historical work would make valuable contributions to contemporary conversations about the future resource development in the Far North. Greater historical study is required, perhaps as outlined below, but this has to be matched by increased engagement by historians in the debates about resource activity in the Arctic.

Gaps in historical research on natural resources and Arctic communities: preliminary themes

The commentary above focuses on the work that historians have undertaken related to resource development in the Far North. The scholarship is spotty and irregular, with areas of real contribution and surprisingly large gaps in the historical work. While the list of possible subjects is long – and other scholars could readily add other items – it provides an indication of the possibilities for further historical research on the history of resource development in the North. More specifically, this list and the discussion above highlight ways in which historians have, could and should contribute to the region-wide debate about the impact of resource projects on Arctic communities. These are the areas that seem to require urgent and sustained historical attention.

Non-Aboriginal communities and resource-dependant societies. We know far too little about the non-Aboriginal people involved in resource-dependant societies. There is very little work to date on labourers involved with resource development, the private business people who worked alongside the mining companies and the social dynamics of the communities themselves. It is vital that we understand more about how these communities functioned, about social and economic relationships among the resource workers and others and about wages, working conditions and other aspects of camp life in the North. There is a particularly urgent need to look into the operations of labour organizations, particularly after World War II, in enhancing the working and social conditions for working people and their families. A surprisingly limited amount of scholar work has been devoted to the people who constituted the majority of the resource work force and who, across much of the North, made up the majority of the total population.

Transiency in northern resource communities. The northern resource sector is notoriously transient. Workers come North to "make a killing, not to make a living," and typically leave on a seasonal basis or after a few years of high-grading the resources of the North. The contours of demographic transiency are poorly understood and need far greater analysis as this pattern is among the most important social characteristics of the North.[15]

Pre-World War II resource communities. While there has been some scholarly work on the post-World War II resource centres, study of the pre-World War II mining and oil camps is extremely thin. The historical amnesia about this topic, save for the Klondike Gold Rush, means that we have a skewed sense of the historical evolution of northern resource communities and, indeed, the whole resource economy in the region.

Value systems (Indigenous, newcomer populations, outside corporations, governments, Southerners generally) related to Arctic Resource Development. Interestingly, few scholars have offered much more than superficial comments about one of the most important aspects of northern resource development, namely the attitudes and assumptions the various participants hold about the sector. The result is recourse to stereotypes and generalities. We need to know much more about how participants, local and distance, viewed the resources and the development processes in the Far North.

Impact of Pre-World War II resource developments on Aboriginal peoples. It is fascinating that the national preoccupation with the impact of the fur trade and contemporary concern about the impact of resource activities has not been matched by an historical concern about the nature of social, economic and cultural changes associated with resource activity in the North. Comments available to date are limited in scope and depth.

Wage economies in the natural resource sector, including the role of unions in Arctic development. Workers, oddly, have attracted very little attention in the Far North and, more generally, outside the industrial and manufacturing sector. There is a great deal to be learned from the operation of the wage economy in the sector, with a particular need to understand the role of unions in post-World War II Arctic development activity. This is one of the most important themes in post-war history but it has attracted little attention.

High Arctic resource developments, focusing on nearby Indigenous communities and temporary workers. To the degree that northern resource activity has attracted any attention it has focused on the middle North regions. The high Arctic activities, such as oil and gas exploration in the 1970s and 1980s, have not yet drawn scholarly study, particularly as this relates to the Aboriginal communities in the North.

Coping with winter and Arctic conditions: the effects on northern communities. One of the great oddities of northern scholarship is that very little attention has been paid to the impact of winter, easily the dominant characteristic of northern life. The specific study of the impact on winter and Arctic conditions on the natural resource sector is a matter of high priority, but has yet to draw serious or systematic research.

Private and public sector investments in infrastructure and the impact of these investments on Arctic communities. Each northern mining project attracted significant private and public infrastructure investment. In most instances, the development of a road, railway, airport of other facilities had a substantial and often negative impact on nearby Aboriginal communities. The nature and extent of that impact remains little known.

Indigenous engagement with the resource economy. Recent research in British Columbia has shown that Aboriginal people have historically been more deeply engaged in resource activities than is generally believed. Working from the apparent assumption that Aboriginals were not engaged in the sector, scholars have so far given little attention to the degree and nature of Indigenous participation.

Environmental impact of /resource activities on northern ecosystems, Indigenous peoples and Arctic communities. The crux of much of the debate about northern resource development is that these initiatives have significant impacts on the environment, communities and peoples of the North. Surprisingly, there has been little historical study to determine the impact of such development projects in the past, which could shed important light on the long-term impact of mining and other resource activity.

Indigenous protests against resource development. Canadians were shocked when Aboriginal people protested the construction of the Mackenzie Valley Pipeline and managed to stop the project. Very little is known about Indigenous responses – protests, support or indifference – to other northern mining project. This is an important theme that warrants much deeper investigation.

Government policies re: community formation, servicing and transitions in Arctic regions. The study of federal government policy (and provincial government policy in the northern provinces) has typically focused on broad questions of strategy and general regional management. Much less attention has been given to community-level strategies and the activities associated with support, serving and sustaining northern communities, specifically those affected by major resource projects.

Mine/Project-specific studies of the impact of resource developments in the Arctic In recent years, specific studies of the Pine Point mine have added dramatically to our understanding of northern mining activity. We need many more such studies, particularly of the smaller and less well-known resource projects and the early stage developments of places such as the Norman Wells field. These are essential building blocks in the attempt to understand the full impact of resource development.

Resource speculation and Arctic communities (particularly related to oil and gas). Over time, a great deal of time and effort is devoted to projects – like High Arctic natural gas the Beaufort Sea oil and gas – that do not result in ongoing resource activity. The period of speculation, investigation, preliminary testing and heavy promotion is vital to the development process and can have sizeable impacts on associated communities.

The military, strategic considerations and major investments in northern natural resource development and infrastructure. In many northern countries, resource activities have vital strategic considerations and, further, military investments can spark expanded resource development. The contrast between the United States and Canada in this regard is notable and is worthy of expanded study. To date, the strategic work on the Arctic has focused on international relations and much less on the practical impact of military engagement on the region.

The advent of environmental remediation in the Arctic. Environmental assessment is an integral part of northern development, and is a key factor in the establishment, monitoring and remediation of mining and other project sites. Little is known about the early stages of remediation and the emergence of environmental assessment and remediation as a political, economic and environmental force in northern life.

Fly in-Fly out camps and the transformation of northern societies. Fly-camps now dominate the operation of the northern resource frontier. The concept of mobile workforces is not particularly new, however. Much more work is required on the origins of the mobile work force and the implications of this labour management approach for the social and economic development of the North, particularly as this relates to northern and Aboriginal communities.

Aboriginal land claims and land claims settlements and the development frontier. Since the 1970s, Aboriginal land claims (and, later, settlements) have reshaped the development process in the North. Sufficient time has passed that historians should expand their studies of the relationship between land claims, land claims agreements and implementation measures and the exploitation of northern resources. This kind of contemporary history could be extremely valuable in understanding the future of resource activity in the Arctic.

Comparative aspects of northern resource development and Arctic communities. Resource development is an international enterprise. Capital, professional expertise and the workforce are truly global in nature. It was not always as international as the sector is in the twenty-first century, but we have a limited sense of how resource development occurred in other Arctic and sub-Arctic environments. Comparative historical study remains quite limited, despite the fact that it is well-suited for such a global industry as resource extraction, with historic and contemporary international connections.

History, historians and the improvement of community benefits

The historical profession has been less than fully engaged in the debates about the current and future challenges of Canada and other nations. The nature of the discipline – backward-looking by definition and generally disconnected from public policy debates – favours informing the public about previous developments rather than focusing specifically on reshaping the present and influencing the future. As the recent contributions to contemporary debates from Niall Ferguson, Simon Schima, Jared Diamond and Margaret MacMillan demonstrate, historians can be of fundamental importance in defining and clarifying public policy issues (Ferguson, 2012 and 2014; Schima, 2003 and 2004; Diamond, 2005; Liu and Diamond, 2005; MacMillan, 2003). In a different vein, historians are playing major roles in helping the Canadian and international legal community understand and redefine Aboriginal legal and treaty rights. The idea of the policy-engaged historians, not commonplace only twenty years ago, is becoming much more familiar.

The situation facing northern, largely Aboriginal, communities seeking to capitalize on the opportunities presented by resource development lends itself to historical engagement. Aboriginal peoples themselves take a strong historical view on contemporary opportunities. Community memories are strong, extending across many generations, and there is a strong tradition of examining current opportunities in light of earlier experiences. There should, as a consequence, be a receptive local and regional audience for historical studies of the impact of resource projects on local communities. In addition, historians are well placed to look at specific or region-wide developments in a detailed, multi-party perspective, utilizing company documents, government materials and interviews with a variety of participants to determine the local impacts of resource activities. Historians have the opportunity, for example, to compare the effects on Aboriginal communities of long-running mining operations, such as those in the Dawson City gold fields, the Elsa-Keno properties and the Yellowknife mine and contrast those with short-term or episodic developments, including Whitehorse Copper, Cantung, Cyprus-Anvil and others. There is now sufficient passage of time to test various interventions, ranging from corporate hiring policies, government training programs, community-based initiatives and corporate financial and other commitments to communities. Very few of these were on offer in the 1950s. Beginning in the 1960s, companies and governments started paying attention to community impacts, both in preventative terms (isolating mining camps from Indigenous settlements) and by promoting participation.

Arctic communities considering resource development in their traditional territories need to know what has worked – and what has not worked in other jurisdictions, with other Indigenous populations, and with various resource projects. At present, information is substantially anecdotal, without the substantial and sustained analysis that an historical investigation would produce. One worries, in fact, that major contemporary decisions are being made in the absence of historical understanding, particularly of post-World War II resource developments and initiatives designed to encourage or support Aboriginal engagement. It is important, therefore, for historians to examine the resource frontier from the community perspective, to determine what has worked and what has been less successful. A detailed evaluation of the community impact of specific resource developments, combined with a Northern Canadian and comparative Arctic study of resource activity generally, could provide crucial to Aboriginal decision-making going forward and could produce a substantial quantity of solid, evidence-based advice on how to evaluate current and future possibilities.

Concluding comments

This overview of historical work on northern resource development and suggestions about areas for future historical studies is, by nature, individual and idiosyncratic. Other scholars, and there are fortunately many fine historians working on northern themes, would undoubtedly have a different set, equally valid and rich with scholastic opportunity. What stands out is that historians have not been particularly active in studying resource development and, in particular, the impact of such projects on Arctic residents, communities and environments. Historians believe that historical understanding, while worthy on its own, contributes substantially to contemporary awareness and forward planning. If the Arctic is to better understand the likely implications for the North from expanded resource development, it follows that a greater appreciation for historical patterns and processes could and should be invaluable.

Historical research, of course, requires substantial documentary and other records. For many years, the abundant and remarkable records of the Hudson's Bay Company served like a magnet for northern researchers, to the detriment of studies on other areas and topics. There is no comparable set of documentary materials on natural resource developments, although the federal and territorial government collections contain a substantial amount of material. The corporate record, unlike the Hudson's Bay Company's archive, is scattered and far from complete. Many of the pre-World War II projects have precious little documentary materials and even the post-war records are much less complete than historians would like. The richest sources, one that is being tapped by the newest generation of scholars, include oral testimony, photographic records, video materials and physical objects. The passage of time, of course, erodes these sources, creating a sense of urgency about the historians' collective responsibility to gather, preserve and utilize the existing historical records.

Arctic communities approach the prospect of major resource development with nervousness. They want to know how the projects, some of them massive in scale and potential impact, might affect the North and they want and need strategies for mitigating the negative effects and maximizing the positive elements. Understanding the experiences of the past could provide the North with an excellent means of considering contemporary options and planning for the future. It is incumbent on historians, therefore, to both expand their work on northern resource development and find new ways of sharing their scholarship with the affected communities. History matters. To historians falls the task of making history relevant and meaningful to Arctic peoples wrestling with the prospect of rapid resource-driven change.

Notes

1 For overview histories of the North, see Morris Zaslow, *The Opening of the Canadian North, 1870–1914* (Toronto and Montreal, McClelland and Stewart, 1971), 339; Zaslow, *The Northward Expansion of Canada, 1914–1967* (Toronto, McClelland and Stewart, 1988), 423; Ken Coates, *Canada's Colonies: A History of the Yukon and Northwest Territories* (Toronto, James Lorimer & Company, 1985), 251; William Morrison, *True North: The Yukon and Northwest Territories* (Toronto, Oxford University Press, 1998), 202; Shelagh Grant, *Polar Imperative: A History of Arctic*

Sovereignty in North America (Vancouver, Douglas and McIntyre, 2010), 540. For more recent developments, see John D. Hamilton, *Arctic Revolution: Social Change in the Northwest Territories, 1935–1994* (Toronto, Dundurn Press, 1994), 304; and Ken Coates and Judith Powell, *The Modern North: People, Politics and the Rejection of Colonialism* (Toronto, James Lorimer & Company, 1989), 168.

2 See for example Kerry Abel's excellent study *Changing Places*. See also Ken Coates and Bill Morrison, *The Forgotten North: The Forgotten North: A History of Canada's Provincial Norths* (Toronto, Lorimer, 1992), 144; and Ken Coates and W. R. Morrison, *The Historiography of the Provincial Norths* (Thunder Bay, Centre for Northern Studies, 1996), 335.

3 See the excellent contributions of Robert Bone, *The Canadian North: Issues and Challenges* (Toronto, Oxford University Press, 2009), 310.

4 There is superb work available on the fur trade. See, for example A. J. Ray, *I Have Lived Here Since the World Began: An Illustrated History of Canada's Native people* (Toronto, Key Porter Books, 2005), 422. See also Frank Tough, *As Their Natural Resources Fail: Native Peoples and the Economic History of Northern Manitoba, 1870–1930* (Vancouver, UBC Press, 1996), 292; James Waldrum, *As Long as the Rivers Run: Hydroelectric Development and Native Communities in Western Canada* (Winnipeg, University of Manitoba Press, 1988), 253.

5 One significant exception relates to Cassiar, British Columbia. When the mining company was in the final stages of wrapping up their Cassiar operations, the University of Northern British Columbia arranged for a large volume of company documents to be transferred to the university.

6 See the work by Arn Keeling, Sandlos and Piper.

7 There is not a great deal of writing on northern mining developers, but see Lewis Green, *The Gold Hustlers* (Alaska, Northwest Books, 1977), 339.

8 This issue is examined in Coates and Morrison, "Northern Visions: Recent Writing in Northern Canadian History", *Manitoba History 10*(Autumn 1985): 2–8; and Coates and Morrison, "Writing the North," in Sherrill Grace, ed., *Essays in Canadian Writing* (1997).

9 The work cited earlier by Neeling, Piper, Sandlos are excellent examples of this approach.

10 Many examples of this scholarly approach can be found in the *Northern Review*, the only Canadian scholarly journal based North of 60. See "The Northern Review", Yukon College website, accessed October 15, 2015, http://yukoncollege.yk.ca/index.php/research/pages/the_northern_review.

11 One of the best works of this type, albeit with little coverage of resource issues, is Julie Cruikshank, *Life Lived Like a Story: Life Stories of Three Yukon Native Elders* (Lincoln, University of Nebraska, 1990), 428.

12 See Thomas Berger, *Northern Frontier, Northern Homeland* (Douglas & Macintyre, 1971), 280.

13 On the Klondike, see the excellent study by Charlene Porsild, *Gamblers and Dreamers: Women, Men, and Community in the Klondike* (University of British Columbia Press, 1998), 264. See also Ken Coates and William R. Morrison (eds.), *The Historiography of the Provincial Norths* (Thunder Bay, Ont. Centre for Northern Studies, Lakehead University, 1996), 335.

14 See the excellent work associated with the international project, "Network in Canadian History & Environment (Niche)", accessed October 15, 2015, http://niche-canada.org/about/.

15 An early effort on this theme is Coates and Morrison, *Sinking of the Princess Sophia: Taking the North Down With Her* (Fairbanks, University of Alaska Press, 1991), 220.

References

Atkins, F. and A. MacFadyen, 2008. "A Resource Whose Time Has Come? The Alberta Oil Sands as an Economic Resource." *The Energy Journal 29* (Special Issue): 77–98.

Barman, J. 1991. *The West Beyond the West: A History of British Columbia*. Toronto: University of Toronto Press.

Berger, T. 1971. *Northern Frontier, Northern Homeland*. Toronto: Douglas & Macintyre.

Berkes, F. 1988. "The Intrinsic Difficulty of Predicting Impacts: Lessons from the James Bay Hydro Project." *Environmental Impact Assessment Review 8*(3): 201–220.

Blum, H. 2011. *The Floor of Heaven: A True Tale of the Last Frontier and the Yukon Gold Rush*. New York: Crown/Archetype.

Bone, R. 2009. *The Canadian North: Issues and Challenges*. Toronto: Oxford University Press.

Bothwell, R. 2011. *Eldorado: Canada's National Uranium Company*. Toronto: University of Toronto Press.

Boutet, J.-S., A. Keeling and J. Sandlos, 2015. "Mining and the Aboriginal Social Economy." In C. Southcott (ed.) *Northern Communities Working Together: The Social Economy of Canada's North*. Toronto: University of Toronto Press, 198–227.

Brigham, L. 2013. "Arctic Marine Transport Driven by Natural Resource Development." In S. Majuri (ed.) *Baltic Rim Economies Expert Articles 2013*. Turku: Pan – European Institute.

Cater, T. and A. Keeling. 2013. "'That's Where Our Future Came From': Mining, Landscape, and Memory in Rankin Inlet, Nunavut." *Etudes/Inuit/Studies 37*(2): 59–82.

Coates, K. 1985. *Canada's Colonies: A History of the Yukon and Northwest Territories*. Toronto: James Lorimer & Company.

Coates, K. and W. Morrison. 1985. "Northern Visions: Recent Writing in Northern Canadian History." *Manitoba History 10*(Autumn): 2–8.

Coates, K. and W. Morrison, 1991. *Sinking of the Princess Sophia: Taking the North Down With Her*. Fairbanks: University of Alaska Press.

Coates, K. and W. Morrison. 1992. *The Alaska Highway in World War II: The US Army of Occupation in Canada's Northwest*. Norman, OK: University of Oklahoma Press.

Coates, K. and W. Morrison. 1992. *The Forgotten North: A History of Canada's Provincial Norths*. Toronto: Lorimer.

Coates K. and W. Morrison, 1994. *Working the North: Labor and the Northwest Defence Projects, 1942–1946*. Fairbanks: University of Alaska Press.

Coates, K. and W. Morrison. 1996. *The Historiography of the Provincial Norths*. Thunder Bay: Centre for Northern Studies.

Coates, K. and W. Morrison. 1996. "Writing the North." In S. Grace (ed.) *Essays in Canadian Writing*, Toronto: ECW, 59.

Coates, K. and W. Morrison, 2005. *The Land of the Midnight Sun: A History of the Yukon*. Montreal: McGill-Queens University Press, 2005.

Coates, K. and J. Powell. 1989. *The Modern North: People, Politics and the Rejection of Colonialism*. Toronto: James Lorimer & Company.

Cowie, B., B. James and B. Mayer, 2015. "Distribution of Total Dissolved Solids in McMurray Formation Water in the Athabasca Oil Sands Region, Alberta, Canada: Implications for Regional Hydrogeology and Resource Development." *AAPG Bulletin 99*(1): 77–90.

Cruikshank, J. 1990. *Life Lived Like a Story: Life Stories of Three Yukon Native Elders*. Lincoln: University of Nebraska.

Diamond, J. 2005. *Collapse: How Societies Choose to Fail or Succeed.* Toronto: Penguin.

Dodds, K. 2015. "Northward ho! Obama, Diefenbaker and the North American Arctic." *Polar Record*: 1–4.

Ferguson, B. 1986. *Athabasca Oil Sands: Northern Resource Exploration, 1875 to 1951.* Regina: Canadian Plains Research Center.

Ferguson, N. 2012. *Civilization: The West and the Rest.* New York: Penguin Books.

Ferguson, N. 2014. *The Great Degeneration: How Institutions Decay and Economies Die.* New York: Penguin Books.

Finnegan, G. and J. Jacobs, 2015. "Canadian Interprovincial Employees in the Canadian Arctic: A Case Study in Fly-in/Fly-out Employment Metrics, 2004–2009." *Polar Geography 38*: 1–19.

Galbraith, L., B. Bradshaw and M. Rutherford, 2007. "Towards a New Supraregulatory Approach to Environmental Assessment in Northern Canada." *Impact Assessment and Project Appraisal 25*(1): 27–41.

Gibson, R. 2006. "Sustainability Assessment and Conflict Resolution: Reaching Agreement to Proceed with the Voisey's Bay Nickel Mine." *Journal of Cleaner Production 14*(3–4): 334–348.

Grant, S. 2010. *Polar Imperative: A History of Arctic Sovereignty in North America.* Vancouver: Douglas and McIntyre, 173–188.

Grant. S. and B. Hodgins. 1986. "The Canadian North: Trends in Recent Historiography." *Acadiensis 16*(1): 173–188.

Gray, C. 2010. *Gold Diggers: Striking it Rich in the Klondike.* Toronto: HarperCollins.

Green, L. 1977. *The Gold Hustlers.* Alaska: Northwest Books.

Guest, H. 1985. *A Socioeconomic History of the Klondike Goldfields, 1896–1966,* Microfiche Report Series, no. 181. Ottawa: Parks Canada.

Hall, R. 2013. "Diamond Mining in Canada's Northwest Territories: A Colonial Continuity." *Antipode 45*(2): 376–393.

Hamilton, J. 1994. *Arctic Revolution: Social Change in the Northwest Territories, 1935–1994.* Toronto: Dundurn Press.

Hood, B. and G. Raikie, 1998. "Mineral Resource Development, Archaeology and Aboriginal Rights in Northern Labrador." *Études/Inuit/Studies 22*(2): 7–29.

Hunt, W. 1974. *North of 53°: The Wild Days of the Alaska–Yukon Mining Frontier, 1870–1914.* New York: Macmillan.

Isard, P. 2010. "Northern Vision: Northern Development during the Diefenbaker Era." Master of Arts Thesis, History. Waterloo: University of Waterloo.

Keeling, A. 2012. "Mining Waste" and "Sewage." In C. Zimring (ed.), *SAGE Encyclopedia of Consumption and Waste.* London: Sage, 553–556, 799–801.

Keeling, A. and J. Sandlos. 2008. "Environmental Justice Goes Underground? Historical Notes from Canada's Northern Mining Frontier." *Environmental Justice 2*(3): 117–125.

Kerry, A. 1993. *Changing Places. History, Community, and Identity in Northeastern Ontario.* Montreal: McGill-Queen's University Press.

Lackenbauer, W. 2009. *From Polar Race to Polar Saga: An Integrated Strategy for Canada and the Circumpolar World.* Toronto: Canadian International Council.

Lithwick, H. 1987. "Regional Development Policies: Context and Consequences." In W. Coffey and M. Polese (eds.), *Still Living Together. Recent Trends and Future Directions in Canadian Regional Development.* Montreal: Institute for Research on Public Policy, 121–155.

Liu, J. and J. Diamond. 2005. "China's Environment in a Globalizing World." *Nature 435*(7046): 1179–1186.

MacMillan, M. 2003. *Paris 1919*. Toronto: Random House.

Manore, J. 1999. *Cross-Current: Hydroelectricity and the Engineering of Northern Ontario*. Waterloo: Wilfrid Laurier University Press.

McGhee, R. 2001. *The Arctic Voyages of Martin Frobisher: An Elizabethan Adventure*. Montreal and Kingston: Canadian Museum of Civilization and McGill-Queen's University Press.

Morrison, W. 1998. *True North: The Yukon and Northwest Territories*. Toronto: Oxford University Press, 1998.

Murphy, R. 1948. "Eldorado mine." *Structural Geology of Canadian Ore Deposits. The Canadian Institute of Mining and Metallurgy, Geology Division*: 259–268.

Nassichuk, W. W. 1987. "Forty Years of Northern Non-renewable Natural Resource Development." *Arctic 40*(4): 274–284.

Neufeld D. and F. Norris. 1996. *Chilkoot Trail: Heritage Route to The Klondike*. Whitehorse: Lost Moose Publishing Company and Parks Canada.

Page, R. 1986. *Northern Development: The Canadian Dilemma*. Toronto: McClelland & Stewart.

Piper, L. 2004. "Backward Seasons and Remarkable Cold: The Weather over Long Reach, New Brunswick, 1812–1821." *Acadiensis XXXIV*(1): 31–55.

Piper, L. 2007. "Subterranean Bodies: Mining the Large Lakes of Northwest Canada, 1921–1960." *Environment and History 13*(2): 155–186.

Piper, L. 2009. *The Industrial Transformation of Subarctic Canada*. Vancouver: UBC Press.

Piper, L. and J. Sandlos, 2007. "A Broken Frontier: Ecological Imperialism in the Canadian North." *Environmental History 12*(4): 759–795.

Porsild, C. 1998. *Gamblers and Dreamers: Women, Men, and Community in the Klondike*. Vancouver: University of British Columbia Press.

Ray, A. J. 2005. *I Have Lived Here Since the World Began: An Illustrated History of Canada's Native People*. Toronto: Key Porter Books.

Richardson, B. 1975. *Strangers Devour the Land*. Toronto: Macmillan.

Rodon, T. and F. Lévesque. 2015. "Understanding the Social and Economic Impacts of Mining Development in Inuit Communities: Experiences with Past and Present Mines in Inuit Nunangat." *Northern Review 41*: 13–39.

Rynard, P. 2000. "'Welcome In, But Check Your Rights at the Door': The James Bay and Nisga'a Agreements in Canada." *Canadian Journal of Political Science/Revue canadienne de science politique 33*(2): 211–243.

Sandlos J. and A. Keeling. 2012. "Claiming the New North: Mining and Colonialism at the Pine Point Mine, Northwest Territories, Canada." *Environment and History 18*(1): 5–34.

Sandlos, J. and A. Keeling, "Aboriginal Communities, Traditional Knowledge, and the Environmental Legacies of Extractive Development in Canada." *Extractive Industries and Society* (accepted June 2015 as part of special issue on Arctic extractive industries).

Sandlos, J. and A. Keeling. 2012. "Claiming the New North: Mining and Colonialism at the Pine Point Mine, Northwest Territories, Canada." *Environment and History 18*(1): 5–34.

Schama. S. 2003. *A History of Britain: The British Wars 1603–1776*. Vol. 2. New York: Random House.

Schama, S. 2005. *Citizens: A Chronicle of the French Revolution*. London: Penguin.

Stuart, R. 1980. "The Underdevelopment of Yukon." Paper presented to the Canadian Historical Association annual meeting, Montreal, June 1980.

Suluk, T. and S. Blakney. 2008. "Land Claims and Resistance to the Management of Harvester Activities in Nunavut." *Arctic 61*(1): 62–70.

Tough, F. 1996. *As Their Natural Resources Fail: Native Peoples and the Economic History of Northern Manitoba, 1870–1930.* Vancouver: UBC Press.

Waldrum, J. 1988. *As Long as the Rivers Run: Hydroelectric Development and Native Communities in Western Canada.* Winnipeg: University of Manitoba Press.

Webb, M. 1993. *Yukon: The Last Frontier.* Vancouver: UBC Press.

Wismer, S. 1996. "The Nasty Game: How Environmental Assessment is Failing Aboriginal Communities in Canada's North." *Alternatives Journal 22*(4): 10–17.

Wonders, W. (ed.). 2003. *Canada's Changing North.* Montreal: McGill-Queen's Press.

Wright, A. 1976. *Prelude to Bonanza: The Discovery and Exploration of the Yukon.* Sydney: Gray's Publishing Ltd.

Zaslow, M. 1971. *The Opening of the Canadian North, 1870–1914.* Toronto and Montreal: McClelland & Stewart.

Zaslow, M. 1988. *The Northward Expansion of Canada, 1914–1967.* Toronto: McClelland & Stewart.

3 Social impacts of non-renewable resource development on indigenous communities in Alaska, Greenland, and Russia[1]

Peter Schweitzer, Florian Stammler, Cecilie Ebsen, Aytalina Ivanova, and Irina Litvina

In July 2009, a public meeting was held in the village of Kivalina, Alaska. It was an opportunity for community members to voice their opinions and concerns regarding the Red Dog Mine's development plans and especially the mine's new waste management permit. Community members were concerned about the mine's impacts on their land, their health, and their community as a whole. A female community member said the following: "For the health and safety of our people and our subsistence way of life, our concerns should be heard and taken seriously, taken into consideration very seriously, for the benefit of our future generations to come" (Public Comments 2009: 19).

A couple of days later a corresponding meeting was held in Kotzebue. A male community member stated:

> I've always liked working in mine companies, you know. I worked for a mining company in Nome in the '50s and I worked for a mining company in Deering . . . I got along real well with most of them. One of the reasons why Deering – it's one of the only villages in the region that don't speak Inupiaq because of the influence we have. We talk English all the time because of the mining. . . . Because of the fact that we were influenced by these people, you know, we figured out something better than what we're doing. Consequently, they lost a lot of us for a while, a lot of our – at that time, we didn't have high schools in our villages.
>
> (Public Comments 2009: 21–22)

These statements are indications of how resource development projects have shaped and are shaping Indigenous communities and Indigenous ways of life. They show that it is of high priority for community members to protect a subsistence lifestyle; that values are changing because of resource development projects; and that people are migrating from rural areas to pursue education, jobs, and new lives because of resource development. Arctic Indigenous people are communicating with resource development industries and governments to maintain a voice in the debate about how to develop and how it might impact community sustainability, culture, and health (Nuttall 2010: 14).

Impacts of resource development are often explored through socio-economic measurements of population influx and outflux, community involvement, previous impact events, occupational composition, local benefits, presence of outside agency, and mitigation measures (Burdge, Field and Wells 1988: 43). What are missing from such explorations are cultural impacts of resource development. These are important to take into account, posing questions such as: what are the cultural values of the community and how are these affected? What cultural traditions will change and how? How are economy and culture interconnected? How can impacts be identified and measured without knowledge of the cultural context of the community?

The following analysis was conducted on the basis of available and relevant scholarly publications in English, Russian, and Danish. Temporally the analysis will focus on impacts of resource development from the 1970s to the present in the Alaskan and Greenlandic cases, and on post-Soviet developments regarding Russia. Data on impacts of resource development on Indigenous communities is primarily found in "grey literature" such as Environmental Impact Statements and Social Impact Assessments (EIS, SIA). Data can also be found in academic and governmental reports and analyses. Transcripts and summaries of public meetings are also a source of data. A small amount of data has been retrieved from newspaper articles.

The remainder of the chapter is organized in the following way. First, a short presentation of the three regional contexts will be provided. Second, an overview over the Alaskan and Greenlandic specifics will proceed along the lines of three topics: i) formal and informal economic impacts, ii) subsistence and cultural values, and iii) human mobility. The Russian specifics will be presented under the following three sub-headings: 1) ethnic demography, health and socio-economic issues, ii) processes in Indigenous cultures and traditional economies, and iii) Indigenous adaptations and future scenarios. The final section of the chapter will bring the different regional contexts together again, to provide overall conclusions and identify research gaps for individual regions and beyond.

Regional contexts: Alaska, Greenland, and Russia

Mining has been a part of Alaska's history since the 1800s while oil was discovered in the 1960s and drilling began on the North Slope in 1971. The contemporary Alaska we are concerned with is about much more than the white male miner from the south looking for gold. Of the 710,231 people who lived in Alaska in 2010, approximately 104,871 are Alaska Native/American Indian (Laborstats.alaska.gov 2010). There are more than 220 Indigenous communities in Alaska all relying to some extent on subsistence from the land for their livelihoods.

Most villages were established where traditional subsistence camps once were as schools and churches were built and children were required to attend school (McClintock 2009: 120–121). Development of natural resources often takes place in remote areas of Alaska on land that Alaska Natives own, use for hunting and fishing, and consider as part of their cultural heritage. In 1971 the Alaska Native Claim Settlement Act (ANCSA) was passed awarding Alaska Natives 44 million acres of land and 962.5 million for giving up claims to the rest (Hensley 2009: 159).

Greenland has seen mining activities since the early 1900s (Haley et al. 2011: 48). Approximately 56,500 people live in Greenland and of these 50,000 are born in Greenland (Nanoq 2012). There are 17 villages/settlements in Greenland, all of which are predominantly Indigenous. A great part of Greenland's modern history cannot be understood without mentioning that Denmark colonized Greenland from 1721 to 1953. In 1953 Greenland was accepted as part of the Danish kingdom but special conditions continued to keep Greenland in a subordinate position (Thisted 2005). In 1979 Greenland was granted home rule, giving Greenlanders more power to make independent decisions about their country and in 2009 Greenland was given self-government. Self-government is an important step towards Greenland becoming independent but they continue to rely on annual subsidies from Denmark. One way to become economically independent is to develop the natural resource industry in Greenland, which is why there has recently been a boom in resource development (Haley et al. 2011: 56–57).

Alaska Natives and Greenlanders have gone from being nomadic to settlers, from hunters to businessmen, from wearing skin clothing to wearing industrial produced clothing, and from eating off the land to, at least partly, eating store-bought foods (Kruse 2011: 10). Over the past 60 years Arctic Indigenous people have undergone tremendous changes affecting their cultures and social organization, but a great threat to their lifestyle still remain: resource development.

The Russian Arctic is a strategically important region which produces about a half of the country's energy resources – oil and gas, creating 12–15% of country's GDP, which accounts for more than a quarter of all Russia's export. In the economic structure of the Russian Arctic gas is the most important industry (estimated more than 80% of Russian gas), second to mining, which is primarily focused of non-ferrous metals, for example the copper-nickel industry of Norilsk in North Central Siberia. The Arctic zone also produces most of the Russian diamonds, 100% of antimony, apatite, phlogopite, vermiculite, barite, rare metals, and more than 95% of platinum, nickel, cobalt and 60% of copper. Moreover, fishery is also very important, counting for more than a third of Russian fishing and about 20% of country's canned fish production. The total value of mineral resources in the Arctic exceeds 30 trillion US$. Two-thirds of this amount belongs to the revenues from fuel and energy resources.

While this contribution refers to examples all over the Russian Arctic, there is a particular emphasis on the West Siberian oil and gas province. This is because many of the social impacts and developments in the field of Indigenous peoples and extractive industries can be best studied there, or only studied there. The reason is the geographical and cultural setting in which extractive industries have developed within this region: within the Russian Arctic. Only in Yamal and the Khanty-Mansi areas of West Siberia have Indigenous livelihoods continued to thrive in direct coexistence with the industry. In other Russian Arctic areas the two groups have often less contact or are spatially further apart. Just west of the Ural Mountains, the European Nenets and Komi also herd their reindeer in direct vicinity of the oil industry, but on a smaller scale than in neighboring Yamal (Habeck 2002; Stuvøy 2011; Wilson 2016; Stammler and Peskov 2008). In Taimyr, the large-scale mining industry (Norilsk Nickel) is concentrated around one big city, where reindeer herding is not practiced any more, and Nganasan and Dolgans engage more in hunting (Ziker 2002; Klokov 1997). In Yakutiya, most of the contacts between Indigenous people and industry evolve around discussions of infrastructure projects (electric lines, railroads, pipelines) (Ivanova 2007; Fondahl and Sirina 2006; Yakovleva 2014; Sidortsov et al. 2016).

Specifics of the Alaskan and Greenlandic situations

A. Formal and informal economic impacts

As noted in the introduction, the consideration of social impacts of extractive resource development was heavily influenced by developments in both Alaska and the Canadian North. The Alaska pipeline and the debate on the Mackenzie Valley pipeline played a major role in institutionalizing the consideration of these impacts in what are now called social impact assessments (Burdge, Field and Wells 1988). Chapter 4 of this volume describes how these SIAs are integrated in the varying versions of Environmental Impact Assessments. What is notable is the emphasis on economic impacts. Formal economic impacts of resource development have been researched extensively focusing on consequences for a state/country as a whole and for local communities. Economic impacts are often the most obvious and most attractive impacts of development and as such a large body of literature focuses on them. Rural communities are often investigated through a rural-urban continuum, where development towards urban/modern living standards is interpreted as positive and desirable (Burdge, Field and Wells 1988: 37). The urban economic model has become the standardized measure for prosperity, which a small community can reach with the help of an economic boost. Literature on formal economic impacts of resource development was analyzed in order to show the kind of research that has been done in impacts of resource development.

In literature from Alaska, the many benefits of resource development are being emphasized and the continued development of mines, oil extraction, and industrial fisheries is being argued for (Baring-Gould and Bennett 1975; Colt and Schwoerer 2010; Goldsmith 2008; Huskey and Nebesky 1979, McDowell Group 2002, 2011, 2012; Rogers 2006; Trenholm 2012; Vining 1974). SIAs prepared for resource development contractors are showing how the mining and oil industries will provide new and more jobs, higher average salaries, and through taxes contribute to local economies (Baring-Gould and Bennett 1975; McDowell Group 2002, 2011, 2012; Rogers 2006; Trenholm 2012). In 1974, Aidan Vining wrote that for the state the only "ideal impact" of the development of the pipeline was an increase in employment opportunities (Vining 1974: 3). Employment is emphasized in all

SIAs; listing how many direct and indirect jobs a development will create and what kind of wages they will have (Colt and Schwoerer 2010; Goldsmith 2008; McDowell Group 2002, 2011, 2012). These reports and analyses do not consider how the informal economies of rural Alaska are being affected by development and how that relates to their overall economies. Subsistence is investigated in terms of access to, abundance of, and quality of subsistence resources (Rogers 2006), but not as an economic factor. Resource development is primarily estimated in terms of economic, and monetary values because this is expected to be the main reason why a local community might accept and promote resource development (Haley et al. 2011: 57).

There is limited evidence that resource development can have positive impacts on the amount of subsistence activities (for the Alaskan North Slope, see various publications by Kruse and Haley, as well as Schweitzer et al. 2014). In a variety of older reports, subsistence harvesting had not been considered an economic system, which might have resulted in downplaying the impacts of resource extraction on it. A single report compares the economic value of a wild salmon ecosystem to the value of the Pebble mine proposal (Trenholm 2012). The report argues that although the economic value of the Pebble mine seems greater now, the long-term value is questionable, while the value of a wild salmon ecosystem will continue, as it is a renewable resource (Trenholm 2012: 81). The wild salmon ecosystem will also maintain and promote social and cultural relationships that a mine will only disrupt. The report emphasizes the contrast between a mine's boom-and-bust cycles and the wild salmon ecosystem's reliability (Trenholm 2012:81). Greenlandic SIAs and other literature regarding impacts of resource cite the main impact of resource development as a positive formal economic impact: jobs will be created and good working relations will be provided for the workers (Watkinson 2009: 7; Grontmij 2012: vii). The recent study "To the Benefit of Greenland" (The Commission 2014) takes an approach that puts local and regional benefits at the center, no matter whether they are economic, cultural or ecological.

Formal and informal economic activities can be difficult to distinguish in Greenland where commercial fisheries are the most important export industry (Frederiksen et al. 2012: 95), and fishing at the same time is an important informal economic source for the individual Greenlandic family (Nuttall 1998: 113). Private catches are both shared and sold on local markets while the government simultaneously encourages people to sell their meat to the national fishing export company Royal Greenland (Nuttall 1998: 118–119).

As described in greater detail in Chapter 4, assessments from Greenland are somewhat differently organized than EISs and SIAs from Alaska. One reason for this might be found in differing governmental and political structures. Assessments from Greenland are more formal in their design. They are written as reports on how already established, homogenous, governmental structures on both local and national levels might be affected by the opening of a mine or oil drilling on the coast. In fact, the cultural implications of resource development are completely left out in some SIAs. Even more problematic is the interpretation of cultural values as cultural heritage sites (Watkinson 2009). In the SIA for the Nalunaq goldmine a paragraph on cultural values states that the only cultural value in the area of the mine is a Norse settlement that they will maintain the access to (Watkinson 2009: 54–55). Culture is exclusively interpreted and reduced to historical material. One reason for the lack of an advocate perspective might be that Greenlanders have already achieved self-government and the need to establish themselves as an oppressed minority in a dominating nation-state has been abandoned (Nuttall 1998: 109).

In many ways Greenland finds itself in the developing phase that Alaska saw in the 1970s. Everyone wants to develop resources in Greenland and the new independent government is working hard to achieve contracts that will benefit Greenlanders as much as possible. They are also just starting to estimate the consequences of resource developments such as the social, economic, and cultural impacts. They are posing questions such as "What are the social consequences of large in-migration of Chinese mine workers on small Indigenous communities?" (Harvey 2012), which is a question similar to that posed by researchers in the 1970s in Alaska asking: "What will happen when a large group of workers from mainland U.S. comes up and live in or near rural Alaska Native villages?" (Vining 1974).

Informal economic activities are generally not considered in the SIAs or other analyses. Indigenous communities in Alaska and Greenland are constantly adapting to the economic, political, social, and cultural environments they are a part of and their small-scale economies depend on their abilities to utilize informal economic means such as subsistence harvesting (Nuttall 1998: 97). They need to adapt to economic boom-and-bust cycles, continue to fight for self-determination, and stimulate local empowerment; all of which is done through informal economic activities (Nuttall 1998: 97). One informal economic activity is subsistence. However, subsistence is much more than just an economic value.

B. Subsistence and cultural values

Chapter 12 of this volume details the need to better understand the importance of subsistence activities in order for communities in the north to properly benefit from resource development. In the context Alaska and Greenland subsistence can have many meanings and values attached to it, but the lack of literature on subsistence as impacted by resource development creates a narrow idea of what subsistence is.

Subsistence is closely connected to and can be considered a cultural value. In his book, "A Yupiaq Worldview" (2006), Oscar Kawagley explains how Alaska Natives share certain worldview characteristics resulting in certain shared values. Alaska Natives all try to live in harmony with their environment: "This has required the construction of an intricate, subsistence-based worldview, a complex way of life with specific cultural mandates regarding the ways in which the human being is to relate to other human relatives and the natural and spiritual worlds" (Kawagley 2006: 8). The shared emphasis on harmony, balance and reciprocity has resulted in some common Alaska Native values (Kawagley 2006: 9).

Among these shared values is the importance of sharing, the importance of cooperation within an extended family and respecting and thanking the universe for what have been given (Kawagley 2006: 10). The emphasis on sharing is also a value among Greenlanders (Nuttall 1998: 83). Sharing among hunter-gatherers has been studied extensively in anthropology. Nurit Bird-David (1992) refers to them as having an economy of cosmic sharing between people, animals, and the environment. However, this is a kind of economy that is not acknowledged in EISs and SIAs.

Similar to Haley and Magdanz (2008), Natuk Lund Olsen writes that Greenlandic culture can be understood through the saying, "To live is to survive" and the only thing keeping death at bay is food (Olsen 2011: 409). Food is not just nutrition; it is also customs, norms, and identity shaping (Olsen 2011: 412; Jeppesen 2008: 96). Quoting Lotte Holm, Olsen explains: "We incorporate our environment in our bodies when we eat – we let the environment pass through us" (Olsen 2011: 413).

Many Arctic Indigenous people live in subsistence dependent communities with an economy based on hunting, fishing, and gathering as well as supplemental wage employment (Haley and Magdanz 2008: 25). Sharing of subsistence foods help reinforce and maintain social relationships, while participating in subsistence activities teach new generations about values and identity. Olsen writes that: "Through the ages gastronomy has proved to be a stronger cultural force among the peoples of the world than linguistic or other influences" (Olsen 2011: 425). The two only things that have survived are the Greenlandic language and Greenlandic food. However, many Greenlanders, especially in the capital Nuuk, do not speak Greenlandic anymore and so Greenlandic food is really the main cultural activity (Olsen 2011: 426). When subsistence is threatened by resource development, whole cultures and economic systems are threatened.

Resource development is impacting subsistence in major ways: by threatening the natural environment and by increasing the cash flow to Indigenous people resulting in changing lifestyles, priorities, and values (Jorgensen 1990: 14–19). EIAs describe how resource development will impact land and animals, which in turn will have impacts on local residents' subsistence hunting, fishing, and gathering. These are described as *potential* problems that continue to be monitored while the actual effects have not been proven (Boertman et al. 1998; ConocoPhilips 2008; Frederiksen et al. 2012; LGL 2012; Perry, Jones and Ward 2010, 2011; Tetra Tech 2009).

In Greenland, noise from oil drilling is defined as an issue as the noise will disturb animals and scare them away (Boertman et al. 1998: 11). Oil drilling and oil spills have an effect on the environment and will therefore impact subsistence and access to subsistence (Bortman et al. 1998: 25, 33). In the EIS for the National Petroleum Reserve in Alaska, ConocoPhilips states that drilling, construction of ice roads, and overland moves might affect subsistence in the area, but that these are expected to only have short-term effects (ConocoPhilips 2008: 26). Tetra Tech has much the same arguments in the 2009 EIS for the Red Dog Mine (Tetra Tech 2009). What is interesting here is that both EISs and presumably a large number of other EISs have held public meetings with residents of affected Indigenous communities. The summaries and comments from these meetings are publically available and they show that local residents are very much voicing concerns about changes to lands and animals. They are worried about the changes they see and want their livelihoods protected as shown in the introduction to this analysis.

In 2011, the Institute of Social and Economic Research (ISER) at the University of Alaska Anchorage published a number of articles discussing the development of an Arctic Subsistence Observation System under the Arctic Observing Network Social Indicators Project (AON-SIP). This project aims at investigating and mapping Arctic change (Kruse 2011). The scope of the project goes beyond changes associated with resource development and includes climate change, political change, and more. However, some of the changes to subsistence that the project has recorded include birds dying from poisoning; foxes with sore-like spots; and changing caribou migrations patterns (Kruse 2011: 16). Phenomena that could be related to resource development and that are threatening subsistence but are not mentioned in impact assessments.

The Exxon Valdez oil spill is an example of the negative impacts resource development can have on Indigenous communities. The oil spill occurring in 1989 in the Prince William Sound of Alaska caused economic, cultural, spiritual, and social disruptions to Alaska Native communities (Gill and Picou 1997; Ritchie 2004). Liesel Ritchie describes how the oil spill had both short-term and long-term consequences for subsistence in South Alaska (Ritchie 2004: 392).

Another consequence of the spill was changes to the Indigenous diet as people were forced to replace subsistence foods with store-bought food. Changing diets can result in a change in values, as subsistence lifestyles are increasingly abandoned or downscaled to part time. Assessments from the 1970s raise concern over how Indigenous communities that have not had much contact with the Western world might change as a result of that contact established through the construction of the pipeline in Alaska. It is clear that an in-depth analysis of contemporary understandings, values, and roles of subsistence in Greenlandic and Alaskan Indigenous communities is necessary in order to understand what impacts resource development will have on such communities. Still, as had been mentioned above, it is evident by now that resource development can also have positive impacts. Or, to put it differently, the prevention of local resource development can be seen as another assault by outsiders, this time in the pursuit of the resource "wilderness" (Brower 2015).

C. Human mobility

Alaska holds a special place in history as the place where people first migrated to settle in America approximately 50,000–15,000 years ago (Langdon 2002: 6). People migrated from Canada to Greenland approximately 4,500 years ago and have been studied extensively in connection to past human migration. Studies of past migration differ widely from the migration that an analysis like the present is investigating, but the focus on past migration in Alaska and Greenland might have overshadowed the importance of present-day migration in connection to resource development.

As noted in Chapter 5 of this volume, in Canada, migration is closely connected to resource development, as workers move to an area being developed and then move away again when the development is downscaled, leaving behind struggling local communities, businesses, and sometimes ghost towns. In Alaska, migration as connected to resource development is not dealt with in-depth. Only one recent social-economic impact assessment mentions migration and here it is only to say that migration is low (Rogers 2006: 77). Looking further back the construction of the pipeline caused some migration as workers followed the construction further north. It did not, however, create ghost towns as in Canada because only temporary camps were established and these were located away from Indigenous communities in the hope that workers would not disturb the ways of life in the villages (Vining 1974). No actual studies of the effects of pipeline work camps on Indigenous communities are available. One study of the effects of pipeline work camps on a non-Indigenous community (Valdez) describes how workers are hard-working, hard drinking, and from outside the region (Baring-Gould and Bennett 1975: 23). Most of them do not participate in local community life aside from buying liquor, books, and clothes. Relationships with locals were minimal but considered friendly or civil and most social problems were found within the camps, not involving residents of Valdez (Baring-Gould and Bennett 1974: 23–25).

In the EIS for the Red Dog Mine extension in 2009 the historical context of the Red Dog Mine was briefly discussed. The following was said about migration:

> Conversely, another potential impact on the cultural integrity of the region is in-migration due to the mine. One of the key concerns brought up at the 1984 public hearings regarding the development of the Red Dog Mine site was the possibility of an influx of people from outside into communities and villages (Public Hearing Transcripts 1984). These individuals have the potential to bring stronger influences of western culture, which could further alter local culture. With this concern in mind, the mine and its related infrastructure was purposefully located away from any established villages to avoid directly impacting any one community. These interactions, however may still take place at the mine site between local and non-local employees.
>
> (Tetra Tech 2009: G-17)

A follow-up to this potential impact has not been identified. As the Red Dog Mine is still active, it would be highly relevant to investigate the impacts of the in-migration to the surrounding villages and hub-town: Kotzebue.

Martin, Killorin and Colt (2008: 7) mention that while Alaska has experienced large in- and outflows of people in connection to the construction of the pipeline, today migration has changed to a one-sided out-migration. The question is: why are people leaving the villages? How big an effect has resource development had on this trend of out-migration? Building on SLICA and the work by Hamilton and

Seyfrit (1994), Martin Killorin and Colt (2008: 8) write that people hope to enhance their well-being by moving to places with more jobs and better education. People are migrating from rural to urban areas in both Greenland and Alaska, and one of the open questions is whether resource development can slow down, or revert, the process. Since the 1950s, people have been leaving Alaskan villages for hub towns and bigger cities (Martin, Killorin and Colt 2008: 3). Between 1990 and 2000, the net migration rate out of the Northwest Arctic Region was -4.7% (Tetra Tech 2009: G-17). In Greenland, a similar trend is found in more recent years as people move from smaller villages to bigger ones (SMV 2010: 10). It is repeatedly indicated that young people leave the villages to return later either because they were unsuccessful in creating new lives elsewhere or because they want to give back to the community they belong to (Hamilton and Seyfrit 1994: 191; Gram-Hanssen 2012: 60). In Alaska, more young women than men have been leaving the villages in pursuit of higher education and different lifestyles (Hamilton and Seyfrit 1994: 191). A gap-analysis of management of living resources in Greenland (Müller-Wille et al. 2005) names migration as a gap that should be analyzed. It is emphasized that the role of in- and out-migration is not clear (Müller-Wille et al. 2005: 17). Today workers in Alaska and Greenland are flown in on a bi-weekly basis, working two-three week shifts and are then flown out again. There are no studies investigating the impacts of this traffic on Indigenous as well as non-Indigenous communities. In proposed projects in Greenland workers are expected to live in camps while working but it has not been decided whether these camps will accommodate families and welcome women and children (Agersnap 1997).

Specifics of the Russian situation

A. Ethnic demography, health, and socio-economic issues

With perestroika, the disastrous situation of Russia's Arctic minorities came to light, mostly in areas where intensive extractive industrial development had devastated the environment that they relied on for their traditional livelihood. Rather than scholarly discussions, an article in the communist newspaper entitled "the big problems of small peoples" (Pika and Prokhorov 1988) generated the discussion (Fondahl and Poelzer 1997). Since the early 1990s, there has existed a significant number of scientific publications dedicated to the issues of Indigenous living conditions in the Russian Arctic, which analyze the causes of the current situation (Sokolova 1990, 1995, 2003; Karlov 1991). Significant contributions to the subject of ethnic and demographic processes have been made by Pika and Bogoyavlenskiy (1995), Pika (1999), Volzhanina (2003, 2004, 2009, 2010), Kvashnin (2003, 2007, 2009, 2011), Tuisku (2002), Zen'ko (2001), and Balzer (1999). Most studies were conducted with community-based field research. Ethno-sociological monitoring and socio-economic situation in Indigenous settlements and the tundra were analyzed by Kharamzin and Khayrullina (2003), Ezyngi (2004), and Stammler (2005). In the following we shall highlight some examples of such research, focusing on areas where extractive industries have had a particularly strong social impact.

Volzhanina (2009) summarized the major dynamics and demographic characteristics of the Yamal Nenets, describing the influence of traditional life and knowledge on time, family, and relations with the Russian population. The calculated demographic indices testified Aboriginal stability. Referring to census data and land management expeditions, the author identified a high absolute population growth. Comparing trends in demographic structures of nomadic and the settled industrial population of Yamal she revealed differences of traditional and modern types of reproduction. Artyukhova and Pirig (2004) put demography of Indigenous peoples in a broader context of the basic trends of the national ethnographic and historical science. In this light, specific demographic trends among the Yamal Nenets lie in their high natality and the general increase in the Indigenous share of the entire population, in an area that is exceptional in the Russian Arctic for its general population stability, and even slight increase since the end of the Soviet Union (Heleniak 2009).

B. Processes in Indigenous cultures and traditional economies

There are two overlapping economies in the north: a native rural traditional economy, which is most important for Indigenous livelihoods, and the main extractive industry run by the urban incomer population as well as fly-in/fly-out population in the Russian Arctic. Some hunting territories of the native peoples have become accessible for newcomers due to the development of transportation facilities (Ziker 2003), built closer to extractive industrial cities.

The complexity of the interrelationship between Indigenous cultures and extractive industries is seen in a very broad range of phenomena related to Indigenous societies in the North. In discussing social impacts of resource development on northern communities, researchers in Russia often highlight notions of Indigenous well-being, the loss of cultural continuity, linguistic assimilation, alienation from power, the adaptation to a new social environment, and the threat of industry to traditional activity.

Ezyngi (2004) examined the ethnocultural characteristics of the current situation in Yamal, including such parameters as Indigenous well-being and interrelationship with the processes in the district during the past 100 years. He conducted an analysis of the socio-cultural situation assessing the impacts of factors linked to resource development on Indigenous lifestyle and culture. The following tendencies in the Nenets culture were observed: a) the native language has been preserved as a spoken language, however, there are tendencies of transition to another language (Russian); b) there is an increase in total number of ethnically mixed families; and c) there are cases of duality in consciousness between the Indigenous and national identity. The author argues that the intensive industrial development in the region has affected all the cultural aspects of the tundra population.

Klokov (1997) called the decrease of land available for Indigenous livelihoods "territorial ethnocide", even when it was linked to a stable or even growing Indigenous population. According to the author, territorial ethnocide happens due to the gradual transition to the more efficient economic system of the dominant society. Klokov further identifies that the main problem for Indigenous adaptation to a rapidly changing society and environment around them lies in the passiveness of their approach. He argues that Indigenous peoples make few attempts to actively shape their relations with the industrial population.

Golovnev and Osherenko (1999) analyzed the Nenets history up to the start of gas industry development as a dialogue of cultures in which the major attention was focused on ethnocultural traditions and motivations. In their research the Nenets appear as an exceptionally resilient group with a phenomenal capacity to preserve their culture. Stammler (2005) challenged this notion and showed that rather than in the exceptionality of Nenets culture of resisting assimilation, the reason for the continued thriving of Nenets nomadism lies in the savviness of the Nenets to identify their niche in a dominant Soviet society, but also not less in the tolerance of the Soviet representatives to grant them that niche. As a result, Nenets society adopted and adapted many innovations from incomers, but without therefore abandoning their own traditions based on their herding-based mobility (Stammler 2013).

Many scholars have addressed the issue of Indigenous cultural survival and the development of their traditional economies. Reindeer herding, hunting, fishing, plants gathering, and processing of natural materials form the basis of life-support, language, culture, and national psychology of the Indigenous peoples. Aleksandr Pika was among the first Russian anthropologists to address existing cultural and political issues facing the Arctic Indigenous people. In his concept of "neotraditionalism" (Pika 1999), he related to earlier (Soviet) ideas of paternalistic governance of the Indigenous North, identifying a new approach for combining old and new that respects the traditions and emphasizes the role of preservation of the natural and social environment in development of the northern Indigenous peoples.

Along these lines, Yuzhakov and Mukhachev (2001) argued for a special status and support of what they called "ethnic reindeer herding". They acknowledge that during the Soviet and post-Soviet periods, preoccupation with quantitative performance indexes for meat production misses the most important point in Russian Arctic livelihoods: the importance of this lifestyle for sustaining the peoples themselves, allowing them to live with less welfare, more independence, and consequently also more prestige and pride. Rapid industrial development of natural resources and new market relations has resulted in further loss of the significance of traditional economies as a source of livelihood for the Indigenous population. However, it was mostly during the years of transition to a market economy that negative processes linked to resource-based economic development destroyed the traditional economic bases for Indigenous lifestyles and led to a catastrophic decrease in living standards (Vitebsky 1990; Osherenko 1995; Pika and Bogoyavlensky 1995).

C. Indigenous adaptations and future scenarios

Since the beginning of the 1990s, researchers have proposed several scenarios for Indigenous adaptation to the social impacts of industrial development in traditional lands. Dryagin (1995) describes one of them as a positive adaptation, based on preserved social ties of the tribe and introduction of a collective bargaining with relatives, which could promote social and economic self-government in the tundra. In fact, it is this idea of creating clan communities and collective agreements with oil and gas companies, which currently is taking place (Stammler and Ivanova 2016). According to the second scenario, the emphasis is on converting the young tundra-based Indigenous population to individual employment within the dominant Russian population, as well as promoting interethnic marriages. In the scenario's case site, this development would result in a complete assimilation of Indigenous youth in the Pur district – Russia's main gas province in Yamal, West Siberia.

Kvashnin (2009) observed that even if present development would develop according to the first scenario, it is not likely to be sustainable. The present condition reminds him of the Soviet Era's paternalistic policy towards the Indigenous people, where oil and gas companies had taken on the role of the party and business leaders, providing Aboriginals with subsidies and other benefits. Such an approach, according to Kvashnin, creates dependent attitudes among the Aboriginal population and stimulates constant expectations of care. Evaluating the situation in one community (*obshchina*), he predicts a gradual reduction of traditional herding on the territory and gradual resettlement of the Nenets to villages, with a seasonal fishery in the last non-contaminated rivers and lakes.

However, not all the scenario predictions are grim. Some researchers note the Indigenous ability to adapt to the changing social and environmental conditions (e.g., see Crate 2002; Hovelsrud and Smit 2010). In research on the Forest Nenets, Karapetova (2012) acknowledged that despite some negative changes, contracts with industry have made it possible to purchase new freezers and construct new fish processing facilities, which have greatly contributed to the export and barter of Indigenous products. This has improved the quality of life for some Indigenous families. Karapetova noted that the traditional economy no longer exists in the same form as in the 1980s. Hunting has become unprofitable, as fur is not in demand any more. It is a sign of resilience that Indigenous people are still able to focus on traditional activities such as fishing and herding which still serve as the basis for preservation of their ethnic cultural traditions and enables them to adapt to environmental and social transformations.

Conclusions and research gaps

This last section identifies some research gaps relating to a better understanding of social impacts of resource development in Alaska, Greenland, and Russia. Work on these gaps is necessary to ensure that the benefits of resource development are maximized, and negative impacts are minimized, as far as communities in these regions are concerned.

There is little research of a more applied nature that takes into consideration social-cultural aspects of the Indigenous population with a focus on their real everyday problems. Only in its infancy is research that analyses practices and the scope of Indigenous participation in the development process. Works by Novikova (2013a, 2013b, 214), and by Wilson (2016; Novikova and Wilson 2013) are important steps in this direction. Such research would not necessarily put a price-tag to answering the above-mentioned points and how much more benefit Indigenous or local people would get in such cases from industrial development. However, it would show ways for Indigenous cultures and livelihoods to become more resilient to the damages of industrial development, and become more active participants in the development process. Results of such research could be also used to recreate and evaluate the outcome of different development scenarios, where Indigenous communities have a different level of involvement, which directly influence their integration into the process.

Another topic, with a future research potential is an in-depth study of Indigenous population in the context of monitoring the socio-cultural and environmental conditions, which can determine the effects from anthropogenic pollution on the environment, creating a greater danger for the present and future generations of its Indigenous inhabitants (Dallmann et al. 2011). Above all natural and ecological factors, a clean environment is first and foremost the basis for Indigenous livelihoods and cultures. Much more than urbanized industrial lifestyles do Indigenous land users depend on the state of their environment. In other words, a polluted lake is dangerous for those who drink its water, eat its fish, and worship its spirits. It is not dangerous for inhabitants of an apartment bloc built on its shores, where the only "use" of the lake for humans is as a site of visual beauty. More than that, environmental pollution and health consequences happen anyway in all cases of industrial development. These topics are therefore the prime occasion for building partnership relations between the parties involved in industrial development (Sovet Federatsii 2009).

The current socio-economic and environmental situation in the Arctic has been thoroughly studied in recent decades. We identify, however, significant gaps in the practical implementation of development plans, providing better participatory decision-making. Academic articles on the practice of Indigenous participation in the development process and its problem solving (e.g. Novikova and Wilson 2013) are still rare, although various attempts to implement participatory mechanisms designed in the West for Russian cases are already ongoing (Muraskho and Yakel' 2009), as well as Russian domestic initiatives (Stammler and Ivanova 2016b).

Future research could place more importance on the influence of culturally specific relations to the resources among the different land users. It makes a big difference if the land is seen as a treasure house void of any other meaning except its potential of extraction, or seen as an animated landscape inhabited by people, animals, and spirits that are all closely cooperating in supporting survival and reproduction – physically and culturally – in a harsh environment. Recent research has just started looking into that more closely (Wilson and Stammler 2016; Stammler and Ivanova 2016).

As mentioned in several other chapters of this volume there is a research gap in regards to our understanding of the cumulative impacts of extractive industrial development on local people. It would be very useful to these communities to better understand how industrial development generates these cumulative impact chains. To name some examples: new construction leading to a new immigration of workers, who then bring their own social and cultural norms into the Arctic, who then start interacting with local and Indigenous cultures. This can lead to significant changes in ethno-social structures, for example through intermarriages between Indigenous (often women) and recent incomers from central Asia (often men). This in turn leads to gender shifts, which, and noted in Chapter 14, is a poorly studied topic. In the few available sources on gender in Russia (Povoroznyuk, Habeck and Vaté 2010), the impacts of extractive industries have so far not been studied. Related to this topic is also the role of domestic violence, sexual, alcohol, and drug abuse, and prostitution in places where industry workers meet Indigenous locals. While the authors of Chapter 14 note a lack of research in this area, this topic is far better studied in the Canadian Arctic than in the Russian Arctic.

The same is true with the cumulative effects of industrial migration on religion in the Arctic. Many of the more recent migrants are Muslims, and almost every industrial Russian Arctic city features a mosque these days. At the same time, Alaskan and Greenlandic cultural and religious landscapes are being altered through migration processes triggered by resource development (and visions of potentially thousands of Chinese workers on the island of Greenland circulating in the press). Still, there are hardly any studies about how such changes are being perceived by Indigenous peoples some of whom are in the process of reviving their own pre-Christian religious traditions.

Another gap can be identified in the absence of specific studies on Indigenous industrial workers. Studies are still rare (Dudeck 2008), and the topic is important also in applied research: how do Indigenous industry workers deal with labor regimes that they are not used to? How do they cope with discipline at work? How do they adapt their traditional harvesting practices to employment schedules? How can they make best use of their extensive knowledge of the environment for making extractive industries environmentally and socially as viable as possible?

Finally, we argue that the informal economic impacts of resource development are generally left out of SIAs and other impact analyses, which create a gap in literature that leads to a number of impact aspects being neglected. Indigenous communities in Alaska and Greenland are constantly adapting to the economic, political, social, and cultural environments they are a part of and their small-scale economies depend on their abilities to utilize informal economic means such as subsistence harvesting (Nuttall 1998: 97). They need to adapt to economic boom-and-bust cycles, continue to fight for self-determination and stimulate local empowerment; all of which is done through informal economic activities (Nuttall 1998: 97). One informal economic activity is subsistence. However, subsistence is of much more than economic value at this analysis has shown.

Note

1 This chapter was prepared by two different teams of authors over a span of almost four years. Cecilie Ebsen at UAF (under the supervision of Peter Schweitzer) conducted an initial literature review regarding Alaska and Greenland in 2012 and 2013, while Irina Litvina did a similar review for Russia at the Arctic Centre in Rovaniemi under the supervision of Florian Stammler. Aytalina Ivanova and Stammler completed the Russian part in 2016 as part of the NRC project 'Challenges in Arctic Governance: Indigenous territorial rights in the Russian Federation', number 257644/H30.

References

Agersnap, T. 1997. Råstofudvinding – sociale virkninger af fire mulige projekter. Kamikposten. Retrieved at: www.kamikposten.dk/global/maskinrum/rutine/leksikon.aspx ?tag=periode&folder=hvadermeningen&sprog=da&punkt=199710&soegestreng= &udvalgt=19971028c1

Artyukhova and Pirig Артюхова И.Д., Пириг Г.Р. 2004. "О демографической ситуации и сохранении традиционного образа жизни коренного населения Ямало-Ненецкого автономного округа" Вестн. Тюмен. нефтегаз. ун-та № 3: Региональные социальные процессы. С. 64–70.

Balzer, M. M. 1999. *The Tenacity of Ethnicity: A Siberian Saga in Global Perspective*. Princeton, NJ: Princeton University Press.

Baring-Gould, M. and M. Bennett. 1975. *Social Impact of the Trans-Alaska Pipeline Construction in Valdez, Alaska 1974–1975*. Anchorage: Department of Sociology, UAA. Prepared for the Mackenzie Valley Pipeline Inquiry.

Berger, T. R. 1985. *Village Journey: The Report of the Alaskan Native Review Commission*. New York: Hill and Wang.

Berry, M. C. 1975. *The Alaska Pipeline: The Politics of Oil and Native Land Claims*. Bloomington: Indiana University Press.

Bird-David, N. 1992. "Beyond 'The Original Affluent Society': A Culturalist Reform- ulation." *Current Anthropology* 33(1): 25–47.

Djenegaard, P. and I. K. Dahl-Petersen. 2008. "Befolkningsundersøgelse i Grønland 2005–2007: Levevilkår, livsstil og helbred." *SIF's Grønlandstidsskrifter nr. 18*, Køben- havn: Statens Institut for folkesundhed.

Boertman, D., A. Mosbech, P. Johansen, and H. Petersen. 1998. *Olieefterforskning og miljø i Vestgrønland*. Miljø og Energi Ministeriet, Danmarks Miljøundersøgelser.

Braund, S., J. Kruse, and F. Andrews. 1985. *A Social Indicators System for OCS Impact Monitoring*. U.S. Department of the Interior Minerals Management Service.

Brower, C. 2015. "ANWR Worshippers Fail to Consider the Inupiat." *Alaska Dispatch News*, February 10, 2015. www.adn.com/article/20150210/anwr-worshippers-fail- consider-i-upiat (accessed on March 15, 2015).

Burch, E. 1985. *Subsistence Production in Kivalina, Alaska. A Twenty-Year Perspective*. Juneau: Division of Subsistence, Alaska Department of Fish and Game. Technical Paper No. 128.

Burdge, R., D. Field, and S. Wells. 1988. "Utilizing Social History to Identify Impacts of Resource Development on Isolated Communities: The Case of Skagway, Alaska." *Impact Assessment* 6(2): 37–54.

Burdge, R. and F. Vanclay. 1996. "Social Impact Assessment: A Contribution to the State of the Arts Series." *Impact Assessment* 14(1): 59–86.

Bureau of Minerals and Petroleum, Greenland 2009. *Guidelines for Social Impact Assessments for Mining Projects in Greenland*, Bureau of Minerals and Petroleum.

Cerveny, L. 2005. *Tourism and Its Effects on Southeast Alaska Communities and Resources: Case Studies from Haines, Craig, and Hoonah, Alaska*. Res. Pap. PNW-RP-566. Portland, OR: U.S. Department of Agriculture, Forest Service, Pacific Northwest Research Station.

Colt, S. and T. Schwoerer. 2010. "Socioeconomic Impacts of Potential Wishbone Hill Coal Mine Activity." Anchorage: Institute of Social and Economic Research, University of Alaska.

The Committee for Greenlandic Mineral Resources to the Benefit of Society. 2014. *For the Benefit of Greenland*. Nuussuaq, Greenland: Ilisimatusarfik, University of Greenland.

ConocoPhilips. 2008. National Petroleum Reserve-Alaska (NPR-A) 5-Year Winter Exploration Drilling/Well Testing Program. ConocoPhilips Alaska Inc.

Crate, S. 2002. "Co-option in Siberia: The Case of Diamonds and the Vilyuy Sakha." *Polar Geography* 26(4): 418–435.

Dallmann, W. K., V. Peskov, O. A. Murashko and E. Khmeleva. 2011. "Reindeer herders in the Timan-Pechora oil province of Northwest Russia: an assessment of interacting environmental, social, and legal challenges." *Polar Geography* 34: 229–247.

Delaney, A., R. Becker Jacobsen, and K. Hendriksen. 2012. Greenland Halibut in Upernavik: a preliminary study of the importance of the stock for the fishing populace, Aalborg University: Innovative Fisheries Management.

Dixon, M. 1978. *What Happened to Fairbanks? The Effects of the Trans-Alaska Oil Pipeline on the Community of Fairbanks, Alaska*. Boulder, CO: Westview Press.

Dryagin Дрягин, B.B. 1995. "Проблемы адаптации коренных народов Тюменского Севера к производственным структурам." Народы Сибири и сопредельных территорий. Томск, 95–104.

Dudeck, S. 2008. "Indigenous Oil Workers between the Oil Town of Kogalym and Reindeer Herder's Camps in the Surrounding Area." In Stammler, F. and Eilmsteiner-Saxinger, G. (eds). *Biography, Shift-labour and Socialisation in a Northern Industrial City – The Far North: Particularities of Labour and Human Socialization*. Rovaniemi and Vienna: Arctic Centre, University of Lapland and University of Vienna, pp. 141–144.

Ezyngi Езынги, X. M. 2004. Проблемы сохранения традиционной культуры ненцев Ямала. Докт. дисер. Тюмень: ТГУ

Fondahl, G. and G. Poelzer. 1997. "Indigenous Peoples of the Russian North." *Cultural Survival Quarterly* 21(3).

Fondahl, G. and A. A. Sirina. 2006. "Rights and Risks: Evenki Concerns Regarding the Proposed Eastern Siberia–Pacific Ocean Pipeline." *Sibirica* (5)2: 115–138.

Frederiksen, M., B. David, U. Fernando, and A. Mosbech. 2012. South Greenland: A Preliminary Strategic Environmental Impact Assessment of Hydrocarbon Activities in the Greenland Sector of the Labrador Sea and the Southeast Davis Strait, Aarhus University: Danish Center for Environment and Energy & Greenland Institute of Natural Resources.

Gill, D. and J. S. Picou. 1997. "The Day the Water Died: Cultural Impacts of the Exxon Valdez Oil Spill." In Steven Picou, J. et al. (eds.) *The Exxon Valdez Disaster: Readings on a Modern Social Problem*. Dubuque, IA: IA Books, pp. 167–187.

Goldsmith, S. 2008. How North Slope Oil Has Transformed Alaska's Economy. Anchorage: Institute of Social and Economic Research, University of Alaska Anchorage.

Golovnev, A. V. and G. Osherenko. 1999. *Siberian Survival: The Nenets and their Story*. Ithaca and London: Cornell University Press.

Gram-Hanssen, I. 2012. *Youth Creating Sustainable Communities in Rural Alaska.* Thesis, Northern Studies, Fairbanks: University of Alaska Fairbanks.

Greenland Bureau of Minerals and Petroleum. 2012. Report to Inatsisartut, the Parliament of Greenland, Concerning Mineral Resources Activities in Greenland, Greenland Bureau of Minerals and Petroleum, Greenlandic Self Government.

Grontmij. 2012. Executive Summary on the Social Impact Assessment Aappaluttoq Ruby Project for True North Gems Inc. Prepared for True North Gems.

Habeck, J. O. 2002. "How to Turn a Reindeer Pasture into an Oil Well, and Vice Versa: Transfer of Land, Compensation and Reclamation in the Komi Republic." In Kasten, E. (ed.), *People and the Land. Pathways to Reform in Post-Soviet Siberia.* Berlin: Reimer, pp. 125–147.

Haley, S. 2004. "Institutional Assets for Negotiating the Terms of Development: Indigenous Collective Action and Oil in Ecuador and Alaska." *Economic Development and Cultural Change* 53(1): 191–213.

Haley, S., M. Klick, N. Szymoniak, and A. Crow. 2011. "Observing Trends and Assessing Data for Arctic Mining." *Polar Geography* 34(1–2): 37–61.

Haley, S. and J. Magdanz. 2008. "The Impacts of Resource Development on Social Ties: Theories and Methods for Assessment." In O'Faircheallaigh, C. and Ali, S. (eds.), *Earth Matters: Indigenous Peoples, the Extractive Industries and Corporate Social Responsibility.* Sheffield, UK: Green Leaf Publishing.

Hamilton, L. and C. L. Seyfrit. 1994. "Female Flight? Gender Balance and Outmigration by Native Alaskan Villagers." *Arctic Medical Research* 53(2): 189–193.

Harvey, F. 2012. "Europe looks to open up Greenland for natural resource extraction." *The Guardian.* Retrieved at: www.guardian.co.uk/environment/2012/jul/31/europe-greenland-natural-resources

Heleniak, T. E. 2009. "The Role of Attachment to Place in Migration Decisions of the Population of the Russian North." *Polar Geography* 32(1–2): 31–60. http://doi.org/10.1080/10889370903000398.

Hensley, W. 2009. *Fifty Miles from Tomorrow.* New York: Sarah Crichton Books.

Hjemmestyret. 1997. Socio-økonomiske virkninger af råstofudvikling. Kamikposten. Retrieved at: www.kamikposten.dk/global/maskinrum/rutine/leksikon.aspx?tag=periode&folder=hvadermeningen&sprog=da&punkt=199710&soegestreng=&udvalgt=1997102800

Hovelsrud, G. K. and B. Smit (eds.) 2010. *Community Adaptation and Vulnerability in Arctic Regions.* Berlin: Springer.

Huskey, L. and W. Nebesky. 1979. Northern Gulf of Alaska Petroleum Development Scenarios: Economic and Demographic Impacts. Technical Report, 34. Alaska OCS Socioeconomic Studies Program.

Information Insights. 2006. Alaska–Canada Rail Link Strategic Environmental Analysis: Socio-economic Impact Assessment – Alaska. Final Report. Fairbanks: Information Insights, Inc.

Ivanova, A. 2007. "The Price of Progress in Eastern Siberia: Problems of Ecological Legislation and Political Agency in a Russian Region." In Kankaanpää, P., Ovaskainen, S., and Pekkala, L. (eds) *Knowledge and Power in the Arctic: Proceedings at a Conference in Rovaniemi 16–18 April 2007.* Rovaniemi: Arctic Centre Report 48, pp. 61–69.

Jeppesen, C. 2008. "Kost." In Bjerregaard, P. and Dahl-Petersen, I. K. (eds), *Befolkningsundersøgelse i Grønland 2005–2007: Levevilkår, livsstil og helbred.* SIF's Grønlandstidsskrifter nr. 18, København: Statens Institut for folkesundhed.

Jorgensen, J. 1990. *Oil Age Eskimos*. Berkeley, CA: University of California Press.

Karapetova Карапетова, И.А. 2012. Пуровские лесные ненцы: некоторые аспекты хозяйственно-культурной и социальной адаптации // Вестник археологии, антропологии и этнографии. Тюмень: Изд-во ИПОС СО РАН, № 1(16). С. 92–101.

Карлов, В. В. 1991. "Народности Севера Сибири: особенности воспроизводства и альтернативы развития." *Советская этнография* 5: 3–15.

Kawagley, A. O. 2006. *A Yupiaq Worldview: A Pathway to Ecology and Spirit*. 2nd Edition. Long Grove, IL: Waveland Press.

Kharamzin and Khayrullina Харамзин Т.Г., Хайруллина Н.Г. 2003. *Обские угры (социологические исследования материальной и духовной культуры)*. Тюмень: Изд-во ТюмГНГУ.

Клоков К. В. 1997. Традиционное природопользование народов Севера: концепция сохранения и развития в современных условиях// Этнографические и этноэкологические исследования. Вып. 5. СПб

Knapp, G., and M. Lowe. 2007. *Economic and Social Impacts of BSAI Crab Rationalization on the Communities of King Cove, Akutan, and False Pass*. Anchorage: Institute of Social and Economic Research, University of Alaska.

Kruse, J. 1991. "Alaska Inupiat Subsistence and Wage Employment Patterns: Understanding Individual Choice." *Human Organization* 50(4): 317–326.

Kruse, J. 2006. Indicators of Social, Economic, and Cultural Cumulative Effects Resulting from Petroleum Development in Alaska: A Review. Available at www.ceaa-acee.gc.ca/155701CE-docs/Jack_Kruse-eng.pdf (accessed between July 2014 and July 2018).

Kruse, J. 2010. "Sustainability from a Local Point of View: Alaska's North Slope and Oil Development." In Winther, G. (ed.). *The Political Economy of Northern Regional Development – Yearbook 2008*. TemaNord: Nordic Council of Ministers, pp. 55–72.

Kruse, J. 2011 "Developing an Arctic Subsistence Observation System." *Polar Geography* 34(1–2): 9–35.

Kruse, J., M. Lowe, S. Haley, V. Fay, L. Hamilton, and M. Berman. 2011. "Arctic Observing Network Social Indicators: An Overview." *Polar Geography* 34(1–2): 1–8.

Квашнин Ю. Н. et al. 2003. *Этнография и антропология Ямала*. Новосибирск: Наука.

Квашнин Ю. Н. 2007. Лесные ненцы п. Ханымей // Вестник археологии, антропологии и этнографии. Тюмень: Изд-во ИПОС СО РАН, № 8. С. 252–256.

Квашнин Ю. Н. 2009. Пяко-Пуровская община лесных ненцев: проблемы сохранения традиционного хозяйства // Человек и север: антропология, археология, экология. Материалы всероссийской научной конференции. Вып. 1. Тюмень: Изд-во Ин-та проблем освоения Севера СО РАН, С. 9–11.

Квашнин Ю. Н. 2011. Енисейские ненцы: Этнографические исследования спустя полвека // «Не любопытства ради, а познания для…»: К 75-летию Ю.Б. Симченко / Ред. Н.А. Дубова, Ю.Н. Квашнин. М.: Старый сад, С. 283–313.

Labourstats.alaska.gov. 2010. Demographic Profile for Alaska. Retrieved at: http://live.laborstats.alaska.gov/cen/dp.cfm

Langdon, S. 2002. *The Native People of Alaska: Traditional Living in a Northern Land*. Anchorage: Greatland Graphics.

Larsen, J. N., P. Schweitzer, and G. Fondahl (eds.). 2010. *Arctic Social Indicators*. Copenhagen: Nordic Council of Ministers.

LGL Ltd., environmental research associates and Grontmij. 2012. Environmental Impact Assessment Shallow Coring in Baffin Bay, Northwest Greenland, prepared for Shell Kanumas A/S.

Lowe, M. 2011. "Arctic Observing Network Social Indicators and Northern Commercial Fisheries." *Polar Geography* 34(1): 87–105.

Magdanz, J., N. Braem, B. Robbins, and D. Koster. 2010. Subsistence Harvest in Northwest Alaska, Kivalina and Noatak. Juneau, AK: Division of Subsistence, Department of Fish and Game.

Martin, S., M. Killorin, and S. Colt. 2008. Fuel Costs, Migration, and Community Viability, Final Report, prepared for Denali Commission. Anchorage: Institute of Social and Economic Research, University of Alaska Anchorage.

McClintock, S. 2009. Coastal and Riverine Erosion Challenges: Alaskan Villages' Sustainability. In UNESCO 2009 Climate Change and Arctic Sustainable Development: Scientific, Social, Cultural and Educational Challenges. Paris: UNESCO.

McDowell Group. 2002. ANWR and the Alaska Economy: An Economic Impact Assessment. Prepared for Supporting Alaska Free Enterprise. Anchorage: McDowell Group.

McDowell Group. 2011. Socioeconomic Impacts of the Fort Knox Mine. Anchorage: McDowell Group.

McDowell Group. 2012. The Economic Impacts of Alaska's Mining Industry. Prepared for Alaska Miners Association. Anchorage: McDowell Group.

Müller-Wille, L., M. C. S. Kingsley, and S. S. Nielsen (eds.). 2005. Socio-Economic Research on Management Systems of Living Resources: Strategies, Recommendations and Examples. (Proceedings of the Workshop on "Social and economic research related to the management of marine resources in West Greenland", Greenland Institute of Natural Resources, Nuuk (Greenland), 18–20 November 2003). Inussuk – Arctic Research Journal 1/2005. Nuuk: Greenland Home Rule.

Мурашко О. А., and Якель Ю. Я. 2009. "Экологическое соуправление — основа взаимодействия коренных малочисленных народов, органов власти и промышленных компаний." In Federatsii, S. (ed.). *Федеральное Собрание Российской Федерации Совет Федерации Вопросы взаимоотношений коренных малочисленных народов с промышленными компаниями.* Moscow, pp. 75–93.

Nanoq. 2012. Befolkning i Grønland. Retrieved at: http://dk.nanoq.gl/Emner/Om%20 Groenland/Befolkning_i_Groenland.aspx

Northern Economics. 2002. *An Assessment of the Socioeconomic Impacts of the Western Alaska Community Development Quota Program.* Anchorage, AK: Northern Economics.

Новикова Н. 2013a. "Защита культурного наследия коренных малочисленных народов Севера в контексте промышленного освоения. Вестник НГУ. Серия: история, филология. Том 12, выпуск 3: Археология и этнография. Новосибирск", pp. 93–101.

Новикова Н. 2013b. "Коренные народы российского Севера и нефтегазовые компании: преодоление рисков." *Арктика: экология и экономика* 3(11): 102–111.

Новикова Н. 2014. *Охотники и нефтяники: исследование по юридической антропологии.* Москва: Наука.

Novikova, N. and E. Wilson. 2013. "The Sakhalin2 Project Grievance Mechanism, Russia." In Wilson, E. and Blackmore, E. (eds.) *Dispute or Dialogue? Community Perspectives on Company-led Grievance Mechanisms.* London: IIED, pp. 84–109.

Nuttall, M. 1998. *Protecting the Arctic: Indigenous Peoples and Cultural Survival.* Amsterdam: Harwood Academic Publishers.

Nuttall, M. 2010. *Pipeline Dreams: People, Environment and the Arctic Energy Frontier.* Copenhagen: International Work Group for Indigenous Affairs.

Olsen, N. L. 2011. "Uden grønlandsk mad er jeg intet." In Høiris, O. and Marquardt, O. (eds.) *Fra Vild Til Verdensborger: Grønlandsk Identitet Fra Kolonitiden Til Nutidens Globalitet.* Aarhus: Aarhus Universitetsforlag, pp. 409–431.

Osherenko, G. 1995. "Property Rights and Transformation in Russia: Institutional Change in the Far North (Yamal Peninsula)." *Europe–Asia Studies* 47: 1077–1108.

Payne, J. T. 1985. *Our Way of Life is Threatened and Nobody Seems to Give A Damn: The Cordova District Fisheries Union and the Trans-Alaska Pipeline.* The Cordova District Fisheries Union.

Perry, J. and R. Bright. 2010. Capricorn Drilling Exploration: Drilling EIA, London: Environmental Resources Management.

Perry, J., C. Jones, and J. Ward. 2011. Environmental Impact Assessment for High Resolution Seismic Site Survey off West Greenland, prepared for Shell Kanumas.

Petterson, J., B. Harris, L. Palinkas, and S. Langdon. 1983. Cold Bay: Ethnographic Study and Impacts Analysis. Technical Report, 93. Alaska OCS Socioeconomic Studies Program.

Pika, A. 1999. *Neotraditionalism in the Russian North: Indigenous Peoples and the Legacy of Perestroika.* Edmonton: Canadian Circumpolar Institute Press.

Pika, A. and D. Bogoyavlenskiy. 1995. "Yamal Peninsula: Oil and Gas Development and Problems of Demography and Health among Indigenous Populations." *Arctic Anthropology* 32(2): 61–74.

/Пика А., and Prokhorov Прохоров Б. 1988. Большие проблемы малых народов. *Коммунист* № 16: 76–83.

Poppel, B., J. Kruse, G. Duhaime, and L. Abryutina. 2007. SLiCA results. Anchorage: Institute of Social and Economic Research, University of Alaska Anchorage.

Povoroznyuk, O., J. O. Habeck, and V. Vaté. 2010. "Introduction: On the Definition, Theory, and Practice of Gender Shift in the North of Russia." *Anthropology of East Europe Review* 28(2): 1–37.

Public Comments. 2009. Red Dog Mine, Draft Authorization, Department of Natural Resources, State of Alaska.

Public Meeting. 2012. NPR-A Subsistence Advisory Panel, Point Lay, 5 June.

Rasmussen, R. O. 2009. "Resource Changes and Community Response: The Case of Northern Fisheries." In Nuttall, A. D. and Nuttall, M. (eds), *Canada's and Europe's Northern Dimension.* Oulu: Thule Institute, University of Oulu.

Ritchie, L. 2004. Voices of Cordova: Social capital in the wake of the Exxon Valdex oil spill. Ph.D. Dissertation, Mississippi State University.

Schweitzer, P. P., R. Barnhardt, M. Berman, and L. Kaplan. 2014. "Inuit Regions of Alaska." In Larsen, J. N., Schweitzer, P. P., and Petrov, A. (eds.), *Arctic Social Indicators. ASI II: Implementation.* Copenhagen: Nordic Council of Ministers: pp. 183–224.

Segal, B. (principal author). 1999. Alaska Native Combating Substance Abuse and Related Violence through Self-Healing: A Report for the People. Anchorage, AK: The Center for Alcohol and Addiction and The Institute for Circumpolar Health Studies.

Sidortsov, R., Ivanova, A., and Stammler, F. 2016. "Localizing governance of systemic risks: A case study of the Power of Siberia pipeline in Russia." *Energy Research & Social Science,* special issue Arctic Energy: Views from the Social Sciences, ed. by Roman Sidortsov, 16: 54–68.

SMV (Strategisk Miljøvurdering). 2010. Kapitel 5: Regionaludvikling og migration. Nuuk: Grønlands Selvstyre.

Соколова З. П.1990. Проблемы этнокультурного развития народов Севера// Расы и народы. М.: Наука, Вып. 20. С. 128–150.

Соколова З. П. 1995. Концептуальные подходы к развитию малочисленных народов Севера // Социально-экономическое и культурное развитие народов Сибири и Севера: традиции и современность. М.: Наука, С. 5–42.

Соколова З. П. 2003. Перспективы социально-экономического и культурного развития коренных малочисленных народов Севера // ВРГНФ,№1. С. 47–62.

Federatsii Совет Федерации. (ed.) 2009. Федеральное Собрание Российской Федерации Совет Федерации. Вопросы взаимоотношений коренных малочисленных народов с промышленными компаниями. Moscow.

Stammler, F. 2005. *Reindeer Nomads Meet the Market: Culture, Property and Globalisation at the End of the Land*. Berlin: Lit Verlag.

Stammler, F. 2013. "Narratives of Adaptation and Innovation: Ways of Being Mobile and Mobile Technologies among Reindeer Nomads in the Russian Arctic." In Miggelbrink, J., Habeck, J. O., Mazzullo, N., and Koch, P. *Nomadic and Indigenous Spaces: Productions and Cognitions*. Farnham, UK: Ashgate, pp. 221–245.

Stammler, F., and Vladislav P. 2008. "Building a 'Culture of dialogue' among stakeholders in North-West Russian oil extraction." *Europe-Asia Studies* 60(5): 831–849.

Stammler, F. and Ivanova, A. 2016a. "Confrontation, Coexistence or Co-ignorance? Negotiating Human–resource Relations in Two Russian Regions." *The Extractive Industries and Society* 3(1): 60–72. http://doi.org/10.1016/j.exis.2015.12.003.

Stammler, F. and Ivanova, A. 2016b. "Resources, Rights and Communities: Extractive Mega-projects and Local People in the Russian Arctic." *Europe–Asia Studies* 68(7): 1220–1244.

Stammler, F. and Peskov, V. 2008. "Building a 'Culture of dialogue' among stakeholders in North-West Russian oil extraction." *Europe-Asia Studies*, 60(5): 831–849.

Stenbaek, M. 1987. "Forty Years of Cultural Change among the Inuit in Alaska, Canada, and Greenland: Some Reflections." *Arctic* 40(4): 300–309.

Stuvøy, K. 2011. "Human Security, Oil and People: An Actor-based Security Analysis of the Impacts of Oil Activity in the Komi Republic, Russia." *Journal of Human Security* 7(2): 5.

Tetra Tech. 2009. Red Dog Mine Extension: Aqqaluk Project. Final Supplemental Environmental Impact Statement (SEIS). Anchorage: Tetra Tech.

Thisted, K. 2005. Postkolonialisme i nordisk perspektiv: relationen Danmark-Grønland. In Bech, H. and Scott Sørensen, A. (eds.), *Kultur på kryds og tværs*. Aarhus: Klim.

Thisted, K. 2011. "Nationbuilding – Nationbranding· Identitetspositioner og tilhørsforhold under det selvstyrede Grønland." In Høiris, O. and Marquardt, O. (eds.), *Fra vild til verdensborger*. Aarhus: Aarhus Universitetsforlag.

Trenholm, M. (ed.). 2012. *Bristol Bay's Wild Salmon Ecosystem and the Pebble Mine: Key Considerations for a Large-Scale Mine Proposal*. Portland, OR: Wild Salmon Center & Trout Unlimited.

Tuisku, T. 2002. "Transition Period in the Nenets Autonomous Okrug: Changing and Unchanging Life of Nenets People." In Karsten, E. (ed.), *People and the Land: Pathways to Reform in Post-Soviet Siberia*, Vol. 1, Siberian Studies. Berlin: Reimer, pp. 189–205.

Udvalget for socioøkonomiske virkninger af olie-og gasudvinding samt mineralindustri. 1997. Socio-økonomiske virkninger af råstofudvinding. Retrieved at: www.kamikposten. dk/global/maskinrum/rutine/leksikon.aspx?tag=periode&folder=hvadermeningen& sprog=da&punkt=199710&udvalgt=1997102810&retur=periode|199710|1997102800

U.S. Principals and Guidelines. 2003. "Principles and Guidelines for Social Impact Assessment in the USA." *Impact Assessment and Project Appraisal* 21(3): 231–250.

Vining, A. 1974. *The Socio-economic Impacts of the Trans-Alaska Pipeline: A Strategy for the State of Alaska*. Boulder, CO: Western Interstate Commission for Higher Education.

Vitebsky, P. 1990. "Gas, Environmentalism and Native Anxieties in the Soviet Arctic: The Case of Yamal Peninsula." *Polar Record* 26:19–26.

Volzhanina Волжанина Е.А. 2003. Динамика фамильного состава коми и ненцев Аксарковской сельской администрации (Приуральский район Ямало-Ненецкого автономного округа) // Словцовские чтения-2003: Матер. докл. и сообщ. XV Всерос. науч.-практ. краевед. конф. Тюмень, pp. 106–109.

Volzhanina Волжанина Е.А. 2004. Современные демографические процессы у коренного населения низовьев реки Ны- да // Словцовские чтения: Матер. докл. и сообщ. XVI Всерос. науч.-практ. краевед. конф. Тюмень, ч. 1. с. 132–133.

Volzhanina Волжанина Е.А. 2009. Проблемы демографии ненцев Ямала в первой трети XX века // Вестник археологии, антропологии и этнографии, No. 9.

Volzhanina Волжанина Е.А. 2010. Численность, расселение и традиционное хозяйство вынгапуровских ненцев в условиях промышленного освоения // Вестн. археологии, антропологии и этнографии. Тюмень: ИПОС СО РАН, № 1(12). С. 174–184.

Watkinson, P. 2009. *Nalunaq Goldmine: Social Impact Assessment*. St Ives, UK: P. I. Watkinson.

Willis, R. 2006. "A New Game in the North: Alaska Native Reindeer Herding 1890–1940." *The Western Historical Quarterly* 37(3): 277–301.

Wilson, E. 2016. "What is the Social Licence to Operate? Local Perceptions of Oil and Gas Projects in Russia's Komi Republic and Sakhalin Island." *The Extractive Industries and Society* 3(1): 73–81. http://doi.org/10.1016/j.exis.2015.09.001.

Wilson, E., and F. Stammler. 2016. "Beyond Extractivism and Alternative Cosmologies: Arctic Communities and Extractive Industries in Uncertain Times." *The Extractive Industries and Society* 3(1): 1–8. http://doi.org/10.1016/j.exis.2015.12.001.

Wilson, R. R., J. R. Liebezeit, and W. M. Loya. 2013. "Accounting for Uncertainty in Oil and Gas Development Impacts to Wildlife in Alaska." *Conservation Letters* 6(5): 350–358.

Yakovleva, N. 2014. "Land, Oil and Indigenous People in the Russian North: A Case Study of the Oil Pipeline and Evenki in Aldan." In Gilberthorpe, E. and Hilson, G. (eds.), *Natural Resource Extraction and Indigenous Livelihoods*. Aldershot, UK: Ashgate, pp. 147–178.

Yarie, S. (ed.). 1982. *Alaska Symposium on the Social, Economic, and Cultural Impacts of Natural Resource Development*. Anchorage: Alaska Pacific University.

Южаков А. and А. Мухачев. 2001. *Etnicheskoe olenevodstvo Zapadnoi Sibiri: nenetskii tip*. Novosibirsk: Agricultural Science Publishers.

Зенько М. А. 2001. Современный Ямал: этноэкологические и этносоциальные проблемы. *Исследования по прикладной и неотложной этнологии* 139.

Ziker, J. 2002. *Peoples of the Tundra. Northern Siberians in the Post-Communist Transition*. Long Grove, IL: Waveland Press.

Ziker, J. 2003. "Assigned Territories, Family/Clan/Communal Holdings, and Common-Pool Resources in the Taimyr Autonomous Region, Northern Russia." *Human Ecology* 31(3): 331–368.

4 Northern environmental assessment

A gap analysis and research agenda

*Bram Noble, Kevin Hanna, and
Jill Blakly*

Arctic environments are rapidly changing under the pressures of human development, resource extraction, and global climate change. The United Nations Environment Programme Globio Report indicates that by 2050, even at modest rates of economic growth, approximately 50 to 80% of the Arctic may reach critical levels of anthropogenic disturbance (Nellemann et al. 2001). Environmental assessment (EA) is amongst the most widely practiced environmental management tools in the world and is applied across the Arctic to assess the impacts of a diverse range of anthropogenic development activities. The 1991 Strategy for the Protection of the Arctic Environment emphasized the importance of EA, leading to the establishment of the Arctic Monitoring and Assessment Program (AMAP) and, subsequently, the adoption of Arctic EA guidelines under the 1997 Alta Declaration by the ministers of Arctic Countries. AMAP (2007) recommends that EA and related planning tools, including strategic EA, be rigorously applied in the Arctic and that attention is directed toward increasing EA's relevance and usefulness.

The effectiveness of EA has been questioned. In a 2005 newsletter of the International Association for Impact Assessment, Richard Fuggle described a "disillusionment" about EA "and skepticism that impact assessments are contributing to better decisions" (Fuggle 2005: 1). In the Finnish context, Pölönen, Hokkanen and Jalava (2011) described the linkages between EA and decision-making processes as a major deficiency and, in Russia, Cherp and Golubeva (2004) identified several challenges in national-level approaches to EA implementation. In Canada's western Arctic, Noble et al. (2013) and BSStRPA (2008) identify challenges to current EA processes in capturing the cumulative effects of energy developments, whilst Harrison (2006) and Voutier et al. (2008) report the challenges to industry of an increasingly complex EA regulatory environment. Morgan (2012), in his review of the international state-of-the-art of EA, warned of weakening EA processes as governments seek more expedited approvals to support economic development initiatives.

Researchers, practitioners, communities, and governments understandably want to know if the time and resources spent on EA is leading to improved environmental management and quality. In a brief to a Canadian parliamentary committee, Hanna and Noble (2011) emphasized the importance of EA as an essential public policy instrument for environmental decision-making, but also noted the need to understand the effectiveness of EA and to improve EA rather than abandon it. A major challenge, however, is that the majority of EA research, particularly research on EA effectiveness, has occurred outside the Arctic.

This chapter examines what work has been done to understand the effectiveness of EA in the Arctic and identifies gaps in research that need to be addressed to understand and improve EA and its relevance to Arctic communities. It provides a gap analysis of EA research across the Arctic by identifying what research exists and what research needs to be done to help increase the effectiveness and benefits of EA. We first provide a brief overview of EA in the Arctic, followed by the gaps in Arctic EA research and priority areas for future research.

Environmental assessment

Environmental assessment is broadly defined as a process for identifying, predicting, evaluating, and mitigating the biophysical, social, and other relevant effects of development proposals prior to major decisions being taken and commitments made (IAIA and IEA 1999). It can also be viewed as a means of strengthening environmental management processes (Morrison-Saunders and Bailey 1999). The underlying intent of EA is to enable project proponents, communities, and decision-makers to enhance the benefits and to minimize the environmental costs of development actions. EA is both a planning and management tool for choosing and designing developments wisely. It can also be viewed in a much broader context – as a means to shape decisions and provide an opportunity for public debate about the merits of a proposed development.

Environmental assessment in the Arctic emerged with the United States' *National Environmental Policy Act* (1969/70), with Canada following in 1972 with the Federal Environmental Assessment Review Office policy that all new federally initiated projects and those under federal jurisdiction be screened for potential pollution effects. It was the Mackenzie Valley Pipeline Inquiry (Berger 1977), however, that "set an international standard for critical and cross-cultural public assessment" and created expectations "about what an assessment process should be" (Gibson and Hanna 2009: 22). Environmental assessment across the Arctic is now a vast subject, capturing biophysical, social, economic, and legal dimensions, and ranging in scope of application from individual project assessments to regional plan assessments and broad policy assessments at the strategic level.

Table 4.1 provides a brief overview of the EA setting in the eight Arctic nations: the United States (Alaska), Canada, Russia, Finland, Sweden, Greenland, Iceland, and Norway. The history, provisions, scope, and requirements of EA vary considerably across the Arctic. Three of the Arctic nations, the US, Canada, and Russia, have national EA systems but also sub-administrative EA (e.g. territorial, state) jurisdictions, including specific provisions and rights granted to Indigenous peoples. With the exceptions of certain sub-jurisdictions (e.g., Norway's Svalbard

Table 4.1 Select northern EA systems and provisions

EA system	EA provision(s)
United States (Alaska)	National Environmental Policy Act (NEPA 1970)
	The state of Alaska applies the US NEPA procedures, and the Alaska Department of Natural Resources is responsible for coordinating the state permitting process for NEPA assessments. In the offshore regions, including the Beaufort and Chukchi Seas, the Bering Sea, Cook Inlet and the Gulf of Alaska, the federal Bureau of Ocean Energy Management is responsible for NEPA analysis and related EA studies. There is no formal strategic EA system in Alaska, but the US NEPA does provide for programmatic EAs of offshore multi-project activities, programs or offshore areas – such as the programmatic EA of the Arctic Ocean outer continental shelf seismic surveys (US DOI MMS, 2006).
	See http://ceq.hss.doe.gov/
Canada (federal)	Canadian Environmental Assessment Act (CEAA 2012); Cabinet Directive for the Environmental Assessment of Policy Plan and Program Proposals (2004)
	Federal EA in Canada is required by way of CEAA (2012), replacing CEAA (1992). EA under CEAA (2012) applies only to physical undertakings (designated projects). EA is the responsibility of three federal authorities: Canadian Nuclear Safety Commission (for nuclear projects); National Energy Board (for international and interprovincial pipelines and transmission lines); Canadian Environmental Assessment Agency (for all other designated projects). Strategic environmental assessment is not legislated; it is required by way of a federal Cabinet Directive and applies only to the policy, plan, or program proposals of a federal department or agency that are submitted to an individual minister or Cabinet for approval and the implementation of the proposal may result in important environmental effects. In an effort to extend the principles of strategic environmental assessment beyond the federal level, the Canadian Council of Ministers of the Environment (2009) released principles and guidelines for regional strategic environmental assessment.
	See www.ceaa-acee.gc.ca/
Mackenzie Valley	Mackenzie Valley Resource Management Act (MVRMA 1998); CEAA (2012)
	The MVRMA emerged from the resolution of three comprehensive land claims in the Northwest Territories, requiring the coordinated management of land and water in the Mackenzie Valley. The MVRMA includes all of the Northwest Territories, with the exception of the Inuvialuit Settlement Region and Wood Buffalo National Park. The Mackenzie Valley Environmental Impact Review Board (MVEIRB), a quasi-judicial co-management board, is responsible for EA at a valley-wide level. The MVEIRB's (2004) guidelines describes EA as "a process, which examines the potential impacts of proposed developments to promote sustainability and avoid costly mistakes . . . to anticipate and avoid environmental problems, rather than reacting and fixing them after they occur." EA is to ensure protection of the environment from the significant adverse impacts; and ensure protection

continued

Table 4.1 continued

EA system	EA provision(s)
	of the social, cultural, and economic wellbeing of residents and communities of the Mackenzie Valley.
	See www.reviewboard.ca/
Inuvialuit Settlement Region	Inuvialuit Final Agreement (IFA 1984); CEAA (2012)
	The IFA established the Inuvialuit Settlement Region (ISR) and a framework for co-management between the Inuvialuit and the federal government. The ISR includes the Beaufort Sea/Mackenzie Delta and Yukon North Slope Region. Projects in the ISR are subject to EA by the Environmental Impact Screening Committee (EISC) and, depending on the potential for significant impacts, the Environment Impact Review Board (EIRB). The EISC is responsible for determining whether a proposed development could have significant environmental effects or affect present or future wildlife harvesting, and making recommendations as to whether the development can proceed or whether a more comprehensive assessment is required under the EIRB. The federal government has jurisdiction over and authority to manage the marine portion of the ISR.
	See www.eirb.ca/
Yukon	Yukon Environmental and Socio-economic Assessment Act (YESAA 2003); CEAA (2012)
	YESSA gives effect to provisions under the Umbrella Final Agreement (1993), between Canada, Government of Yukon, and Yukon First Nations, respecting the assessment of environmental and socio-economic effects. The stated purposes of YESA are, among others, to: require that, before projects are undertaken, their environmental and socio-economic effects are considered; protect and maintain environmental quality and heritage resources; ensure that projects are undertaken in accordance with principles that foster beneficial socio-economic change without undermining the ecological and social systems on which communities and their residents, and societies in general, depend; and provide opportunities for public participation in the assessment process. The Yukon Environmental and Socio-economic Assessment Board administers EA under YESAA.
	See www.yesab.ca/
Nunavut	Nunavut Land Claims Agreement (NLCA 1993); CEAA (2012)
	Article 12 of the NLCA sets out the EA process for the Nunavut Settlement Area and Outer Land Fast Ice Zone. The NLCA established the Nunavut Impact Review Board (NIRB) to conduct EA. In carrying out its EA role, the primary objectives of NIRB are to protect and promote the existing and future wellbeing of the residents and communities of the Nunavut Settlement Area, and to protect the ecosystemic integrity of the Nunavut Settlement Area. The NLCA also allows for the establishment of a Federal EA panel. NIRB's mandate is applied to crown lands, territorial lands, Inuit-owned lands and private lands. The federal government retains an overriding authority on EA.
	See www.nirb.ca/

Table 4.1 continued

EA system	EA provision(s)
Russia	Federal Law on Environmental Protection, 2002; the Federal Law on Ecological Expertise, 1995; Regulation on the Assessment of Environmental Impact, 2000
	EA under the Federal Law on Environmental Protection is defined as a process encouraging an ecologically informed administrative decision on implementation of economic and other activities through identification of possible adverse impacts, assessment of ecological impacts, taking into account public opinions, and developing measures to mitigate and prevent negative impacts. EA is designed for project-level developments and provides a conceptual regulatory approach to review and permitting. Projects subject to mandatory EA include all large-scale energy, industry, and agricultural facilities. There is no strategic EA system in Russia. Environmental assessment in each of Russia's five Arctic regions (Arkhangelsk, Karelia, Komi, Murmansk, and Nenets Autonomous Okrug) is largely influenced by Federal EA regulations.
	See Solodyankina and Koeppel (2009), Cherp and Golubeva (2004); Koivurova, (2008)
Iceland	Environmental Impact Assessment Act (1994); EU EIA Directive (2011/92/EU); EU SEA Directive (2001/42/EC)
	Iceland's Environmental Impact Assessment Act was introduced in 1994 when it joined the European Economic Area (EEA), which legalized the EU Directive on EIA 85/337, and was amended in 2000 and 2005 to align with EU requirements. The Icelandic National Planning Agency is the state authority responsible for the administration of the EA Act. Amongst the purposes of EA under Icelandic state legislation are to minimize the negative environmental impacts of developments and promote the cooperation of stakeholders and other interested parties concerning projects that are subject to assessment. In 2005, the locus of environment-related decision making was shifted to local authorities, allowing the National Planning Agency to focus on assessing impact rather than deciding whether projects should proceed (OECD, 2014). Strategic EA in Iceland has been, informally, part of municipal land use planning since the 1990s. The EU SEA directive (2001/42/EC) was formally transposed in 2006 in the Icelandic Act on Environmental Assessment of Plans and Programmes, which follows the main principles of the EU SEA Directive (OECD, 2014).
	See www.skipulagsstofnun.is/english
Sweden	EU EIA Directive (2011/92/EU); EU SEA Directive (2001/42/EC)
	The EU EIA and SEA Directives set out the principles and procedural requirements for EA in Sweden, leaving it to the discretion of the state as to how the Directives are transformed into national legislation. The Swedish Environmental Code (2000) sets out national EA requirements. The stated purpose of EA under the Code is "to establish and describe the direct and indirect impact of a planned activity or measure on people, animals, plants, land, water, air, the climate, the landscape and the cultural environment, on the management of land, water and the

continued

Table 4.1 continued

EA system	EA provision(s)
	physical environment in general, and on other management of materials, raw materials and energy. Another purpose is to enable an overall assessment to be made of this impact on human health and the environment." See www.swedishepa.se/Legislation/The-Environmental-Code/
Norway	Planning and Building Act (1990); EU EIA Directive (2011/92/EU); EU SEA Directive (2001/42/EC) Norway's first EA legislation was implemented in 1990, as part of the Planning and Building Act. As a member of the EEA, similar to Iceland, the provisions of the EU Directives have also adopted in Norway. However, the Svalbard Islands, which was excluded from the EAA agreement, has its own EA procedure enacted by Norway (Act of 15 June 2001, Ch. VII, Svalbard Environmental Protection Act). Offshore Norway, EA provisions for oil and gas activities are contained in the petroleum legislation, administered by the Ministry of Petroleum and Energy, which also provides for regional EAs prior to offshore areas being made available for licensing. In 2006, Norway introduced an additional framework, an Integrated Management Plan, to guide EA activities and offshore planning process across oil and gas, shipping and the fishing sectors. See www.regjeringen.no/en/dep/md.html?id=668
Finland	EU EIA Directive (2011/92/EU); EU SEA Directive (2001/42/EC) The EU EIA and SEA Directives set out the principles and procedural requirements for EA in Finland, leaving it to the discretion of the state as to how the Directives are transformed into national legislation. Finland's EIA Decree (713/2006) lists further the types of projects that must always be subjected to assessment; however EA may also be required for any projects where adverse environmental impacts are likely, on the basis of decisions made by the regional environment authority. The Directives do not prescribe how EAs should be completed, nor do they contain requirements as to their quality. See Jantunen (2011)
Greenland	Mineral Resources Act (2009) Greenland, as a self-governing territory of Denmark, it is not subject to EU Directives and has established its own EA provisions under its Home Rule government. Under Greenland's Mineral Resources Act, applications for certain mineral exploration and development projects, including offshore hydrocarbon activities, require an EA and a strategic social impact assessment. Prior to opening up new offshore areas for hydrocarbon exploration and licensing, a "strategic environmental impact assessment" is also prepared by the National Environmental Research Institute, the Greenland Institute of Natural Resources and the Mineral License and Safety Authority. See Hansen (2010)

Islands, Canada's Inuvialuit Settlement Region), most EA systems across the Arctic do not contain Arctic-specific provisions (Koivurova 2008). Common to all EA systems in the Arctic, however, is the provision of adequate information about the potential impacts of proposed development actions to support informed decision-making.

The gap analysis was based on a review of recent research on EA in the Arctic. We focused only on literature in the field of "environmental assessment." Our review was a country-specific and Arctic-wide search of the EA literature published in the last 20 years. Papers and reports were identified using Scopus, based on title, keywords, and subject. We also examined, in a much more limited fashion, regulatory reviews and EA system performance to identify key research issues and priorities that may be noted in professional practice literature (e.g., Koivurova, 2002; National Research Council 2003; OECD 2006 2014; Ovind and Sneve 2004; McCrank 2008; Elizabeth et al. 2014).

Results from our review were used to identify gaps in research or understanding of EA in the Arctic. Initial research gaps and priorities were identified based on: knowledge and information missing from Arctic EA research when considering current issues facing the Arctic (e.g., regulatory change, energy development, climate change, adaptation, cumulative effects, social change) (see Prowse et al. 2009; Burkett 2011; Nellemann et al. 2001; Porta and Bankes 2011); and consolidating recommendations from the range of literature reviewed on research needs or priorities.

We also surveyed ten individuals from across the Arctic, from four different countries, involved in EA research or regulation. We asked what they perceived to be the major gaps and issues in EA research in the Arctic that require significant attention. These individuals were purposively identified and either members of our own research networks or recommended by our colleagues, and well published in the field. They were asked to comment on EA research in the Arctic more broadly and not on jurisdictional-specific matters. The sample of experts was not intended to be representative of EA jurisdictions; rather, the intent was to validate Arctic-wide EA research gaps and priorities identified from our review of the literature, and to determine whether there were important research needs not captured in our review.

Survey of Arctic environmental assessment research

We identified 194 papers with a focus on some aspect of EA in the Arctic. From this, we identified eight major research themes, each consisting of sub-themes (Table 4.2). Themes were influenced by the scope of our review and focused on research about formal EA. The themes are not necessarily independent. For example, some research on transboundary EA is set within the context of strategic EA, or project-based EA where multi-jurisdictional impacts are likely to occur. In such cases, we assigned papers to the theme that we perceived as the main representation of its content based on the paper's objectives and recommendations.

Table 4.2 Current EA research themes and sub-themes

EA research theme	Sub-themes	Examples
Collaboration and participation	Collaborative EA; Learning; Participation and deliberative democracy; Community participation and negotiation; Traditional knowledge; Interactive planning; Local capacity for engagement; Conflict and early consultation; Knowledge mobilization; Social equity, empowerment and exclusion	Armitage (2005); Couch (2002); Fitzpatrick et al. (2008); Hildén (2005); Huttunen (1999); Koivurova (2008); Lajoie and Bouchard (2006); Lidskog and Soneryd (2000); Meschtyb et al. (2005); Sasvari (2012); Saarikoski (2000); Wismer (1996)
Cultural, social and health impacts	Integrating spiritual and cultural issues in EA; Social impact follow-up and monitoring; Health integration in EA; Mitigating health impacts; Determinants of health; Scope of health considerations in EA; Planning for community impacts of mega-projects; Strengthening social impact assessment; Consideration place meaning in EA	Ehrlich (2010); Erikstad et al. (2008); Gagnon (2003); Juslén (2005); Kaltenborn (1998); Kwiatkowski and Ooi (2003); Meschtyb et al. (2005); MVEIRB (2008); National Research Council (2003); Noble and Bronson (2005); Noble and Bronson (2006); Sasvari (2012); Storey and Hamilton (2003); Wernham (2007)
Regional environmental assessment	Cumulative environmental effects; Opportunities and constraints to strategic EA in the Arctic; Strategic EA for offshore energy; Need to upstream EA to the strategic level; Regional strategic EA for coordinated marine spatial planning; Strategic EA for marine environmental sustainability; Identifying priorities for Arctic development; Regional assessment for sensitive wildlife	Bruhn-Tysk and Eklund (2002); Doelle et al. (2012); Elvin and Fraser (2012); Fidler and Noble (2012a); Fidler and Noble (2013a); Fidler and Noble (2013b); Hansen and Kørnøv (2010); Johnson et al. (2005); Newton et al. (2002); Noble et al. (2013); Orenstein et al. (2010); Solodyankina and Koeppel (2009)
Procedural effectiveness	Adequacy of EA scoping; Weight of evidence of cultural impacts; Inter-cultural communication; Problems of non-binding decisions; Adaptive approaches; Limitations to highly systematic decision processes; Mitigation effectiveness; Follow-up and monitoring; Timing of application; Institutional evolution; Regulatory complexity	CARC (1996); Cherp and Golubeva (2004); Couch (2002); Ehrlich (2010); Haefele and Cliffe-Phillips (2004); Jalava et al. (2010); MVEIRB (2008); McCrank (2008); Mulvihill and Baker (2001); Nellemann and Vistnes (2003); O'Reilly (1996); Ovind and Sneve (2004); Ross (2004); Voutier et al. (2008)

Category	Description	References
Environmental assessment and decision making	Legality versus legitimacy; Use of EA information in decision process; Action-forcing mechanisms for implementing EA results; Securing knowledge versus influencing decisions; Influence of community consultation on decisions made	Ehrlich (2010); Haefele and Cliffe-Phillips (2004); Hansen (2011); Hildén and Jalonen (2005); Hokkanen (2001); Noble and Birk (2011); Pölönen et al. (2011); Sasvari (2012)
Transboundary assessment and international law and policy	Strategies for Arctic environmental protection; Existence versus uptake of Arctic guidelines for EA; Consultation challenges regarding transboundary issues; Challenges to the variability in EA across the Arctic; Strategic EA to address transboundary impacts; Disregard for Arctic ecosystems in jurisdictional EA processes	Arctic Environment Protection Strategy (1997); Azcarate et al. (2011); Brubaker and Ragner (2010); European Union (2010); Huebert (1998); Koivurova (2002); Koivurova (2008)
Negotiated agreements	Rationale for impact and benefit agreements; Scope and function of impact and benefit agreements; Environmental agreements and regulatory process; Aboriginal engagement in negotiated agreements; Community-based monitoring under environmental agreements; Link between environmental agreements and environmental assessment	Fidler (2009); Fidler and Hitch (2007); Galbraith et al. (2007); Klein et al. (2004); Knotsch and Warda (2009); Noble and Birk (2011); Noble and Fidler (2011); O'Faircheallaigh (2007); Prno and Bradshaw (2008); Sosa and Keenan (2001)
Supporting analytical and evaluation tools and techniques	Scenario analysis; Remote sensing for biodiversity assessment; Geographic Information Systems application; Technical tools for monitoring support; Wildlife assessment	Johnson et al. (2005); Kumpula et al. (2011); Newton et al. (2002); Rigina (2002); Vlassova (2006)

Of the eight themes, "collaboration and participation" and "cultural, social and health impacts" comprised approximately 38% of the papers identified. This was not surprising as such topics have been of long-standing interest in the Arctic. Research on traditional knowledge and community participation, and on the assessment of social, health, and cultural impacts were dominant sub-themes. Research on regional and strategic EA comprised 15% of papers. Much of the research under this theme was within the past five to eight years, and often focused on Arctic energy resource development and cumulative effects. The remaining 47% of the literature addressed various aspects of procedural effectiveness, the influence of EA on decision-making, transboundary assessment and international EA laws and regulations, negotiated agreements (e.g. impact and benefit agreements, environmental agreements), and analytical tools or techniques.

Research gaps and priorities

Below we identify several gaps in research that need to be addressed to advance the effectiveness of EA and ensure its relevance to Arctic communities and stakeholders. Some of these gaps reflect enduring concerns about the efficacy of EA; others reflect emerging issues in Arctic EA (Table 4.3).

Understanding expectations about EA

Notwithstanding the breadth of research on Arctic EA, we found limited research about community and stakeholder experiences and expectations. EA has come under much criticism in recent years; however, Fuggle (2005) warns that EA is not a "magic bullet" that can resolve all environmental and socio-economic issues. Part of the challenge to understanding the efficacy of EA is that the underlying purpose of EA is still much debated. Cashmore (2004) describes EA as a series of nebulous models along a spectrum of philosophies and values concerning the role of science in EA. At one end of this spectrum is the belief that scientific method provides the basis for EA theory and practice; at the other end of the spectrum is the belief that EA is a civic science intended to empower communities, promote social justice, and realize community self-governance (Bond, Morrison-Saunders and Howitt 2012).

Based on research in Sweden, Hilding-Rydevik (2006: 25) suggests that what constitutes "effective" EA can be, "viewed from the various and differing perspectives of the many actor groups that are a part of the EA system and its processes – legislators, proponents, competent authorities, NGOs etc." Different actors have different roles and aspirations and may view EA's effectiveness based on their role in and influence on the decisions that emerge. Sasvari (2012), for example, reports on recent EA experiences of Saami communities and developers and concludes that policies conflict with Saami perceptions and knowledge. In the case of Canada's BHP Diamond mine, O'Reilly (1996) and CARC (1996) argued that the EA was neither rigorous, comprehensive, or fair; but Kwiatkowski and

Table 4.3 Synthesis of northern EA research priorities

Research theme	Research questions
Community and stakeholder expectations about EA	a. What do Arctic communities and stakeholders expect of EA? b. Is EA the right mechanism to meet these expectations?
Efficiency and responsiveness	a. As a process is EA sufficiently expeditious, flexible, and responsive to communities' and proponent's needs in the context of a rapidly changing Arctic economic and biophysical environment? b. What reforms are needed to ensure EA processes are sufficiently expeditious, flexible and responsive to communities' and proponent's needs without compromising its effectiveness?
Impact and influence of EA	a. What influence has EA had on development decisions across the Arctic? b. What lessons can be learned from the decades of EA application to resource-mega-projects across the Arctic?
Capacity for meaningful engagement in EA	a. Has past engagement in EA facilitated learning and capacity building in Arctic communities? b. What is the current capacity of Arctic communities and Aboriginal organizations to be meaningfully engaged in EA, or to use EA as a planning and decision-making tool? c. Given the expected increase in development applications in an ice free Arctic, what are the capacity building requirements to ensure sustained and meaningful engagement in EA? d. What institutional or process reforms are needed to ensure more effective engagement in EA in the face of limited resources and under the time constraints of evolving EA processes?
Strengthening EA through land use planning and science	a. Are current regional planning, science, and monitoring programs in the Arctic responsive to the regulatory and broad governance needs of EA? b. What are the opportunities and mechanisms to improve EA practice through better integration with regional planning? c. How can current Arctic science and monitoring programs be better integrated into EA practices and decision making?
Applied regional and strategic EA	a. As part of developing futures-oriented planning and assessment processes, what frameworks and methods are needed to support EA application at the regional or sub-regional scale? b. What are the lessons, opportunities, and institutional requirements needed to scale-up regional EA to Arctic planning regions and transboundary eco-regions?
Climate change	a. What are the implications of climate change for current EA systems and processes? b. How should EA be used to help anticipate and respond to the impacts of climate change? c. How can climate change adaptation needs be addressed through EA processes?

Ooi (2003) characterized it as an integrated assessment that considered social, cultural, health, and environmental impacts, and resulted in a project that coordinated the concerns of all stakeholders.

In a review of EA in the western Arctic, Noble et al. (2013) identified a range of expectations about EA and what it can and should deliver and to whom. Similarly, in Greenland's offshore oil sector, Olsen and Hansen (2014) report a diversity of expectations concerning what EA should versus what it can deliver regarding participation. They explain that industry often approaches EA to manage expectations and achieve a social license; whilst communities see it as a means to understand a project, prepare for its impacts, and better capture economic opportunities. The requirements for and the practice of EA vary considerably across the Arctic, but there has been limited effort to fully understand the diversity of Indigenous, community, and stakeholder expectations about EA.

Process efficiency and responsiveness

The Arctic Monitoring and Assessment Program (AMAP 2007) recommended that EA and related planning tools be "rigorously applied" in the Arctic, but also emphasized that such tools be "streamlined to increase their relevance and usefulness" (AMAP 2007: viii). Recent literature on EA effectiveness has tended to focus on making the process less cumbersome and more efficient. In Canada's western Arctic, for example, evaluations of EA have largely been from the perspective of industry or regulators and focused on process or regulatory efficiency (e.g., Voutier et al. 2008; Harrison 2006; McCrank 2008). Voutier et al. (2008) note that EA in Canada's Arctic is becoming increasingly complex and that the "regulatory regime will undermine the attractiveness" of Canada's western Arctic to industry investors. In Iceland, a 2014 OECD performance review describes EA as complex and slow and recommends, as a step towards "green growth," streamlining the EA process to reduce administrative costs and delay (OECD 2014). There has been very limited attention to Arctic community views of the EA process, including its complexity and responsiveness to concerns about development, or whether a faster EA process to accommodate regulatory efficiencies compromises its effectiveness.

Impact and influence

Efficiency is a valid concern, but more attention needs to be given to understanding the influence of EA on decisions and its contribution to environmental manage-ment. There has been some limited reporting on the impact of EA on decision outcomes in Canada's Arctic. Ehrlich (2010), for example, addressed the "weighing of evidence" with respect to the assessment of spiritual issues and the role it played in the rejection of three projects and conditional approval of another in Canada's Arctic. However, Hansen (2011) argues that besides securing environ-mental knowledge there is little evidence that EA has influenced decision-making. In the Finnish context, Hildén and Jalonen (2005) report limited influence of

EA on decision-making and development choices, and Pölönen (2006) concludes that Finnish EA legislation does not guarantee that assessment results are transferred to decision-making. Hokkanen (2001) similarly reports that a significant amount of information is generated through Finnish EA, but there is insufficient time to use the resulting knowledge. Impact and influence have also been long-standing issues under Russia's EA system (Cherp and Golubeva 2004). Von Ritter and Tsirkunov (2003) reported that Russian EA has had limited impact on project design, aside from cases of highly visible and internationally financed projects. Amongst the reasons reported for the limited influence are inadequate baseline data, the limited capacities of EA regions to implement regulatory requirements, and the late preparation of EAs – often after important project decisions have been made. This may reflect Russian institutional and political challenges.

Arguably, however, many Arctic EA applications have yielded positive outcomes, and thus lessons in terms of EA's influence and impact. Too often these positive lessons are not shared, leading to lost opportunities for EA learning and improvement (Mulvihill and Baker, 2001).

Capacity for meaningful engagement EA

There is increasing recognition of the need for EA to integrate traditional knowledge and increasing expectations for industry to ensure early and ongoing community engagement (Fitzpatrick, Sinclair and Mitchell 2008; Armitage 2005; Meschtyb et al. 2005; Saarikoski 2000). For the Sierilä hydropower station, northern Finland, Huttunen (1999) reports that community engagement in EA increased mutual learning and understanding, resulting in "a significant awakening to their own empowerment and self-management." At the same time, however, Huttunen cautioned that communities lack capacity to participate effectively in EA. Capacity concerns were similarly raised at the 2011 Alaska Forum on the Environment, reporting to the US Environmental Protection Agency of the need for Arctic communities to not only see the benefits of engaging in EA but also to have the resources to ensure meaningful engagement (Interagency Working Group on Environmental Justice 2011).

Consultation during and after the EA process, combined with increasing EA applications, are raising concerns about the capacity of communities to become meaningfully engaged in EA. In their review of marine planning, assessment, and science programs in Canada's western Arctic, for example, Fidler and Noble (2013a) identified capacity and resources as constraints to meaningful participation. The MVEIRB (2008) also identified the lack of capacity as an ongoing concern, noting community concerns about the constant struggle to retain their capacity to participate in EAs and the increasing workload of Aboriginal groups. McCrank (2008) describes the limited institutional and human resource capacity of impacted Aboriginal organizations as hindering their ability to participate in EA and to document and interpret traditional knowledge to assist in decision-making. That said, not all regions provide an opportunity for engagement. Public participation in Russian EA varies depending on the openness of regional authorities. Russia's EA system has been described as focused on verifying project compliance existing laws and regulations, with limited opportunity for public participation until the very late stages of the process (OECD 2006).

Given the enduring concerns over capacity for engagement in EA, combined with the anticipated growth in EA applications for Arctic development, research is needed on how to facilitate capacity building in Arctic communities and the institutional or process reforms necessary to ensure that such engagement occurs.

Opportunities and mechanisms to link strategic initiatives with EA

Increasingly, scholars, Arctic communities, and environmental organizations are seeking regional and strategic approaches to EA (see IGC 2004; WWF 2005; Cherp and Golubeva 2004; Doelle, Bankes and Porta 2012; Fidler and Noble 2012). There is a need for EA to be more proactive in its approach to planning for future development and to better assess cumulative effects, including climate change and transboundary effects. There are many planning, science, and assessment programs ongoing in the Arctic. How these programs contribute to better EA, and vice versa, remains unclear. In the Canadian Arctic, regional initiatives such as the Integrated Oceans Management Plan (IOMP), Beaufort Regional Environmental Assessment (BREA), Integrated Regional Impact Studies (IRIS), the Cumulative Impact Monitoring Program (CIMP), and various land use plans address issues that may be relevant to EA; however, the challenge is determining how the processes and data and knowledge generated are best translated and used to inform and influence EA and decisions (Fidler and Noble 2013). There is also an assumption that these processes are useful to EA. However, with regard to CIMP, for example, the MVEIRB (2005: 7) reports: "it is unclear to the MVEIRB what information is being collected through CIMP . . . and how this information can be used in the EA process."

The relationships between regional and strategic initiatives and EA is described as "tiering" (Fischer 2007; João 2005), whereby regional or higher-level initiatives (e.g., planning processes, regional studies, monitoring programs) are intended to influence, or provide direction to project EA. How to facilitate tiering remains a significant challenge in practice. Research is required to determine the value added of regional policy, planning, and science initiatives to Arctic EA, and how to better link them with downstream project EA decisions.

Applied research to demonstrate regional strategic EA

In 2008, the Beaufort Sea Strategic Regional Plan of Action (BSStRPA 2008) identified the need for a coordinated and strategic approach to EA in the Beaufort region. The federal response was BREA – a four-year research project to collect data on specific issues related to offshore oil and gas development to identify and fill gaps in baseline data related to offshore activities and the marine environment for supporting project assessments. BREA may prove valuable in these regards, but the strategic need to identify priorities for sustainable development (see Newton, Fast and Henley 2002) is missing from the BREA process.

Work has been done to advance the understanding of, and a generic framework for, regional strategic EA (CCME 2009); and there is a growing volume of research on regional strategic EA opportunities in the Arctic, particularly for the marine environments (Fidler and Noble 2013; Doelle, Bankes and Porta 2012; Kinn 1999). On offshore Norway, for example, regional strategic EA is described as an effective process for determining how to move forward in terms of planning for development and, in the context of offshore hydrocarbon development, where future leasing could occur (Fidler and Noble 2012). However, Ketilson (2011) identified reservations amongst western Arctic communities, regulators, and industry about regional strategic EA due to its "unproven benefits."

The EA community has approached the potential role of regional and strategic EA in the Arctic; but applied research is needed to pilot test regional and strategic EA applications, specifically applications that involve prospective or futures-based assessment, and to learn from those applications.

Adapting EA to climate change

The requirements for, and scope of EA varies considerably across the Arctic, but all regions are faced with the challenges of a rapidly changing climate. The MVEIRB (2005: 7) notes that climate change "has not yet featured prominently in any EA, with the exception of the Mackenzie Gas Project," but "anticipates that climate change issues will play an increasing role in future assessments." We found very little Arctic EA research focused specifically on climate change and impacts. Although recent literature has focused on climate change mitigation in EA (Burdge 2008, Byer et al. 2012), specifically project based GHG emissions assessment (Oshawa and Duinker 2014), the implications for EA processes, and for those Arctic development sectors subject to EA, remains unexplored. Climate change will inevitably impose new demands on EA, but EA may also function as an essential tool for addressing the challenges to new and expanded development under changing climatic conditions.

Other enduring and emerging research needs

There are several other issues that we suggest are important to advancing the effectiveness of EA and its relevance to Arctic communities and stakeholders, but not necessarily captured under the above themes or treated as unique gaps in Arctic EA research. They are persistent needs in EA practice, regardless of political-economic context or location; but they can pose unique challenges for Arctic settings.

The first concerns socio-economic indicators to support EA. The MVEIRB (2005) identifies the lack of baseline information on socio-economic conditions in the Arctic as an ongoing EA concern. This is a concern that extends to other Arctic jurisdictions. Of particular need is to identify indicators for monitoring socio-economic conditions that are not only responsive to regional change, but also help predict and evaluate specific impacts of Arctic resource development. Based on research in Finland's mining sector, Suopajärvi (2013) argues that such indicators should not only be "hard" or "quantitative," but also qualitative, value-based, even hermeneutic in approach.

Second, there is a need to examine the relationship between negotiated agreements (e.g. impact and benefit agreements) and the EA process. In some regions, such as Greenland, negotiated agreements are part of the formal requirements for EA, are arranged between the proponents, affected municipalities, and the national government and are within the realm of public law. In most jurisdictions, however, agreements are negotiated in confidence. The relationship between privately agreements involving communities and proponents, and the public EA process, is poorly understood. Comparative analyses of privatized versus public approaches to negotiated agreements may provide a better understanding of the extent to which negotiated agreements complement or undermine EA.

Third, guidelines for EA under the 1997 Arctic Environmental Protection Strategy for cooperation between the eight Arctic states have not been fully incorporated into national EA systems (Koivurova 2008). The coordination of EA roles and responsibilities between jurisdictions to address transboundary impacts has proven difficult within national EA systems (Fitzpatrick and Sinclair 2009); these challenges are exacerbated when dealing with impacts that involve multiple national Arctic EA systems. Research is needed to explore the institutional, policy, and legal options and opportunities for strengthening international Arctic EA coordination.

Conclusions

This chapter examined the current state of research on Arctic EA and identified gaps in research that should be addressed to advance the effectiveness of EA, enhance its relevance to Arctic communities and stakeholders, and address issues unique to Arctic EA and management. We recommend seven priority research themes for Arctic EA (Table 4.3), and identify several enduring issues that, whilst not unique gaps in Arctic EA research, are important issues to ensuring the effectiveness of EA application in the Arctic. In conclusion, the research gaps and opportunities identified in this chapter are not the *only* research gaps and priorities and others may have different views of what EA is and can or should deliver in Arctic contexts. There is also a diversity of interpretations about what works, what isn't working, and what needs to be done to improve EA across the Arctic. EA should play a key role in planning for the impacts of development and in creating responses that will allow Arctic communities to best respond to new opportunities and to the pressures associated with environmental, social, and cultural change.

Acknowledgements

This chapter was supported by a ReSDA sub-grant, funded by the Social Sciences and Humanities Research Council of Canada. We wish to acknowledge our key informants for contributing their time and expertise to the gap analysis. This paper is an adaptation of a version previously published in Arctic, vol. 86, no. 3; http://dx.doi.org/10.14430/arctic4501.

References

AMAP. 2007. *Arctic Oil and Gas 2007*. Oslo: Arctic Monitoring and Assessment Program.

Arctic Environment Protection Strategy. 1997. *Guidelines for Environmental Impact Assessment (EIA) in the Arctic. Sustainable Development and Utilization*. Finnish Ministry of the Environment.

Armitage, D. 2005. Collaborative Environmental Assessment in the Northwest Territories, Canada. *Environmental Impact Assessment Review*, 25: 239–258.

Azcarate, J., Balfors, B., Destouni, G., and Bring, A. 2011. Shaping a Sustainability Strategy for the Arctic. Paper presented at 2011 International Association of Impact Assessment Annual Conference, Puebla, Mexico, 28 May–4 June.

Beaufort Sea Strategic Regional Plan of Action (BSStRPA). 2008. www.bsstrpa.ca.

Berger, T. 1977. Northern Frontier, Northern Homeland: The Report of the Mackenzie Valley Pipeline Inquiry. Ottawa: Environment Canada.

Bond, A., Morrison-Saunders, A., and Howitt, H. (eds) 2012. *Sustainability Assessment*. London: Routledge.

Burkett, V. 2011. Global Climate Change Implications for Coastal and Offshore Oil and Gas Development. *Energy Policy*, 39: 7719–7725.

Brubaker, D. and Ragner, C. 2010. A Review of the International Northern Sea Route Program (INSROP) – 10 Years On. *Polar Geography*, 33(1–2): 15–38.

Bruhn-Tysk, S. and Mats Eklund, M. 2002. Environmental Impact Assessment A Tool for Sustainable Development? A Case Study of Biofuelled Energy Plants in Sweden. *Environmental Impact Assessment Review*, 22: 129–144.

Burdge, R. 2008. The Focus of Impact Assessment (and IAIA) Must Now Shift to Climate Change! *Environmental Impact Assessment Review*, 28(8): 618–622.

Byer, P., Cestti, R., Croal, P., Fisher, W., Hazell, S., Kolhoff, A., and Kørnøv, L. 2012. *Climate Change in Impact Assessment: International Best Practice Principles*. Special publication series no. 8. Fargo, ND: International Association for Impact Assessment.

Canadian Council of Ministers of the Environment (CCME). 2009. *Regional Strategic Environmental Assessment in Canada: Principles and Guidance*. CCME: Winnipeg, MB.

CARC. 1996. Critique of the BHP Environmental Assessment: Purpose, Structure, and Process. *Northern Perspectives*, 24: 1–4.

Cashmore, M. 2004. The Role of Science in Environmental Impact Assessment: Process and Procedure versus Purpose in the Development of Theory. *Environmental Impact Assessment Review*, 24(4): 403–426.

Cherp, A. and Golubeva, S. 2004. Environmental Assessment in the Russian Federation: Evolution through Capacity Building. *Impact Assessment and Project Appraisal*, 22(2): 121–130.

Couch, W. J. 2002. Strategic Resolution of Policy, Environmental and Socio-economic Impacts in Canadian Arctic Diamond Mining: BHP's NWT Diamond Project. *Impact Assessment and Project Appraisal*, 20(4): 265–278.

Doelle, M., Bankes, N., and Porta, L. 2012. Using Strategic Environmental Assessments to Guide Oil and Gas Exploration Decisions in the Beaufort Sea: Lessons Learned from Atlantic Canada. CIRL Occasional Paper #39. pp. 34.

Ehrlich, A. 2010. Cumulative Cultural Effects and Reasonably Foreseeable Future Developments in the Upper Thelon Basin, Canada. *Impact Assessment and Project Appraisal*, 28(4): 279–286.

Elizabeth, T., Lesser, P., Riedel, A., and Weingartner, K. 2014. *Gap Analysis Report. Preparatory Action: Strategic Environmental Impact Assessment of Development of the Arctic.* Arctic Centre, University of Lapland: European Union.

Elvin, S. and Fraser, G. 2012. Advancing a National Strategic Environmental Assessment for the Canadian Offshore Oil and Gas Industry with Special Emphasis on Cumulative Effects. *Journal of Environmental Assessment Policy and Management*, 14(3): 1–37.

Erikstad, L., Lindblom, I., Jerpåsen, G., Hanssen, M., Bekkby, T., Stabbetorp, O., and Bakkestuen, V. 2008. Environmental Value Assessment in a Multidisciplinary EIA Setting. *Environmental Impact Assessment Review*, 28: 131–143.

European Union. 2010. EU Arctic Footprint and Policy Assessment: Final Report. Berlin: Ecologic Institute.

Fidler, C. 2009. Increasing the Sustainability of a Resource Development: Aboriginal Engagement and Negotiated Agreements. *Environment, Development and Sustainability*, 12(2): 233–244.

Fidler, C. and Hitch, M. 2007. Impact and Benefit Agreements: A Contentious Issue for Environmental and Aboriginal Justice. *Environments Journal*, 35(2): 49–69.

Fidler, C. and Noble, B. 2012. Advancing Strategic Environmental Assessment in the Offshore Oil and Gas Sector: Lessons from Norway, Canada, and the United Kingdom. *Environmental Impact Assessment Review*, 34: 12–21.

Fidler, C. and Noble, B. 2013a. Stakeholder Perceptions of Current Planning, Assessment and Science Initiatives in Canada's Beaufort Sea. *Arctic*, 66(2): 179–190.

Fidler, C. and Noble, B. 2013b. Advancing Regional Strategic Environmental Assessment in Canada's Western Arctic: Implementation Opportunities and Challenges. *Journal of Environmental Assessment Policy and Management*, 15(1): 1–27.

Fischer, T. B. (ed.) 2007. *Theory and Practice of Strategic Environmental Assessment: Towards a More Systematic Approach.* London: Earthscan.

Fitzpatrick, P., Sinclair, J., and Mitchell, B. 2008. Environmental Impact Assessment Under the Mackenzie Valley Resource Management Act: Deliberative Democracy in Canada's North? *Environmental Management*, 42: 1–18.

Fitzpatrick, P. and Sinclair, J. 2009. Multi-jurisdictional Environmental Assessment, in K. Hanna (ed.), *Environmental Impact Assessment: Practice and Participation*. Don Mills ON: Oxford University Press.

Fuggle, R 2005. Have Impact Assessments Passed Their 'Sell By' Date? *International Association for Impact Assessment Newsletter*, 16(1): 1, 6.

Gagnon, C. 2003. Methodology of Social Impact Follow-Up Modeling: The Case Study of a New Aluminium Smelter in Canada (Alma, Alcan), in R. Rasmussen and N. Koroleva (eds.), *Social and Environmental Impacts in the North*. Dordrecht: Kluwer Academic Publishers, pp. 479–489.

Galbraith, L., Bradshaw, B., and Rutherford, M. 2007. Towards a New Supraregulatory Approach to Environmental Assessment in Northern Canada. *Impact Assessment and Project Appraisal*, 25(1): 27–41.

Gibson, R. and Hanna, K. 2009. The Evolution of Federal Environmental Assessment in Canada, in K. Hanna (ed.), *Environmental Impact Assessment Practice and Participation*. Don Mills, ON: Oxford University Press.

Haefele, M. and Cliffe-Phillips, K. 2004. Environmental Impact Assessment Made in the North. Vancouver, BC: Annual conference of the International Association for Impact Assessment, April 2004.

Hanna, K. and Noble, B. F. 2011. The Canadian Environmental Assessment Registry: Promise and Reality. UVP-report. *Journal of the German EIA/SEA Association*, 25(4): 222–225.

Hansen, A. 2011. Strategic Environmental Assessment (SEA) as a Means to Include Environmental Knowledge in Decision Making in the Case of an Aluminum Reduction Plant in Greenland. *Journal of Environmental Planning and Management*, 54(9): 1261–1278.

Hansen, A. and Kørnøv, L. 2010. A Value-Rational View of Impact Assessment of Mega Industry in a Greenland Planning and Policy Context. *Impact Assessment and Project Appraisal*, 28(2): 135–145.

Hansen, H. 2010. SEA Effectiveness and Power in Decision Making – A Case Study of Aluminium Production in Greenland, PhD thesis. Aalborg University.

Harrison, C. 2006. Industry Perspectives on Barriers, Hurdles, and Irritants Preventing Development of Frontier Energy in Canada's Arctic Islands. *Arctic*, 59(2): 238–242.

Hilden, M. 2005. SEA Experience in Finland, in B. Sadler (ed.), *Strategic Environmental Assessment at the Policy Level: Recent Progress, Current Status and Future Prospects*. Ministry of the Environment, Czech Republic, pp. 55–62.

Hildén, M. and Jalonen P. 2005. Implementing SEA in Finland – Further Development of Existing Practice, in M. Schmidt, E. João, and E. Albrecht (eds.), *Implementing Strategic Environmental Assessment*. London: Springer-Verlag.

Hilding-Rydevik, T. 2006. Environmental Assessment – Effectiveness, Quality and Success, in L. Emmelin (ed.), *Effective Environmental Assessment Tools – Critical Reflections in Concepts and Practice*. Research report No. 2006:03, Blekinge Institute of Technology.

Hokkanen, P. 2001. EIA and Decision Making in Search of Each Other: The final Disposal of Nuclear Waste in Finland, in T. Hilding-Rydevik (ed.), *EIA, Large Develop ment Projects and Decision-making in the Nordic Countries*. Stockholm: Nordregio Report 6.

Huebert, R. 1998. New Directions in Circumpolar Cooperation: Canada, the Arctic Environmental Protection Strategy, and the Arctic Council. *Canadian Foreign Policy Journal*, 5(2): 37–57.

Huttunen, A. 1999. The Effectiveness of Public Participation in the Environmental Impact Assessment Process – A Case Study of the Projected Sierilä Hydropower Station at Oikarainen, Northern Finland. *Acta Borealia: A Nordic Journal of Circumpolar Societies*, 16(2): 27–41.

IAIA and IEA. 1999. *Principles of Environmental Impact Assessment Best Practice*. Fargo, ND.

Inuvialuit Game Council (IGC) 2004. Letter to Honourable David Anderson, PC, MP; 21 June.

Jalava, K., Pasanen, S., Saalasti, M., and Kuitunen, M. 2010. Quality of Environmental Impact Assessment: Finnish EISs and the Opinions of EIA Professionals. *Impact Assessment and Project Appraisal*, 28(1): 15–27.

Jantunen, J. 2011. Evaluation of the Finnish EIA System. Conference Proceedings of the International Association for Impact Assessment, Impact Assessment and Responsible Development for Infrastructure, Business and Industry Centro de Convenciones, Puebla, Mexico.

João, E. M. 2005. Key Principles of SEA, in M. Schmidt, E. M. João, and E. Albrecht (eds.), *Implementing Strategic Environmental Assessment*. Berlin: Springer.

Johnson, C., Boyce, M., Case, R., Dean, H., Gau, R., Gunn, A., and Mulders, R. 2005. Cumulative Effects of Human Developments on Arctic Wildlife. *Wildlife Monographs*, 160: 1–36.

Juslén, J. 1995. Social Impact Assessment: A Look at Finnish Experiences. *Project Appraisal*, 10(3): 163–170.

Kaltenborn, B. 1998. Effects of Sense of Place on Responses to Environmental Impacts: A Study among Residents in Svalbard in the Norwegian High Arctic. *Applied Geography*, 18(2): 169–189.

Ketilson, S. 2011. Regional Strategic Environmental Assessment Roles and Stakes in Arctic Oil and Gas Development. MES thesis, University of Saskatchewan, Saskatoon, Saskatchewan.

Kinn, S. J. 1999. Regional Environmental Impact Assessment – Experiences from Norwegian Petroleum Activity. Proceedings from the 3rd Nordic EIA/SEA conference.

Klein, H., Donihee, J., and Stewart, G. 2004. Environmental Impact Assessment and Impact and Benefit Agreements: Creative Tension or Conflict? Presentation at the 2004 International Association for Impact Assessment annual meeting. Vancouver, BC.

Knotsch, C. and Warda, J. 2009. *Impact Benefit Agreements: A Tool for Healthy Inuit Communities?* Ottawa: National Aboriginal Health Organization.

Koivurova, T. 2002. *Environmental Impact Assessment in the Arctic: A Study of International Legal Norms*. Farnham, UK: Ashgate Publishing.

Koivurova, T. 2008. Transboundary Environmental Assessment in the Arctic. *Impact Assessment and Project Appraisal*, 26(4): 265–275.

Kumpula, T., Pajunen, A., Kaarlejärvi, E., Forbes, B., and Stammler, F. 2011. Land Use and Land Change in Arctic Russia: Ecological and Social Implications of Industrial Development. *Global Environmental Change*, 21(2): 550–562.

Kwiatkowski, R. and Ooi, M. 2003. Integrated Environmental Impact Assessment: A Canadian Example. *Bulletin of the World Health Organization*, 81(6): 434–438.

Lajoie, G. and Bouchard, M. 2006. Native Involvement in Strategic Assessment of Natural Resource Development: The Example of the Crees Living in the Canadian Taiga. *Impact Assessment and Project Appraisal*, 24(3): 211–220.

Lidskog, R. and Soneryd, L. 2000. Transport Infrastructure Investment and Environmental Impact Assessment in Sweden: Public Involvement or Exclusion? *Environment and Planning A*, 32(8): 1465–1479.

McCrank, N. 2008. *Road to Improvement: The Review of Regulatory Systems Across the North*. Indian and Northern Affairs Development, Canada.

Meschtyb, N., Forbes, B., and Kankaanpää, P. 2005 Social Impact Assessment along Russia's Northern Sea Route: Petroleum Transport and the Arctic Operational Platform (ARCOP). *Arctic*, 58(3): 322–327.

Morrison-Saunders, A. and Bailey, J. 1999. Exploring the EIA/Environmental Management Relationship. *Environmental Management*, 24(3), 281–295.

Morgan, R. K. 2012. Environmental Impact Assessment: The State of the Art. *Impact Assessment and Project Appraisal*, 30(1): 5–14.

Mulvihill, P. and Baker, D. 2001. Ambitious and Restrictive Scoping: Case Studies from Northern Canada. *Environmental Impact Assessment Review*, 21(4): 363–384.

MVEIRB. 2004. Environmental Impact Assessment Guidelines. Available at www.review board.ca.

MVEIRB. 2005. EIA Made in the North. The Mackenzie Valley Environmental Impact Review Board's Submission to the NWT Environmental Audit. Available at www.review board.ca.

MVEIRB. 2008. Environmental Impact Assessment (EIA) Practitioner's Workshop Report. Available at www.reviewboard.ca/.

MVEIRB. n.d. Issues and Recommendations for Social and Economic Impact Assessment in the Mackenzie Valley. Available at www.reviewboard.ca.

National Research Council. 2003. Cumulative Environmental Effects of Oil and Gas Activities on Alaska's North Slope. Washington, DC: National Academies Press.

Nellemann, C. Kullerud, L., Vistnes, I., Forbes, B. C., Husby, E., Kofinas, G. P. . . . Larsen, T. S. 2001. *GLOBIO Global Methodology for Mapping Human Impacts on the Biosphere: The Arctic 2050 Scenario and Global Application.* UNEP/DEWA Technical Report 3. United Nations Environment Programme, Nairobi.

Nellemann, C. and Vistnes, I. 2003. Large-scale Environmental Dynamics: Assessing Processes and Impacts, in R. Rasmussen and N. Koroleva (eds.), *Social and Environmental Impacts in the North.* Dordrecht: Kluwer Academic Publishers, pp. 3–6.

Newton, S., Fast, H., and Henley, T. 2002. Sustainable Development for Canada's Arctic and Subarctic Communities: A Backcasting Approach to Churchill, Manitoba. *Arctic,* 55(3): 281–290.

Noble, B. and Bronson, J. 2005. Integrating Human Health into Environmental Impact Assessment: Case Studies of Canada's Northern Mining Resource Sector. *Arctic,* 58(4): 395–405.

Noble, B. and Bronson, J. 2006. Practitioner Survey of the State of Health Integration in Environmental Assessment: The Case of Northern Canada. *Environmental Impact Assessment Review,* 26: 410–424.

Noble, B. F. and Birk, J. 2011. Comfort Monitoring? Environmental Assessment Follow-up under Community-Industry Negotiated Environmental Agreements. *Environmental Impact Assessment Review,* 31(1): 17–24.

Noble, B. F. and Fidler, C. 2011. Advancing Indigenous Community – Corporate Agreements: Lessons from Practice in the Canadian Mining Sector. *Oil, Gas and Energy Law Intelligence,* 9(4): 1–30.

Noble, B. F., Ketilson, S., Aitken, A., and Poelzer, G. 2013. Strategic Environmental Assessment Opportunities and Risks for Arctic Offshore Energy Planning and Development. *Marine Policy,* 39: 296–302.

OECD. 2006. Environmental Policy and Regulation in Russia: The Implementation Challenge. Paris, France: Organization for Economic Cooperation and Development.

OECD. 2014. OECD Environmental Performance Reviews: Iceland 2014. Paris, France: Organization for Economic Cooperation and Development.

O'Faircheallaigh, C. 2007. Environmental Agreements, EIA Follow-up and Aboriginal Participation in Environmental Management: the Canadian Experience. *Environmental Impact Assessment Review,* 27(4): 319–342.

Ohsawa, T. and Duinker, P. 2014. Climate-change Mitigation in Canadian Environmental Impact Assessments. *Impact Assessment and Project Appraisal,* 32(3): 222–233.

Orenstein, M., Fossgard-Moser, T., Hindmarch, T., Dowse, S., Kuschminder, J., McCloskey, P., and Mugo, R. 2010. Case Study of an Integrated Assessment: Shell's North Field Test in Alberta, Canada. *Impact Assessment and Project Appraisal,* 28(2): 147–157.

O'Reilly, K. 1996. Diamond Mining and the Demise of Environmental Assessment in the North. *Northern Perspectives,* 24: 1–4.

Ovind, A. and Sneve, M. 2004. Environmental Impact Assessment and Risk Assessment in Northwestern Russia – from a Norwegian Perspective. CEG Workshop March 17–19. Available at: www.iaea.org/OurWork/ST/NE/NEFW/CEG/documents/ws03 2004_Ovind.pdf

Pölönen, I. 2006. Quality Control and the Substantive Influence of Environmental Impact Assessment in Finland. *Environmental Impact Assessment Review*, 26: 481–491.

Pölönen, I., Hokkanen, P., and Jalava, K. 2011. The Effectiveness of the Finnish EIA system – What Works, What Doesn't, and What Could Be Improved? *Environmental Impact Assessment Review*, 31: 120–128.

Porta, L. and Bankes, N. 2011. *Becoming Arctic-Ready: Policy Recommendations for Reforming Canada's Approach to Licensing and Regulating Offshore Oil and Gas in the Arctic.* Washington, DC: The PEW Environmental Group.

Prno, J. and Bradshaw, B. 2008. Program Evaluation in a Northern Aboriginal Setting: Assessing Impact and Benefit Agreements. *Journal of Aboriginal Economic Development*, 6(1): 59–75.

Prowse, T. D., Furgal, C., Chouinard, R., Melling, H., Milburn, D., and Smith, S. L. 2009. Implications of Climate Change for Economic Development in Northern Canada: Energy, Resource, and Transportation Sectors. *Ambio* 38(5): 272–281.

Rigina, O. 2002. Environmental Impact Assessment of the Mining and Concentration Activities in the Kola Peninsula, Russia by Multidate Remote Sensing. *Environmental Monitoring and Assessment*, 75: 11–31.

Ross, B. 2004. The Independent Environmental Watchdog: A Canadian Experience in EIA follow-up, in A. Morrison-Saunders. and J. Arts (eds), *Assessing Impact: Handbook of EIA and SEA Follow-up.* London: Earthscan, pp. 178–196.

Saarikoski, H. 2000. Environmental Impact Assessment (EIA) as Collaborative Learning Process. *Environmental Impact Assessment Review*, 20: 681–700.

Sasvari, A. 2012. Consultation Practices and Assessment of Wind Power Impacts on Indigenous Saami lands. Paper presented at the 32nd Annual Meeting of the International Association for Impact Assessment, 27 May–1 June, Porto, Portugal.

Solodyankina, S. and Koeppel, J. 2009. The Environmental Impact Assessment Process for Oil and Gas Extraction Projects in the Russian Federation: Possibilities for Improvement. *Impact Assessment and Project Appraisal*, 27(1): 77–83.

Sosa, I. and Keenan, K. 2001. *Impact Benefit Agreements between Aboriginal Communities and Mining Companies: Their Use in Canada.* Toronto: Canadian Environmental Law Association.

Storey, K. and Hamilton, L. 2003. Planning for the Impacts of Megaprojects: Two North American Examples, in R. Rasmussen and N. Koroleva (eds.), *Social and Environmental Impacts in the North.* Dordrecht: Kluwer Academic Publishers, pp. 281–302.

Suopajärvi, L. 2013. Social Impact Assessment in Mining Projects in Northern Finland: Comparing Practice to Theory. *Environmental Impact Assessment Review*, 42: 25–30.

Vlassova, T. 2006. Arctic Residents' Observations and Human Impact Assessments in Understanding Environmental Changes in Boreal Forests: Russian Experience and Circumpolar Perspectives. *Mitigation and Adaptation Strategies for Global Change*, 11: 897–909.

von Ritter, K. and Tsirkunov, V. 2003. *How Well is Environmental Assessment Working in Russia? A Pilot Study for the World Bank to Assess the Capacity of Russia's EA System.* Moscow: Alex Publishing House.

Voutier, K., Dixit, B., Millman, P., Reid, J., and Sparkes, A 2008. Sustainable Energy Development in Canada's Mackenzie Delta–Beaufort Sea Coastal Region. *Arctic*, 61(Suppl. 1): 103–110.

Wernham, A. 2007. Inupiat Health and Proposed Alaskan Oil Development: Results of the First Integrated Health Impact Assessment/Environmental Impact Statement for Proposed Oil Development on Alaska's North Slope. *EcoHealth*, 4: 500–513.

Wismer, S. 1996. The Nasty Game: How Environmental Assessment is Failing Aboriginal Communities in Canada's North. *Alternatives Journal*, 22(4):10–17.

World Wildlife Fund (WWF). 2005. *Where Are All the SEAs? Project Finance, and Strategic Environmental Assessment of Major Oil and Gas Developments*. WWF.

5 From narrative to evidence

Socio-economic impacts of mining in Northern Canada

Thierry Rodon and Francis Lévesque[1]

Canadian governments have regularly seen the North as a resource trough that would contribute to the wealth of Canada but also to the social and economic advancement of northern Indigenous peoples. Diefenbaker was the first to articulate this narrative that was reiterated in different fashion by a succession of premier and prime ministers. The last iteration was made by the Harper government that very clearly linked the improvement of living and social conditions in the Arctic to the development of a resource extraction economy in the North. The validity of this powerful yet simplistic narrative has rarely been studied. In this chapter, we show that in Northern Canada, the relation between mining development and social development is at best limited and at worst non-existent.

Historically, mining activities have been responsible for many environmental, health and social impacts, and they will continue to do so in the future. Although many of those impacts have been documented and discussed by researchers, the industry and affected populations, a lot more have not. There is, for example, a significant body of literature that deals with environmental impacts (Duhaime et al. 2003; Natural Resources Canada 2004; MiningWatch Canada 2001; Winfield et al. 2006). Closed mines and abandoned exploration sites also have long-term environmental impacts that have been documented (Laurence 2006; Duhaime et al. 2003; Duhaime, Bernhard and Comtois 2005; Sandlos and Keeling 2013) but that local residents still often underestimate (Kruse 2011: 16).

While they have not been as extensively studied as environmental impacts, health impacts have been the focus of some publications recently (Bronson and Noble 2006; Noble and Bronson 2005; Shandro et al. 2011). It has been found that most health impacts suffered by local populations are indirectly caused by the impacts of mining activities on the environment. For example, by affecting animals and their environment, mining infrastructure (mining pit, roads, ports, etc.) has impacts on food consumption because they may reduce the capacity of local populations to hunt, fish and gather plants (Gibson and Klinck 2005). The lack of quality and fresh food may cause increases in diabetes, heart conditions, obesity and tooth decay (Buell 2006).

However, social impacts are by far the least known, researched and theorized of all the impacts mining activities have in the Canadian North. In order to get a broad understanding of how social impacts of mining activities in the Canadian

North are being discussed, this chapter will explore a wide array of literature: scientific publications (books, articles and reports), governmental guidelines, environmental impact assessments (EIA), impact benefits agreements (IBA) and monitoring reports. The objectives of this chapter are to: present the results of a review of literature about what has been written on the social impacts of mining activities in Northern Canada, highlight knowledge gaps found in the literature and identify priority research for the future.

In order to systematize the analysis of the data found in this body of literature, we have identified three main themes to which are associated different variables. The first theme, economy, was divided in two variables: employment and economic activities. The second theme, people, was made of several variables: demographics, health and well-being, education and training, housing, crime, social cohesion, as well as intragenerational and intergenerational equity. Cultural practices comprised variables like the Inuit language, traditional activities, skills and land use. Each of these variables was in turn divided in a series of indicators that allows us to classify the wide range of social impacts. However, it also has some disadvantages. Some of the indicators, for instance, are not social impacts per se, but indicators that have to be considered in that they may or may not create social impacts depending on a community's unique social, cultural or economic context. Incidentally, what constitutes a social impact and which ones have made it to the list is subject to cultural and economic biases. Moreover, such lists tend to focus on the negative impacts, do not induce analytical thinking about indirect impacts and do not take into consideration the effects of planned interventions (see Vanclay 2002: 184, 188, 189 and 200). We have made a conscious effort to avoid focusing on negative impacts only.

This chapter is not meant to be exhaustive: not all recent EIAs conducted in the Canadian North in the past decade or so have been assessed, and the confidential nature of most IBAs will limit our analysis to a single one (Raglan). Our analysis of monitoring reports will also be limited by the small number of reports that has been released in the past decade. The chapter nevertheless highlights heavy tendencies and very meaningful gaps that will need to be addressed in future researches.

The first section will discuss the legal and political framework surrounding mining in the Canadian North. The Canadian North is not culturally nor politically uniform and mines doing business in various northern regions of the country have varied obligations that will be highlighted here. This section will also be used to define some important concepts employed throughout the chapter, such as environmental impact assessment (EIA), impact and benefit agreements (IBA) and so on. The second section will explore the theoretical social impacts examined in the scientific publications and review the anticipated social impacts discussed in environmental impact assessments. A third section discusses the observed social impacts observed in the few monitoring studies. The final section highlights the knowledge gaps identified in the previous sections and discuss research priorities for the future.

Context: mining in Northern Canada

Mineral exploration and extraction have increased steadily in Northern Canada since World War II, when demand for base minerals amplified for both defense and construction projects (Boutet 2010; Nassichuk 1987; Keeling and Sandlos 2009; Sandlos and Keeling 2012). Mining operations such as Con Mine (1938–2003) and Giant Mine (1948–2004) in Yellowknife, Rankin Inlet (1957–1962), Nanisivik (1979–2001), Polaris (1981–2002) and others, all paved the way for existing and future projects. Currently, several mines are in operation in the Canadian North, such as Raglan Nickel Mine (1999–present) and Canadian Royalties (2012–present) in Nunavik, Diavik (2003–present) and Ekati (1999–present) diamond mines in the Northwest Territories, Voisey's Bay (2005–present) nickel mine in Labrador, Meadowbank (2010–present) gold mine in Nunavut, and Troilus (1996–2010) copper and gold mine in Quebec Subarctic.

In 2011, the growing demand for minerals from China and other emerging economies coupled with changes in policies and laws in Canada (Campbell and Laforce 2010; Carter 2007; Gouvernement du Québec 2008; Isaac and Knox 2004; Laforce 2012; Laforce, Lapointe and Lebuis 2012; Lapointe 2010) has created a short-lived boom that was not sustained. Dozens of mining projects were being proposed all over Northern Canada, however, very few were actually developed.

All these mining projects were set to take place near northern communities, creating a "paradox of development" as many communities are torn between a desire to improve their material conditions and a fear of potential impacts (Bell 2010; Kucera 2009). On the one hand, the development of mining projects is encouraged by the industry, governments, communities and organizations. On the other, many in civil society are concerned with their potential impacts (Bernauer 2010a; Bernauer 2010b).

Meanwhile, the regulatory environment of mining started to change towards the end of the 1980s when the mining industry had to address new environmental and social preoccupations and laws through sustainable development and the increasing acknowledgement of Indigenous rights through EIA process (Laforce, Lapointe and Lebuis 2012). However, these processes rarely fulfill all their expectations.

Although, as noted in Chapter 3, EIAs are designed to support decision-making concerning the potential impacts of mining development (Nunavut Impact Review Board 2012; Noble 2006), they fall short in building trust and capacity among stakeholders (Galbraith, Bradshaw and Rutherford 2007; Dokis 2015; Papillon and Rodon 2017; Canada 2017) and do not always yield expected results (Whitmore 2006). This can be explained by the fact that while EIAs need to incorporate traditional Indigenous knowledge (Canadian Environmental Assessment Agency 2004; Canada 2017), their technical language and scope make them inaccessible to most community members (Shadian 2011; Dokis 2015). This can also be caused by the fact that EIAs only focus on one development project and therefore ignore the cumulative impacts of other development projects in the same area.

Social impacts of mining in the Canadian North: synthesis of scientific and grey literature

Mining activities have both positive and negative impacts. It is essential to consider the relationship between them to get a better overview of how they impact Indigenous communities. So although this part of the chapter focuses on the social impacts identified in the scientific and grey literature, it will use a global and integrated approach and try to put them in relation with one another, as well as try to identify the gaps in published research, grey literature, EIA and monitoring reports.

The different environmental impact assessments (EIAs) reviewed in this chapter are related to projects in the Northwest Territories: Mackenzie Valley pipeline (Berger 1977); in Labrador: Voisey's Bay (Voisey's Bay Nickel Company Ltd 1997); in Nunavut: Jericho (Tahera Corporation 2003; Hornal & Associates Ltd 2003), Kiggavik (AREVA Resources Canada Inc 2011), Mary River (Baffinland Iron Mines Corporation 2012), and in Quebec: Raglan (Makivik Corporation 1995) and Eastmain-1-A (Hydro-Québec 2004).

On local economic activity and employment – scientific literature

The literature indicates that mining has the potential to create important economic benefits for local communities (Hilson 2002), because it increases economic activity, employment (Bowes-Lyon, Richards and McGee 2009; Fidler and Hitch 2007; Duhaime et al. 2003), and collective wealth (O'Reilly and Eacott 1999–2000; Waye et al. 2009). First, the construction and opening of mines offer unique employment opportunities in a sector where salaries are higher than in other extracting resources industries (MiningWatch Canada 2001; Gibson and Klinck 2005: 116–117, 131). Second, mining creates business opportunities for people who are not working at the mine (creation of new business, new jobs, etc.) (Buell 2006: 3). Such opportunities may contribute to the long-term durability and the sustainable development of communities (Duhaime et al. 2003). Finally, royalties obtained through an IBA can contribute to increasing collective wealth (O'Reilly and Eacott 1999–2000; Kennett 1999), plus companies can contribute to the development of infrastructure, housing, etc. (Hilson 2002, Kennett 1999).

An increase of disposable income, however, can create or intensify already existing social issues (Buell 2006). For instance, a higher disposable income increases the chances an individual has to engage in high-risk behavior, such as alcohol or drug abuse, gambling and prostitution (Gibson and Klinck 2005). Such behaviors can have important consequences in Arctic communities because they can lead to violence, women's exploitation, mental illness, suicide (Berger 1977), casualization of labor, poor health and a high level of sexually transmitted infections (Buell 2006). The benefits of employment are often counterbalanced by social and familial perturbations (Noble and Bronson 2005) associated with an increase of disposable income. Also, mining can create an economic dependence to rents and high inflation (Mann 1975) or even hyperinflation (Yakovleva 2005) that can paralyze growth in other sectors (Hilson 2002), especially in poor and underdeveloped countries and areas (Whitmore 2006) like the Arctic.

On local economic activity and employment – EIAs

Rarely quantified in EIAs, job opportunities (Baffinland Iron Mines Corporation 2012: 20; Tahera Corporation 2003: 92; Hornal & Associates Ltd 2003: 4–2; Hydro-Québec 2004: 21–16) and business opportunities (Baffinland Iron Mines Corporation 2012: 161; Tahera Corporation 2003: 95; Hydro-Québec 2004: 21–16; Makivik Corporation 1995: 54) are nevertheless the principal benefits expected from mining projects. The probability that the number of jobs will increase is seen as certain (Hornal & Associates Ltd. 2003: 4–18) and they are also expected to have indirect effects on the economy (AREVA Resources Canada Inc. 2011: 8–18).

The Mackenzie River Pipeline assessment, however, illustrated that most of the job opportunities go to workers from the South (Berger 1977: 120) while Indigenous peoples are "engaged mainly in low paid, unskilled, casual or seasonal employment" (Berger 1977: 133) and that projects tend to benefit the metropolis more than the local economy (Berger 1977: 118). In this 1977 report, Indigenous peoples are looking for ways to conciliate their participation in the growing wage economy without becoming entirely dependent and still earning a living from the land (Berger 1977: 110). Incidentally, a later assessment suggests that the income from wage labor has become essential to the pursuit of the traditional activities associated with the land, since harvesters need disposable income to purchase and maintain equipment (Voisey's Bay Nickel Company Ltd 1997: 22.1.5). Also, beyond the income, the pride of having a job is expected to have beneficial effects for the well-being of those concerned (Hornal & Associates Ltd 2003: 4–11).

Training opportunities, which often come in the form of on-the-job training, are emphasized in nearly every assessment (Makivik Corporation 1995: 46–48; Baffinland Iron Mines Corporation 2012: 32; Berger 1977: 138; Tahera Corporation 2003: 96; Hornal & Associates Ltd 2003: 4–7; AREVA Resources Canada Inc 2011:8–16; Hydro-Québec 2004: 21–11).

The dependence on government transfers was considered to have increased in 1977 (Berger 1977: 151), while it was elsewhere noted to have decreased from 61% of personal income coming from government transfers in 1971, to 28% in 2001, a number higher than that of the general population, but similar to that found in other isolated areas (Hydro-Québec 2004: 16–93). Dependence on the government is thus expected to be lessened (Hornal & Associates Ltd 2003: 4–10). Other anticipated impacts on the economy include an increased need for infrastructure (Baffinland Iron Mines Corporation 2012: 149) and a possible increase in the cost of living during both the construction and operation phases of the project (Voisey's Bay Nickel Company Ltd 1997: 24.2.4; Hornal & Associates Ltd 2003: 4–11).

On health and well-being – scientific literature

Peer-reviewed publications have barely addressed the health impacts of mining on Indigenous communities (Bronson and Noble 2006; Bielawski 2004; Hurtig and San Sebastian 2002; Noble and Bronson 2005; Shandro et al. 2011; Keller 2012). Because the mining impacts on Indigenous health are still misunderstood, a joint SSHRC and CIHR project called *Mining & Aboriginal Community Health* (see: www.impactandbenefit.com/miningandhealth) whose intent it is to bridge our knowledge gaps about health impacts of mining activities has been funded.

As stated, higher income can mean exacerbation of already existing social issues like drug and alcohol consumption, gambling and prostitution because individuals have more money to get involve in those (Gibson and Klinck 2005: 122–125; Buell 2006: 17–19, 22; Natural Resources Canada 2003; Fidler and Hitch 2007; Government of the Northwest Territories 2006; Government of the Northwest Territories 2009). This can have indirect impacts because these behaviors can lead to violence, the exploitation of women, mental illness and suicide, problems to go to work, poor health and sexually transmitted diseases (Archibald and Crkovich 1999; Buell 2006: 18–19; Cyzewski et al. 2014; Goldenberg et al. 2008).

On health and well-being – EIAs

EIA studies found that the already high rates of suicide in the communities (Baffinland Iron Mines Corporation 2012: 102; Voisey's Bay Nickel Company Ltd 1997: 24.1.3.6; Berger 1977: 152; Tahera Corporation 2003: 94) may be increased by the poor management of incomes and stressors which also lead to domestic violence and divorce (AREVA Resources Canada Inc 2011: 10–18).

The high rates of drug and alcohol addictions as well as gambling, which are recognized as already existing, widespread and increasing problems (Baffinland Iron Mines Corporation 2012: 102,103; Makivik Corporation 1995; Voisey's Bay Nickel Company Ltd 1997: 24.1.3.5; Berger 1977: 154; Tahera Corporation 2003: 94; Hydro-Québec 2004: 16–103), are expected to be exacerbated by increased income and lead to a series of related social problems (Hornal & Associates Ltd 2003: 4–10; AREVA Resources Canada Inc 2011: 10–17). Only the most recent EIA implies the creation of "Counselling and healing programs" (Baffinland Iron Mines Corporation 2012: 124).

The high rates of venereal disease (Berger 1977: 153; Tahera Corporation 2003: 94) are expected to increase (AREVA 2011: 10–17), while the high rates of mental illness (Tahera Corporation 2003:94) are expected to decrease thanks to the positive effects of employment and reduced economic stress (Baffinland Iron Mines Corporation 2012: 108; AREVA Resources Canada Inc 2011: 10–17).

On social cohesion – scientific literature

Mining can also affect social cohesion (Labrador West Status of Women Council & Femmes francophones de l'Ouest du Labrador 2004; Barrett-Wood et al. nd.;

Cyzewski et al. 2014). The upsurge of new workers from an environment where economic, social and cultural values are different cause pressures on the cultural identity, social integrity and individual self-esteem which can in turn create or amplify problems such as alcohol abuse or the imitation or unsustainable behaviors (Kennett 1999). Moreover, the loss of social norms and structures regulating people's behavior which is attributed to the speed and extent of changes and to the instability that comes with mining projects can lead to social issues such as anomie and suicide (Parlee and O'Neil 2007).

Indigenous women in communities with growing economies are more likely to be the victims of sexual exploitation, violence and sexually transmitted infections, often through sexual abuse or prostitution (Gibson and Klinck 2005). Familial integrity is also deteriorating, threatened by the demands and stress related to work and the changes in familial roles (Gibson and Klinck 2005). The schedule of a mine worker is indeed long and demanding, leaving the women alone to care for the children, a dynamic that potentially creates additional tensions which can lead to violence and conflicts (Sosa and Keenan 2001).

As highlighted in Chapter 14, the impacts of mining can also hinder the capacity of women to contribute to the community's well-being, especially through their capacity to care for resources and the environment and ensure access to food and other subsistence goods (Kuokkanen 2011). The lack of childcare services limits women's opportunities to work within the mining industry or to get an education (O'Faircheallaigh 1998) and explains why women are often excluded from the decisional process (Natural Resources Canada 2003). Also, Indigenous women are more likely to suffer from poverty (Gibson and Klinck 2005) and households headed by women are more vulnerable to the inflation caused by mining exploitation (Sosa and Keenan 2001).

On social cohesion – EIAs

Social cohesion is not discussed in the EIA reports analyzed here. However, EIA reports mention that although the migration of Indigenous or non-Indigenous in and out of the communities is expected, no numbers are stated (Baffinland Iron Mines Corporation 2012: 14, 15; Tahera Corporation 2003: 99; Hornal & Associates Ltd 2003: 4–12; AREVA Resources Canada Inc 2011: 8–2, 8–19; Hydro-Québec 2004: 16–3). It is expected that mines will remain dependent on workers from the South for many years (AREVA Resources Canada Inc. 2011: 9–17).

On intragenerational and intergenerational equity – scientific literature

Mining can increase intra- and intergenerational inequalities as some people receive benefits or work at the mine while others do not (O'Faircheallaigh 1998; Fidler and Hitch 2007; Irlbacher-Fox and Mills 2007; Davis 2009). Furthermore, regional inequalities can happen between communities of a same region, one that has benefits, the other not. For example, communities not located near the mine are not a concern of mining companies and regional centres that are usually quite impacted by mining development (increase use of health service, influx of passing by workers, etc.) are not taken into account (Gibson and Klinck 2005). There are also inequalities on a cultural level where non-Indigenous workers earn more than Indigenous workers because of their access to better jobs (Gibson and Klinck 2005: 131). All of these lead to an increase in social stratification. Indigenous peoples still feel that they don't benefit as much and that they are disadvantaged (Duhaime et al. 2003). This can also lead to an increase in social tensions (O'Faircheallaigh 1998) and to behaviors such as violence and substance abuse (Buell 2006: 17).

On intragenerational and intergenerational equity – EIAs

The benefits announced for future generations include rent and royalties (Baffinland Iron Mines Corporation 2012: 242) and other benefits stemming from IBAs (Tahera Corporation 2003: 100). Support for programs aiming to facilitate the transfer of traditional knowledge and skills is mentioned in the most recent report (Baffinland Iron Mines Corporation 2012: 225), acknowledging the importance given by the communities to leave something for the next generations to harvest, but also to maintain existing values (AREVA Resources Canada Inc 2011; 9–1).

On traditional activities and land use – scientific literature

Traditional activities are a source of pride that allows the Indigenous not only to face the hardships related to life on the beloved and respected land, but that also allows them to maintain their identity (Buell 2006). Depression and anxiety associated with a loss of identity can in turn lead to violence, drug addiction, exploitation of women and suicide (Berger 1977).

Little is known on the real impacts of mining on traditional economies in the Arctic (Haley et al. 2011). Some say participation in harvesting activities decreases because mining projects use physical space which can cause hostile changes and environmental damages to territories (Bjerregaard and Young 1998; Bjerregaard et al. 2004; Duhaime et al. 2003; Bernauer 2011) and that Indigenous peoples are forced to adapt their practices to the effects of industrialization (Bernauer 2011; Kruse 2011).

However, some research has shown that mining projects in remote communities of the North do not necessarily lead to a decline in traditional values or in local populations' participation in traditional activities since well-paid jobs allow workers to increase their harvesting activities (Koke 2008; Landry, Bouvier and Waaub 2009; Laneuville 2013; LeClerc and Keeling 2015) although these findings are disputed by other researchers who claim that an increased participation in an economy based on salaries can lead to a decreased participation in the traditional subsistence-based economy (Buell 2006; Bernauer 2011; Kruse 2011).

Anticipated impacts on traditional activities and land use – EIAs

The anticipated impacts on traditional activities differ from one assessment to the next. While some expect that the project will have little impact on the hunting, fishing and trapping activities of the local people (Tahera Corporation 2003: 96), some predict that less time, overall, may be available for hunting, trapping and fishing (Hornal & Associates Ltd 2003: 4–10), but point to the increased ability to buy hunting equipment (Hornal & Associates Ltd 2003: 4–8), while others admit not knowing if mining activities could lead to decreased or increased time engaging children in land-based activities (Baffinland Iron Mines Corporation 2012: 50).

On housing – EIAs

If the population living in inadequate housing is growing at an alarming rate in some communities (Baffinland Iron Mines Corporation 2012: 104; Hydro-Québec 2004: 16–14) and stable in others such as Igloolik and Pond Inlet (Nunavut) (Baffinland Iron Mines Corporation 2012: 104), it is expected that new income will increase demand for new properties (Hornal & Associates Ltd 2003: 4–9) or for social housing from people wishing to leave overcrowded housing (AREVA Resources Canada Inc 2011: 11–4). However, mining companies are not involved in social housing and therefore they do not contribute to alleviate the housing shortage in the communities.

On crime – EIAs

Already especially high in Nunavut and the Northwest Territories (Baffinland Iron Mines Corporation 2012: 103; Voisey's Bay Nickel Company Ltd 1997; Tahera Corporation 2003: 94), crime rates increase following the opening of mines (AREVA Resources Canada Inc 2011: 10–19).

The crime most discussed in EIAs is family violence (Baffinland Iron Mines Corporation 2012: 107; Tahera Corporation 2003: 94). The rate of reported spousal abuse is, in Nunavut, for instance, 6.5 times the national rate (Baffinland Iron Mines Corporation 2012: 107; Makivik Corporation 1995: 107). However, while some report an increase (Hornal & Associates Ltd 2003: 4–11; AREVA Resources Canada Inc. 2011: 10–19), others note a decline (Voisey's Bay Nickel Company Ltd 1997: 24.1.3.5).

On Indigenous Languages – EIAs

A loss of the Inuktituk language is expected both because of the immersion in the English language at work and the increased contact of English speakers with the youth (AREVA Resources Canada Inc 2011: 9–19). Only the two most recent assessments address the issue. One plans to provide more services to Inuit in Inuktitut (AREVA Resources Canada Inc 2011: 8–18), while the other stipulates that, at the workplace, employees will be allowed to explain something in Inuktitut to each other, as long as it is then repeated in English, the primary language, for the benefit of the other employees (Baffinland Iron Mines Corporation 2012: 227).

Monitoring social impacts

Very few monitoring reports are available and most consist of qualitative interviews with workers and members of the communities. Their structure does not correspond with that of EIAs and they tend not to monitor anticipated impacts. The monitoring reports reviewed in this chapter are those from the following projects in Labrador: Voisey's Bay (Archibald and Crkovich 1999; Labrador West Status of Women Council & Femmes Francophones de l'Ouest du Labrador 2004); in Nunavut: Nanisivik (Brubacher & Associates 2002; Lim 2013), Nanisivik and Polaris (Bowes-Lyon, Richards and McGee 2009), Jericho (Brubacher Development Strategies Inc. 2009); and in Quebec: Raglan (Salluit, Kangiqsujuaq, Puvurnituq, Quaqtaq, Kangirsuk) (Lanari, Smith and Okituk 1999a; 1999b; 2000a; 2000b; 2000c) and Troilus (Penn and Roquet 2008; Krekshi 2009).

First, job opportunities have not been as numerous as was anticipated in the EIAs. The rate of Inuit or Cree employment promised in the agreements has never been reached (Brubacher & Associates 2002: 17; Bowes-Lyon, Richards and McGee 2009: 319) and has declined after the construction phase (Brubacher & Associates 2002: 17; Bowes-Lyon, Richards and McGee 2009: 319; Penn and Roquet 2008: 29, 30; Krekshi 2009: 59), during which contracts were awarded to local companies in some cases (Bowes-Lyon, Richards and McGee 2009: 383; Penn and Roquet 2008: 29,30), but not others (Brubacher & Associates 2002: 20; Lanari, Smith and Okituk 2000a: 8; Lanari, Smith and Okituk 2000b: 13; Krekshi 2009: 74).

Companies sometimes offer scholarships for students (Bowes-Lyon, Richards and McGee 2009: 385). One even offered $5-million to Nunavut to start a new University (Kennedy et al. 2015). Companies also offer on-the-job training for Inuit workers, although local people are not always interested in the positions available to them (Bowes-Lyon, Richards and McGee. 2009: 379). However, amongst the population, on-the-job training is valued as it offers not only opportunities for employment, but also for promotions (Brubacher Development Strategies 2009: 39–43).

On social cohesion and well-being, many reports mention the numerous family breakdowns occurring as a result of working on the mines (Lanari, Smith and Okituk 1999a: 13; Brubacher Development Strategies 2009: 52; Penn and Roquet 2008: 61, 113; Krekshi 2009: 74). Monitoring reports note an increase both in family violence (Labrador West Status of Women Council & Femmes francophones de l'Ouest du Labrador 2004: 55; Krekshi 2009: 74) and sexually transmitted diseases (Krekshi 2009: 99).

As expected in EIAs, the issue of drug and alcohol addiction arises in a majority of the monitoring reports (Brubacher & Associates 2002: 12, 13; Bowes-Lyon, Richards and McGee 2009: 384; Lanari, Smith and Okituk 1999a: 13; 1999b: 12; 2000a: 13; 2000b: 58, 63; Krekshi 2009: 99).

The quality of housing is said to have improved in at least one case (Labrador West Status of Women Council & Femmes francophones de l'Ouest du Labrador 2004: 28) while the shortage and overcrowding is still a reality elsewhere (Brubacher & Associates 2002: 26; Krekshi 2009: 98). On equity, the reported benefits to the next generations are limited to the money made through the Raglan agreement, which is reportedly not always spent in such a way to actually benefit the whole community (Lanari, Smith and Okituk 1999a: 9; 1999b: 6–7).

Also, as expected in at least some of the EIAs in relation to traditional activities, the increased income is said to help with the purchase of hunting equipment which is perceived as a chance to increase hunting (Bowes-Lyon, Richards and McGee 2009: 382; Lanari, Smith and Okituk 2000c: 10), although concerns over the impacts on wildlife resources were also raised (Lanari, Smith and Okituk 1999a: 5, 6; 1999b: 4; 2000a: 5, 6; 2000b: 7, 8, 37). One report also denotes a loss of words in Inuktituk (Brubacher & Associates 2002: 8).

Finally, often overlooked in the grey and scientific literature as well as in the EIAs, the importance of a mine's closure, be it planned or not, has only recently been acknowledged by mining companies and by regulators. What happened at Nanisivik's closure, for instance, was not agreed in advance. After the closure of the mine was announced, long awaited discussions between the government and the company took place, but financial considerations outweighed any others. In 2004, a plan including the full demolition and reclamation of the town site, which was built next to the mine and approximately 30 km away from the community of Arctic Bay, was approved by the Nunavut Water Board.

Residents of Arctic Bay interviewed after the mine's closure mentioned the financial and emotional struggles of people who lost, along with their jobs, the ability to purchase hunting equipment, or even food, and, in a sense, their hopes for the future. A psychological impact was also experienced by those who were born in Nanisivik and lost their home, the place of birth being particularly important to the Inuit. The demolition of Nanisivik's town site infrastructure was widely seen as wasteful, especially in the case of houses, which the Mayor of Arctic Bay had suggested could have been salvaged whole or for materials, since there was a shortage in his community. Some interviewees were also vexed that the company offered no compensation for the damage done to the environment (see Lim 2013: 52, 53, 54, 65, 77, 80, 81, 86; Tester et al. 2013).

Knowledge gaps

While our understanding of social impacts of mining on northern communities has increased considerably, there are still areas where a better understanding would help communities control negative impacts and improve positive impacts.

It is obvious that many social impacts are poorly measured or not discussed in EIAs, be it by choice or because of the lack of data and resources available. For instance, there are few actual numbers on the employment of Indigenous peoples while numbers on education are completely absent from EIAs. Even if mental health is addressed in most EIAs, other health issues are set aside. Although EIAs mention that workers from the South are likely to migrate to work on the mines, the impact of this situation on northern communities is not considered. Impacts on women also are seldom mentioned even if we know from the scientific literature that women are impacted in many ways by mining development project. Finally, different EIAs contain opposed anticipated impacts on traditional activities, highlighting the poor quality of information on the subject.

The monitoring reports are mostly based on interviews in the affected communities and thus reflect the community perception of mining development, as opposed to actual numbers. For instance, the impacts on economic activity are not any better identified in the monitoring reports than in the EIAs. There are no mentions of the effects on the cost of living or the real numbers of jobs created directly or indirectly for Indigenous peoples by the mines. Moreover, monitoring reports contain no information on changes in life expectancy or reported sicknesses nor can they provide the number of deaths that are related to mining activities. In most cases, the shortage and poor quality of housing is not addressed and neither are changes in crime rates other than family violence. The issue of social cohesion, including migration of Inuit or non-Inuit in and out of the community and its impacts, is also not discussed and the changing gender roles are not addressed either. Finally, no quantitative data on the participation in traditional activities or change in transmission of traditional knowledge and skills is available.

The scientific and grey literature shed some light on these issues but the only reliable and comprehensive research on the economic impact of mining development comes from a team of researchers from the University of Alaska Anchorage that has worked on the Red Dog mine (Haley et al. 2011), but there is nothing similar in Canada.

The question of the impact of mining development on the sustainability of the northern communities is not addressed anywhere in the mining context. However, the concept of sustainability was at the core of the analysis in the Mackenzie Gas Project review (Gibson 2006) but since the project was abandoned, it is impossible the measure the results of this new approach. Finally, the research on the impacts of mining development on Indigenous livelihood has led to contradictory predictions and needs more research.

Conclusion

This survey of the available scientific literature, EIA and monitoring reports shows that many social and economic impacts are considered, however, it is surprising to see that one of the main research gaps is the lack of measurement of the economic impacts of mining development on northern communities in Canada. The little evidence that we have collected from the Raglan case shows that these financial inputs do not necessarily convert into local economic and social development (Rodon et al. 2013, Rodon and Lévesque 2015). As we have seen in the introduction, this is the central argument in the mining development narrative and yet we can not find any data that supports or refutes this claim. There are numerous studies on the impact of mining development on the national and regional economy but no study at the community level.

Other gaps are less surprising and can be explained by the fact that these issues are outside the mining development narrative and are therefore not an important concern for decision makers even if it should be. For example, there is clearly a lack of research, and in some case of concerns, on the impact of mining development on women, on Indigenous cultures and livelihood and on social stratification and social cohesion in northern communities. However, a few academic studies have advocated for more research in these areas.

Many researchers and practitioners are also stressing the need to better account for the cumulative impacts of resource development instead of assessing the projects individually. Cumulative environmental assessments (CEA) and strategic environmental assessment (SEA) have been both promoted, but there are many issues with their implementation (Gunn 2009) and SEA has never been used in the context of northern mining. There has been a resistance both from governments and private corporations to implement such a comprehensive review process. However, in the case of a mining boom, such as the one experienced in 2010, the implementation of a SEA process would be imperative to avoid the compounded negative impacts that such an important development would create in northern communities.

Note

1   The authors would like to acknowledge the valuable research assistance provided by Josianne Grenier and Aude Therrien. The research for this chapter was funded by a grant from the Social Sciences and Humanities Research Council of Canada (Partnership Development Grant 890–2012–0111) and financial support from ReSDA.

References

Archibald, L. and Crkovich, M. (1999). *If Gender Mattered: A Case Study of Inuit Women, Land Claims and the Voisey's Bay Nickel Project.* Ottawa, Status of Woman Canada.
AREVA Resources Canada Inc. (2011). "Kiggavik Project. Draft Environmental Impact Statement." Retrieved July 11, 2012, from Nunavut Impact Review Board: ftp://ftp.nirb.ca/02-REVIEWS/ACTIVE%20REVIEWS/09MN003-AREVA%20KIGGAVIK/2-REVIEW/06-DRAFT%20EIS%20&%20CONFORMITY%20REVIEW/02-DEIS%20SUBMISSION/.
Baffinland Iron Mines Corporation (2012). "Mary River Project – Final Environmental Impact Assessment." Retrieved July 11, 2012, from Nunavut Impact Review Board: ftp://ftp.nirb.ca/02-REVIEWS/ACTIVE%20REVIEWS/08MN053-BAFFINLAND%20MARY%20RIVER/2-REVIEW/08-FINAL%20EIS/FEIS/.
Barrett-Wood, Z., Knotsch, C., Davison, C. and Bradshaw, B. (n. d.). *Translating Knowledge on Impacts of Mining for Aboriginal Community Health: The Issue of Rotational, Two-Week Work Shifts.* National Aboriginal Health Organization. Poster Presentation.
Bell, J. (2010). "A Paradox of Development." Web blog. Retrieved from: http://titiraqti.wordpress.com/2010/01/03/a-paradox-of-development/.

Berger, T. (1977). "Northern Frontiers, Northern Homeland: The Report of the Mackenzie Valley Pipeline Inquiry – Berger Report." Retrieved June 22, 2012, from: http://caid.ca/vol1_mac-pip.html.

Bernauer, W. (2010a). "Mining, Harvesting and Decision Making in Baker Lake, Nunavut: A Case Study of Uranium Mining in Baker Lake." *Journal of Aboriginal Economic Development*, 7(1), 19–33.

Bernauer, W. (2010b). "Uranium Mining, Primitive Accumulation and Resistance in Baker Lake, Nunavut: Recent Changes in Community Perspectives." MA thesis, University of Manitoba, Winnipeg.

Bernauer, W. (2011). *Mining and the Social Economy in Baker Lake, Nunavut.* University of Saskatchewan, Saskatoon: Prepared for the Northern Ontario, Manitoba, and Saskatchewan Regional Node of the Social Economy.

Bielawski, E. (2004). *Rogue Diamonds: Northern Riches on Dene Land.* Seattle, WA: University of Washington Press.

Bjerregaard, P. and Young, T. K. (1998). *The Circumpolar Inuit: Health of a Population in Transition.* Copenhagen: Munksgaard.

Bjerregaard, P., Young, T. K., Dewailly, É. and Ebbensson, S. O. (2004). "Indigenous Health in the Arctic: An Overview of the Circumpolar Inuit Population". *Scandinavian Journal of Public Health*, 32(5), 390–395.

Boutet, J.-S. (2010). "Développement ferrifère et mondes autochtones au Québec subarctique, 1954–1983." *Recherches amérindiennes au Québec*, XL(3), 35–52.

Bowes-Lyon, L.-M., Richards, J. P. and McGee, T. M. (2009). "Socio-economic Impacts of the Nanisivik and Polaris Mines, Nunavut, Canada." In J. P. Richards (Ed.), *Mining, Society, and a Sustainable World* (pp. 371–396). Berlin: Springer.

Bronson, J. and Noble, B. F. (2006). "Health Determinants in Canadian Northern Environmental Impact Assessment." *Polar Record*, 42(223), 315–324.

Brubacher & Associates (2002). *The Nanisivik Legacy in Arctic Bay. A Socio-Economic Impact Study.* Ottawa: Department of Sustainable Development, Government of Nunavut.

Brubacher Development Strategies Inc. (2009). *Kitikmeot Socio-Economic Monitoring Committee. Jericho Diamond Mine.* 2007 Socio-Economic Monitoring Report. Ottawa, Brubacher Development Strategies Inc.

Buell, M. (2006). *Resource Extraction Development and Well-Being in the North. A Scan of the Unique Challenges of Development in Inuit Communities*, Ajunnginiq Centre, National Aboriginal Health Organization.

Campbell, B. and Laforce, M. (2010). "La réforme des cadres réglementaires dans le secteur minier: Les expériences canadienne et africaine mises en perspective." *Recherches amérindiennes au Québec*, XL(3), 69–84.

Canada: Expert Panel for the Review of Environmental Assessment Processes (2017). *Building Common Ground: A New Vision for Impact Assessment in Canada*, The Final Report of the Expert Panel for the Review of Environmental Assessment Processes. Ottawa: Canadian Environmental Assessment Agency. Retrieved May 2018, from: www.canada.ca/en/services/environment/conservation/assessments/environmental-reviews/environmental-assessment-processes/building-common-ground.html.

Canadian Environmental Assessment Agency (2004). "Considering Aboriginal Traditional Knowledge in Environmental Assessments Conducted Under the Canadian Environmental Assessment Act: Interim Principles." Retrieved June 7, 2012, from: www.ceaa.gc.ca/default.asp?lang=En&n=4A795E76–1.

Carter, R. A. (2007). "Canadian Mining: A Crowded House." *Engineering & Mining Journal*, 208(7), 68–77.

Cyzewski, K., Tester, F., Aaruaq, N. and Blangy, S. (2014). *The Impact of Resource Extraction on Inuit Women and Families in Qamani'tuaq, Nunavut Territory. A Qualitative Assessment.* Pauktuutit Inuit Women of Canada and UBC School of Social Work.

Davis, G. A. (2009). "Extractive Economies, Growth, and the Poor." In J. P. Richards (Ed.), *Mining, Society, and a Sustainable World* (pp. 37–60). Berlin: Springer.

Dokis, C. A. (2015). *Where the Rivers Meet: Pipelines, Participatory Resource Management, and Aboriginal-State Relations in the Northwest Territories.* Vancouver: UBC Press.

Duhaime, G., Bernard, N. and Comtois, R. (2005). "An Inventory of Abandoned Mining Exploration Sites in Nunavik, Canada." *The Canadian Geographer*, 43(3), 260–271.

Duhaime, G., Bernard, N., Fréchette, P., Maillé, M.-A., Morin, A. and Caron, A. (2003). *The Mining Industry and the Social Stakes of Development in the Arctic.* Québec: Chaire de recherche du Canada sur la condition autochtone comparée.

Fidler, C. and Hitch, M. (2007). "Impact and Benefit Agreements: A Contentious Issue for Environmental and Aboriginal Justice." *Environments Journal Volume*, 35(2), 49–69.

Galbraith, L., Bradshaw, B. and Rutherford, M. B. (2007). "Towards a New Supraregulatory Approach to Environmental Assessment in Northern Canada." *Impact Assessment and Project Appraisal*, 25(1), 27–41.

Gibson, G. and Klinck, J. (2005). "Canada's Resilient North: The Impact of Mining on Aboriginal Communities." *Pimatisiwin: A Journal of Aboriginal and Indigenous Community Health*, 3(1), 115–139.

Gibson, R. B. (2006). *Sustainability-Based Assessment Criteria and Associated Frameworks for Evaluations and Decisions.* A report prepared for the Joint Review Panel for the Mackenzie Gas Project.

Goldenberg, S., Shoveller, J., Ostry, A. and Koehoorn, M. (2008). "Youth Sexual Behaviour in a Boomtown: Implications for the Control of Sexually Transmitted Infections." *Sexually Transmitted Infections*, 84(3), 220–223.

Gouvernement du Québec (2008). *Amérindiens et Inuits du Québec. Guide Intérimaire et matière de consultation des communautés autochtones.* Québec: Groupe interministériel de soutien sur la consultation des Autochtones.

Government of the Northwest Territories (2006). *Communities and Diamonds: 2005 Annual Report of the Government of the Northwest Territories under the BHP Billiton, Diavik and De Beers Socio-Economic Agreements.* Yellowknife: Government of the Northwest Territories.

Government of the Northwest Territories (2009). *Communities and Diamonds. 2008 Annual Report of the Government of the Northwest Territories Under the BHP Billiton, Diavik and De Beers Socio-Economic Agreements.* Yellowknife: Government of the Northwest Territories.

Gunn, J. H. (2009). "Integrating Strategic Environmental Assessment and Cumulative Effects Assessment in Canada." Ph.D. thesis, Geography, University of Saskatchewan.

Haley, S., Klick, M., Szymoniak, N. and Crow, A. (2011). "Observing Trends and Assessing Data for Arctic Mining." *Polar Geography*, 34(1–2), 37–61.

Haley, S., Szymoniak, N., Crow, A. and Schwoerer, T. (2011). "Social Indicators for Arctic Mining," *ISER Working Paper*. University of Alaska Anchorage: Institute of Social and Economic Research.

Hilson, G. (2002). "An Overview of Land Use Conflicts in Mining Communities." *Land Use Policy*, 19, 65–73.

Hornal, R. & Associates Ltd. (2003). *Socio-economic Baseline Study of the Kitikmeot Communities, Nunavut and Yellowknife, Northwest Territories.* Prepared for Tahera Corporation's Jericho Diamond Project.

Hurtig, A. and San Sebastian, M. (2002). "Geographical Differences in Cancer Incidence in the Amazon Basin of Ecuador in Relation to Residence Near Oil Fields." *International Journal of Epidemiology,* 31(5), 1021–1027.

Hydro Québec (2004). "Projet de l'Eastmain-1-A-Sarcelle-Rupert. Étude d'impact sur l'environnement." Retrieved July 11, 2012, from: www.hydroquebec.com/rupert/fr/etudes.html.

Irlbacher-Fox, S. and Mills, S. J. (2007). "Devolution and Resource Revenue Sharing in the Canadian North: Achieving Fairness Across Generations," Walter and Duncan Foundation Discussion Paper.

Isaac, T. and Knox, A. (2004). "Canadian Aboriginal Law: Creating Certainty in Resource Development." *University of New Brunswick Law Journal,* 53(3), 3–42.

Keeling, A. and Sandlos, J. (2009). "Environmental Justice Goes Underground? Historical Notes from Canada's Northern Mining Frontier." *Environmental Justice,* 2(3), 117–125.

Keller, J. (2012). *Les impacts socio-économiques de l'exploitation minière sur les communautés autochtones de l'Arctique.* Québec: Ministère des Ressources naturelles et de la Faune du Québec et Géologie Québec.

Kennedy, S., Black, K. and Rodon, T. (2015). *Northern University Futures: Working Together to Develop a University in Inuit Nunangat.* Report prepared for the Inuit Nunangat University Working Group. Iqaluit, Nunavut.

Kennett, S. A. (1999). *A Guide to Impact and Benefits Agreements.* Calgary: Canadian Institute of Resources Law.

Koke, P. E. (2008). *The Impact of Mining Development on Subsistence Practices of Indigenous Peoples: Lessons Learned from Northern Quebec and Alaska.* MA thesis, The University of Northern British-Columbia.

Krekshi, L. E. N. P. F. C. (2009). *Indigenous Peoples' Perspectives on Participation in Mining. The Case of James Bay Cree First Nation in Canada.* Department of Urban Planning and Environment. Division of Urban and Regional Studies, Kungliga Tekniska högskolan (Royal Institute of Technology, Sweden).

Kruse, P. (2011). "Developing an Arctic Subsistence Observation System." *Polar Geography,* 34(1–2), 9–35.

Kucera, J. (2009). "Oil on Ice." *The Atlantic* (November). Retrieved May 4, 2018, from: www.theatlantic.com/magazine/archive/2009/11/oil-on-ice/7716/.

Kuokkanen, R. (2011). "Indigenous Economies, Theories of Subsistence and Women." *American Indian Quarterly,* 35(2).

Labrador West Status of Women Council & Femmes Francophones de l'Ouest du Labrador. (2004). *Effects of Mining on Women's Health in Labrador West.* In collaboration with MiningWatch Canada and the Steelworkers Humanity Fund, with generous assistance from the Lupina Foundation.

Laforce, M. (2012). "Régulation du projet minier de Voisey's Bay au Labrador. Vers un rééquilibrage des pouvoirs dans certains contextes politiques et institutionnels." In M. Laforce, B. Campbell and B. Sarrazin (Eds.), *Pouvoir et régulation dans le secteur minier. Leçons à partir de l'expérience canadienne* (pp. 157–189). Québec: Presses de l'Université du Québec.

Laforce, M., Lapointe, U. and Lebuis, V. (2012). "Régulation du secteur minier au Québec et au Canada. Une redéfinition des rapports asymétriques est-elle possible?"

In M. Laforce, B. Campbell and B. Sarrazin (Eds.), *Pouvoir et régulation dans le secteur minier. Leçons à partir de l'expérience canadienne* (pp. 9–50). Québec: Presses de l'Université du Québec.

Lanari, R., Smith, S. and Okituk, P. (1999a). *A Report to the Community of Salluit.* Raglan Mine: Action-oriented social research program: Makivik Corporation.

Lanari, R., Smith, S. and Okituk, P. (1999b). *A Report to the Community of Kangiqsujuaq.* Raglan Mine: Action-oriented social research program: Makivik Corporation.

Lanari, R., Smith, S. and Okituk, P. (2000a). *A Report to the Community of Puvirnituq.* Raglan Mine: Action-oriented social research program: Makivik Corporation.

Lanari, R., Smith, S. and Okituk, P. (2000b). *A Report to the Community of Quaqtaq.* Raglan Mine: Action-oriented social research program: Makivik Corporation.

Lanari, R., Smith, S. and Okituk, P. (2000c). *A Report to the Community of Kangirsuk.* Raglan Mine: Action-oriented social research program: Makivik Corporation.

Landry, V., Bouvier, A.-L. and Waaub, J.-P. (2009). *La planification territoriale autochtone au Canada : Le rôle de l'évaluation environnementale stratégique dans la cogestion adaptative.* Montréal: GEIGER (Groupe d'études interdisciplinaires en géographie et environnement régional).

Laneuville, P. (2013). *Chasse et exploitation minière au Nunavut : une expérience inuit du territoire à Qamani'tuaq.* MA thesis. Université Laval.

Lapointe, U. (2010). "L'héritage du principe de free mining au Québec et au Canada." *Recherches amérindiennes au Québec,* XL(3), 9–25.

Laurence, D. (2006). "Optimisation of the Mine Closure Process," *Journal of Cleaner Production,* 14, 285–298.

LeClerc, E. and Keeling, A. (2015). "From Cutlines to Traplines: Post-industrial Land Use at the Pine Point Mine." *The Extractive Industries and Society,* 2(1), 7–18.

Lim, T. W. (2013). *Inuit Encounters with Colonial Capital: Nanisivik – Canada's First High Arctic Mine.* Retrieved February 5, 2013 from: circle.ubc.ca.

Makivik Corporation (1995). *The Raglan Agreement Entered into between Makivik Corporation, Qarqalik Landholding Corporation of Salluit, Northern Village Corporation of Salluit, Nunatulik Lanholding Corporation of Kangiqsujuaq, Northern Village Corporation of Kangiqsujuaq and Société Minière Raglan du Québec Ltée to which intervened Falconbridge Limited.*

Mann, D. (1975). *The Socio-Economic Impact of Non-Renewable Resource Development on the Inuit of Northern Canada.* Renewable Resources Studies, Waterloo, University of Western Ontario / University of Waterloo.

MiningWatch Canada (2001). *Mining in Remote Areas – Issues and Impacts,* MiningWatch Canada.

Nassichuk, W. W. (1987). "Forty Years of Northern Non-Renewable Natural Resource Development." *Arctic,* 40(4), 274–284.

Natural Resources Canada. (2003). *The Social Dimension of Sustainable Development and the Mining Industry.* A Background Paper. Ottawa: Natural Resources Canada, Minerals and Metals Sector.

Natural Resources Canada. (2004). *CSR Case Study: Teck Cominco Building a Culture of Responsibility.* Ottawa: Natural Resources Canada.

Noble, B. F. (2006). *Introduction to Environmental Impact Assessment: Guide to Principles and Practice.* Toronto: Oxford University Press Canada.

Noble, B. F. and Bronson, J. (2005). "Integrating Human Health into Environmental Impact Assessment: Case Studies of Canada's Northern Mining Resource Sector." *Arctic,* 58(4), 395–405.

Nunavut Impact Review Board. (2012). *Draft Guidelines for the Preparation of an Environmental Impact Statement for Hope Bay Mining Ltd.'s Phase 2 Hope Bay Belt Project* (NIRB File No. 12MN001). Cambridge Bay: Nunavut Impact Review Board.

O'Faircheallaigh, C. (1998). "Resource Development and Inequality in Indigenous Societies: Aboriginal Politics and Public Sector Management." *World Development* 26(3), 381–394.

O'Reilly, K. and Eacott, E. (1999–2000). "Aboriginal Peoples and Impact and Benefit Agreement: Summary of the Report of a National Workshop." *Northern Perspectives*, 25(4). Retrieved from: www.carc.org/pubs/v25no4/2.htm.

Papillon, M. and Rodon, T. (2017). "Proponent-Indigenous Agreements and the Implementation of the Right to Free, Prior, and Informed Consent in Canada." *Environmental Impact Assessment Review*, 62(January), 216–224.

Parlee, B. and O'Neil, J. (2007). "'The Dene Way of Life': Perspectives on Health from Canada's North." *Journal of Canadian Studies*, 41(3).

Penn, A. and Roquet, V. N. P. C. (2008). *Implementing the Troilus Agreement*. A Joint Study of Cree Employment and Services Contracts in the Mining Sector. Montreal, Cree Nation of Mistissini, Cree Regional Authority and Inmet Mining Corporation.

Rodon, T. and Lévesque, F. (2015). "Understanding the Social and Economic Impacts of Mining Development in Inuit Communities: Evidence from Past and Present Mines in Inuit Nunangat." *Northern Review*, 41, 13–39.

Rodon, T., Lévesque, F. and J. Blais (2013) "De Rankin Inlet à Raglan, le développement minier et les communautés inuit." *Études Inuit Studies*, 37(2).

Sandlos, J. and Keeling, A. (2012). "Claiming the New North: Mining and Colonialism at the Pine Point Mine, Northwest Territories, Canada." *Environment and History*, 18(1), 5–34.

Sandlos, J. and Keeling, A. (2013). "Zombie Mines and the (Over)burden of History." *Solutions Journal*, 4(3).

Shadian, J. (2011). *Post-Sovereign Governance: An Inuit Framework for Arctic Resource Management*. Paper presented at the ICASS VII: Circumpolar Perspectives in Global Dialogue: Social Sciences beyond the International Polar Year, Akureyri, Iceland.

Shandro, J. A., Veiga, M. M., Shoveller, J., Scoble, M. and Koehoorn, M. (2011). "Perspectives on Community Health Issues and the Mining Boom–Bust Cycle." *Resources Policy*, 36, 178–186.

Sosa, I. and Keenan, K. (2001). *Impact Benefit Agreements Between Aboriginal Communities and Mining Companies: Their Use in Canada*. Toronto: Environmental Mining Council of British Columbia, Canadian Environmental Law Association and CooperAcción: Acción Solidaria para el Desarrollo.

Tahera Corporation (2003). *Jericho Project – Final Environmental Impact Statement*. Retrieved July 11, from: Nunavut Impact Review Board: ftp://ftp.nirb.ca/02-REVIEWS/COMPLETED%20REVIEWS/00MN059-JERICHO%20DIAMOND%20MINE/2-REVIEW/09-FINAL%20EIS/030121–00MN059-Jericho%20Final%20EIS-167a-IT2E.pdf.

Tester, F. J., Drummond, E., Lambert, J. and Lim T. W. (2013). "Wistful Thinking: Making Inuit Labour and the Nanisivik Mine Near Ikpiarjuk (Arctic Bay), Northern Baffin Island." *Études/Inuit/Studies*, 37(2), 15–36.

Vanclay, F. (2002). "Conceptualizing Social Impacts." *Environmental Impact Assessment Review*, 22, 193–211.

Voisey's Bay Nickel Company Ltd. (1997). "Voisey's Bay Mine/Mill Project Environmental Impact Statement." Retrieved July 11, 2012, from: www.vbnc.com/eis/index.htm.

Waye, A., Young, D., Richards, J. P. and Doucet, J. A. (2009). "Sustainable Development and Mining: An Exploratory Examination of the Roles of Government and Industry." In J. P. Richards (Ed.), *Mining, Society, and a Sustainable World* (pp. 151–182). Berlin: Springer.

Whitmore, A. (2006). "The Emperors' New Clothes: Sustainable Mining?" *Journal of Cleaner Production*, 14, 309–314.

Winfield, M. S., Jamison, A., Wong, R. and Czajkowski, P. (2006). *Nuclear Power in Canada: An Examination of Risks, Impacts and Sustainability*. Calgary: The Pembina Institute.

Yakovleva, N. (2005). *Corporate Social Responsibility in the Mining Industries*, Farnham, UK: Ashgate.

6 Measuring impacts

A review of frameworks, methodologies, and indicators for assessing socio-economic impacts of resource activity in the Arctic

Andrey N. Petrov, Jessica Graybill, Matthew Berman, Philip Cavin, Vera Kuklina, Rasmus Ole Rasmussen, and Matthew Cooney

This chapter summarizes and critically reviews existing scholarship, knowledge, and practices regarding the measurement of socio-economic impacts of resource development on Northern communities. In this chapter, we review frameworks and systems of indicators in multiple Arctic jurisdictions, providing description of key studies and sites. Focusing on socio-economic impact assessment procedures, namely measurement systems and methodologies, our guiding question is: *What work has been done to allow the measurement of social and economic impacts of resource development on Northern communities?* To answer this, we outline existing practices in Arctic regions, identify successes and barriers to sustainable development, and reflect on the future of impact measurement and indicators.

This chapter responds directly to key concerns of this book by answering questions about how to measure and better understand resource development impacts. Clearly, measurement of impacts is essential for understanding how communities may benefit better from resource development and how negative impacts can be mitigated. For communities to be able to make good choices about resource development agreements, they must have access to suitable information about the positive and negative consequences of resource development. Our review of knowledge about measuring impacts via indicators will provide Northern communities with the kind of knowledge needed to develop better futures.

Types of impacts and socio-economic impact measurement

Two important impacts of resource development are environmental and socio-economic (Therivel and Ross, 2007). While research on the cumulative impacts of resource development is becoming standard procedure in pre- and post-development impact assessment (Duinker and Greig, 2006; Frank, 2012), environmental impacts assessment (EIA) is more widely conducted and implemented than socio-economic impact assessment (SEIA; Mitchell and Parkins, 2011). Vanclay defines SEIA as "the process of assessing or estimating, in advance, the social consequences that are likely to follow from specific policy actions or project development, particularly in the context of appropriate national, state or provincial environmental policy legislation" (2002: 190). Social impacts, thus, incorporate measurements of

> all social and cultural consequences to human populations of any public or private actions that alter the ways in which people live, work, play, relate to one another, organise to meet their needs, and generally cope as members of society. Cultural impacts involve changes to the norms, values, and beliefs of individuals that guide and rationalise their cognition of themselves and their society.
>
> (Vanclay, 2002: 190)

Because SEIA is relatively new, measurements of such impact are nascent. However, SEIA is required for resource development permits in some jurisdictions; in others, formal or informal rules (e.g., impact, benefit agreements) exist. The history of SEIA is closely related to development of Arctic regions. In the US, the need for SEIA became apparent during the 1970s Alaska oil boom and pipeline planning (Burge and Vanclay, 1995). In Canada the first SEIA occurred in the Mackenzie Valley where social impacts considered during development decision-making processes (Berger, 1977). Legislation efforts continued in other northern places throughout the decades: US (1980s), Europe (1970s–90s) and in Russia (1990s–2000s). The fundamental concern of Arctic SEIAs is the definition and scope of socio-economic impacts regarding resource activities.

Prior attempts to define the scope of SEIA have resulted in questions asked that identify the socio-economic *domains* to be measured and *indicators* to be utilized. Domains, typically linked with specific impacts, are spheres of human life associated with distinct activities, behaviors, norms, locations, relationships, and institutions. For example, Armour (1990) distinguishes among impacts affecting ways of life, culture, and community. Further elaboration of this classification include up to eight impact types corresponding to various domains (Burge, 1994, 1999; Vanclay, 1999, 2002; Juslen, 1995; Interorganizational . . ., 1994; 2003; Taylor et al., 1995; BOR, 2002). The Arctic Social Indicators Report (Larsen et al., 2010) considers six domains important to Arctic residents. Scholars recognize eight domains as highly relevant for measuring Arctic SEIs in places of resource development: human health, material well-being (incomes and wages), employment and participation, social well-being and cohesion, population dynamics (demographics and migration), cultural vitality (including traditional activities and closeness to nature), empowerment and fate control, and education and human capital (AHDR, 2004; Burge, 1994; Vanclay, 1999; 2002; Juslen, 1995;

Interorganizational . . ., 2003; Larsen et al., 2010; GNWT, 2003). Thus, SEI are domain-specific or integrated data series that allow comparison of human well-being and its change over time and over space (Mitchell and Parkins, 2011; Force and Machlis, 1997). Indicators are quantitative or qualitative measurements designed to measure direct or indirect impacts of resource development on human lives. Indicators should be clearly defined, reproducible, unambiguous, understandable, and practical (Larsen et al., 2010).

Based on sociological and economic methods (Burge, 2004), SEIAs include data about social processes and impacts that are often gathered from surveys, workshops, focus groups, community meetings, and ethnographic studies. Quantitative and qualitative data assist analysis of community concerns, capacities, and aspirations. Economic data are subject to standard (e.g., cost benefit, input-output, fiscal) analyses. Risk-benefit analysis may be used in projects that bear significant risk, say, to human health (MVEIRB, 2007). Additionally, most SEIAs include anticipatory assessments, which involve scenario building, baseline comparisons, and case study analysis. One concern about SEIA methodologies is their reliance on quantitative indicators. Recent increase in qualitative and community-based methods requires SEIA frameworks to consider other types of information (e.g., gathered from town hall meetings, focus groups) as involving community members in information collection before the start of a project for long-term monitoring purposes continues to emerge as a research practice (Knopp et al., 2013; Petrov et al., 2013).

Identifying specific indicators for SEI monitoring is challenging for two main reasons. First, the eclectic approach of early indicator research often organized multiple variables into loose categories (Burge, 1994; Vanclay, 1999; 2002; Juslen, 1995; Interorganizational . . ., 1994; Kusel, 2001; Force and Machlis, 1997). The inclusion of dozens of loosely connected variables (Mitchell and Parkins, 2011) has been critiqued, citing lack of focus, inclusion of variables only indirectly connected to resource activity, and confusion between impacts and contributing social processes (Vanclay, 2002). Limitations have been addressed by improving the clarity of the system by differentiating social process (demographic, economic, legal, socio-cultural, and emancipatory) from social impact (health, quality of life, social cohesion, cultural well-being, livability of place) and from specific outcomes experienced by humans (Slootweg, Vanclay and van Schooten, 2001; Vanclay, 2002). Second, indicators must be narrowed to a small, representative set of qualitative or quantitative variables. This pragmatic approach identifies relevant, valid, and reliable measures among all available indicators and are based on expert assessment of indicator quality (Mitchell and Parkins, 2011). For example, the ASI reports use seven indicators to characterize six domains of human well-being, which were selected based on definitional clarity, data availability, robustness, scalability, and inclusiveness (Larsen et al., 2010; 2014). This approach may "condense real-life complexity into a manageable amount of meaningful information" (Larsen et al., 2010: 23).

Arctic regional case studies and SEIA applications

SEIA in Alaska, USA

The US has an SEIA framework due to the National Environmental Policy Act (NEPA) of 1969, which requires assessment of impacts of development on human environments. While NEPA does not explicitly mention SEI, federal agencies must include social sciences in assessing the effect of federal actions on the human environment (Sec. 102 [42 USC § 4332(a)]). Government agencies, thus, must consider social impacts when preparing Environmental Impact Statements. Since 1978, social impacts assessments are required if biophysical and social impacts are interrelated (U.S. Council on Environmental Quality, 1978), but only in the 1980s did federal courts require social impacts to be considered (Freudenberg, 1986). Since then, consideration of SEI has become more systematic (Becker, 1997; Galisteo Consulting, 2002) and the EIA process is now the dominant process at the federal level for assessing social impacts.

Various federal agencies publish and update SEIA guidelines and manuals (US Forest Service, 1982; Interorganizational . . ., 1994, 2003; BOR, 2002). The Interorganizational Committee on Guidelines and Principles for Social Impact Assessment (1994) provides detailed guidelines for anticipatory SEIA and contains definitions of social impacts, principles of assessment planning and implementation, and lists suggested domains and variables. An updated version (Interorganizational, 2003) suggests 32 indicators (Table 6.1), consistent with other recommendations (see Burge, 1998, Galisteo Consulting, 2002). The recommended methodological principle is a baseline comparative analysis of indicators using existing case studies as scenario-building instruments where indicators cover mixed variables describing social change processes and development impacts (Vanclay, 2002).

The US's anticipatory SEIA framework is often used to generate scenarios based on assumptions informed by case studies of communities in post-development stages. Many indicators in anticipatory frameworks could be utilized for post-development SEI monitoring, but a potential shortcoming of this approach is the lack of variables designed to monitor post-development conditions. Measurements appropriate for identifying baseline conditions are not always suitable for detecting change, and inability to account for the actual consequences of resource development is a recurring problem for SEIAs (Schweitzer et al., 2013).

Alaska's specific history of SEIA began with a consideration of the effects on Inuit of the Trans-Alaska Pipeline planning in 1973. With introduction of federal offshore oil and gas lease sales in 1976, the US Bureau of Land Management (later the Minerals Management Service, now Bureau of Ocean Energy Management) began a Social and Economic Studies Program to assemble social and cultural research on coastal communities relevant to impact assessment and planning (Peat Marwick and Mitchell, et al. 1977). Over 100 technical reports, many focused on economic and demographic change, have provided content for NEPA documents related to oil and gas leasing.

Table 6.1 U.S. Interorganizational Committee on Guidelines and Principles for Social Impact Assessment list of recommended variables

Population change	Community resources
Population size density & change	Change in community infrastructure
Ethnic & racial composition & distribution	Indigenous populations
Relocating people	Changing land use patterns
Influx & outflows of temporaries	Effects on cultural, historical, sacred &
Presence of seasonal residents	archaeological resources

Community & institutional structures	Community and family changes
Voluntary associations	Perceptions of risk, health & safety
Interest group activity	Displacement/relocation concerns
Size & structure of local government	Trust in political & social institutions
Historical experience with change	Residential stability
Employment/income characteristics	Density of acquaintanceships
Employment equity of disadvantaged groups	Attitudes toward proposed action
	Family & friendship networks
Local/regional/national linkages	Concerns about social well-being
Industrial/commercial diversity	
Presence of planning & zoning	

Political & social resources
Distribution of power & authority
Conflict newcomers & old-timers
Identification of stakeholders
Interested and affected parties
Leadership capability & characteristics
Interorganizational cooperation

The University of Alaska's Institute of Social and Economic Research developed the Man in the Arctic Program (MAP) in the 1970s, providing an econometric model of the state's economy. The model projected economic and demographic impacts of outer continental shelf leasing in 1979 (Huskey, 1979) and when the Minerals Management Service began selling offshore oil and gas leases in the 1980s, projections of population, employment, and income change from the model were incorporated into EISs for each sale. Use of this model in subsequent EISs has helped project the economic effects of other projects (e.g., Alaska Natural Gas Pipeline, Susitna hydropower project) on Alaska communities.

In the 1990s, economic growth in urban Alaska created concern for the well-being of rural Indigenous communities. Studies at this time focused on local-regional social, economic, and cultural effects of resource development, especially the impacts of the 1989 Exxon Valdez oil spill and the North Pacific Fishery

Table 6.2 Emerging gaps and research priorities

General gaps/barriers	Priorities
Lack of theoretical framework for indicators development	Continue development of theoretical, interdisciplinary frameworks for SEIAs
Not integrated, fragmentary measurement and monitoring systems	Develop integrated systems that address multiple socio-economic well-being domains
Lack of integration among assessment indicators, monitoring, and management	Develop monitoring systems that allow long-term observation of well-being. Develop integrated SEIA frameworks that include multiple monitoring and management components
Narrowly defined indicators	Broaden the scope of SEIA where appropriate to include broader impacts such as such as institutional changes, policies, political issues, as well as territorially extended impacts
Limited ability to collect/analyze data and resultant predictive modeling based on inadequate baseline information	Enhance data collection, implement data collection mandate within business-community agreements, develop frameworks for community self-monitoring
Dependency on "imported" (southern) indicators	Develop Arctic-specific indicators
Lack of understanding of impact significance and thresholds	Focus research efforts on understanding social and economic thresholds, resilience, and adaptation capacities
Quantitative tilt	Improve and integrate non-quantitative methodologies, develop community-based monitoring programs (CBMP)
Limited follow-up and motoring during and after the project.	Enhance and implement follow-up and monitoring SEIA components, develop methodologies for long-term monitoring during and after project's completion
Lack of community participation in the measurement/monitoring/ follow-up stages	Develop principles of community involvement after project's commencement. Direct targeted efforts to developing CBM frameworks
Developer-driven assessment	Advance co-management and oversight institutions, improve understanding of co-management processes and implementation
Negotiated nature of indicators	Develop flexible but firm frameworks, standards for SE impact monitoring
Lack of tools and techniques for Addressing Smaller Projects	Develop cost-effective methods of SEIA; ensure SEIA is conducted regardless of the scale of the project

Management Council's supervision of offshore fisheries. As NEPA processes evolve, SEIAs have become more comprehensive and sophisticated. For example, the EIS for the Red Dog Mine expansion EIS (Tetra Tech, 2009) describes projected impacts of the mine and port expansion, outlining incremental and cumulative effects on land use, recreation, subsistence, public health, cultural resources, transportation, jobs, income, cost of living, and environmental justice. However, little attempt has been made to address potential impacts related to control over destiny and health and to discuss the distribution of material well-being within communities and regions. Impacts on cultural continuity are addressed only qualitatively.

While the NEPA EIS represents the main vehicle for SEIA in Alaska, the state must also determine whether use of state lands and resources is in the "best interest" of the people of Alaska (AS 38.05.035). The *Best Interest Findings* (BIFs), issued by the Department of Natural Resources often contain helpful qualitative analysis of potential social impacts. Comprehensive BIFs include those of North Slope state oil lease sales (ADNR, 2008; ADNR, 2011), containing job estimates by community, qualitative discussion of potential public health effects, and possible mitigation measures to reduce impacts on subsistence harvesting, job training programs for local residents, and requirements for consultation with affected communities.

SEIA in Arctic Canada

In Canada, the federal EIA process is regulated by the Canadian Environmental Assessment Act (CEAA, 2012). Effects to be considered include impacts on the health and socio-economic conditions of Aboriginal peoples; physical and cultural heritage; current use of lands and resources for traditional purposes; and structures, sites or things of historical, archaeological, paleontological, or architectural significance. EIAs can be conducted either by a responsible authority or review panel to understand the cumulative effects of resource activity by the federal government in cooperation with other jurisdiction(s).

Canada's federal government occasionally passes special legislation imposing EIAs or similar procedures on specific regions and projects. For example, the Mackenzie Valley Resource Management Act (MVRMA, 1998) manages the EIA for the Mackenzie Valley Gas Project EIA, creating an integrated co-management structure for public and private lands and waters. Canadian provinces and territories have their own legislative acts and guidelines for conducting EIAs, often creating additional complications for processing project submissions. Jurisdictions attempt to harmonize EIA requirements for efficient and effective environmental assessment that provides high quality information and conclusions about environmental effects to satisfy their respective legislative requirements (e.g., GNL & GC, 2011).

We describe three SEIAs conducted in the Canadian North, emphasizing domain and variable selection, measurement, and monitoring: (1) a traditional anticipatory SEIA (Mackenzie Valley Gas Project), (2) a monitoring and follow-up program established to detect post-SEIA impacts (Communities and Diamonds reports in the Northwest Territories, hereafter NWT), and (3) an integrated inter-jurisdictional measurement framework with anticipatory and follow-up components (Labrador-Island Transmission Link project).

Case study: Mackenzie Valley

The Mackenzie Valley Gas Project EIA presents information about communities and individuals who might be affected by the project's proposed activities (Imperial Oil, 2004). Mackenzie Valley Environmental Impact Review Board provided guidance for the assessment conducted under the Mackenzie Valley Resource

I. People and the Economy
1) Population composition and dynamics
    i) Census counts and population estimates
    ii) Ethnicity in region
2) Economic activity
    i) Labor force by SOC
3) Labor force
    i) Adult education attainment: high school graduation and some postsecondary education
    ii) Population, particiption, employment and unemployment rates
    iii) Profile of the working-age population
4) Income sources and amounts
    i) Corporate tax status
    ii) Employment income and income support beneficiaries
5) Cost of living
    i) Cost of living
    ii) Estimated food price index

II. Infrastructure and Community Services
1) Transportation infrastructure
2) Utilities, energy, and communications
3) Housing and recreation
    i) Housing and repairs needed
4) Governance

III. Individual, Family and Community Wellness
1) Health conditions
    i) Cases of respiratory disease treated by physicians
    ii) Cases of infectious and parasitic diseases treated by physicians
    iii) Cases of sexually transmitted infections
    iv) Cases of accidents, injuries, and poisoning treated by physicians
    v) Death from injuries
    vi) Case of mental disorders treated by physician
2) Health care facilities and services
    i) Number and type of health care facilities
3) Family concerns and community conditions
    i) Hospitalization and alcohol-related illnesses
    ii) Spousal assaults
    iii) Children taken into care/services
    iv) Young Offenders Act offenses
    v) Violent crimes and property crimes

III. Individual, Family and Community Wellness (cont.)
4) Social services facilities and services
    i) Protection service features
5) Education and training
    i) Levels of education training
    ii) Education facilities

IV. Traditional Culture
1) Participation in traditional harvesting
    i) Harvesters/number of people who harvest
    ii) Adults who hunted or fished
    iii) Harvest data
    iv) Total and per capita edible weights of harvested wildlife
    v) Replacement values for wildlife harvesters
    vi) Country food consumption
2) Trapping
    i) Active trappers and average income
3) Aboriginal language
    i) Aboriginal language speakers

V. Nontraditional Land and Resource Use
1) Land ownership
2) Granular resources
3) Timber resources
4) Mineral resources
5) Oil and gas resources
6) Nontraditional resources harvesting
7) Tourism and recreation
8) Other commercial active marine operations
9) Environmentally protected areas
10) Visual and aesthetic resources

VI. Heritage Resources
1) Prehistoric period
2) Historic period
3) Paleontological sites
4) Environmental setting

Figure 6.1 Mackenzie Valley socio-economic baseline indicators

Management Act (MVEIRB, 2005; 2007). The SEIA was completed in August of 2004 by Imperial Oil Resource Ventures Limited, ConocoPhillips Canada Limited, Shell Canada Limited ExxonMobil Canada Properties, and Aboriginal Pipeline Group and focused on how the project affected community wellness. An issues-focused approach to baseline information examined community wellness, including physical, emotional, social, cultural, and economic well-being of individuals, families, and the community. The objective was to develop community well-being indicators from available data for 32 communities in NWT and Alberta (Figure 6.1). Five regions—Inuvialuit, Gwich'in, Sahtu, Deh Cho, and the Industrial and Commercial center—comprised the study area and six domains—people and the economy; infrastructure and community services; individual, family, and community wellness; traditional culture; non-traditional land and resource use; and heritage resources—provided socio-economic measurements. Data was collected from the 1991, 1996, and 2001 censuses, Canadian federal and NWT surveys NWT's Health and Social Services, and RCMP records. The extensive indicators provide socio-economic status data about individuals, communities, and the NWT. The MVEIRB process for impacts assessment resulted in extensive input from Aboriginal stakeholders and a relatively comprehensive measurement system (Galbraith, Bradshaw and Rutherford, 2007) but failed to provide suggestions for monitoring impacts after commencement of the proposed resource development or consider impacts outside the eight communities, a common problem of narrowly focused SEIAs.

Case study: Communities and Diamonds monitoring

Established in 1997, the government of NWT required the Communities and Diamonds (CAD) report via a socio-economic agreement (SEA) to reflect the commitments and predictions made by the company during its environmental assessment. SEAs are follow-up programs for EIAs required under territorial jurisdiction. The CAD report is published annually as a follow-up to environmental assessments and includes three diamond mines: BHP Billiton Ekati Diamond Mine, Diavik Diamond Mine, and De Beers Snap Lake Mine. The Department of Industry, Tourism and Investment leads the assessment program and negotiates agreements on behalf of the NWT. CAD identifies the impacts of resource development and community responses to changes associated with mining activities and helps verifying assumptions made during the first assessment or project proposal. Increasingly, CAD is a monitoring tool to identify socio-economic trends in local communities. It is an exemplary post-SEIA follow-up and monitoring framework that helps communities, government and the companies to assess development effects and devise response and mitigations strategies. Recent CADs provide overviews of 20 socio-economic areas in NWT (in Yellowknife, Bechoko, Detah, Gameti, Lutsel k'e, Ndilo, Wekweeti, and Whati).

The NWT government report examines sustainable economic development through five domains—community, family and individual well-being; cultural well-being and traditional economy; non-traditional economy; net effect on

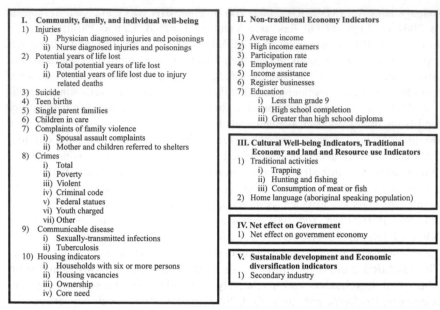

| I.   Community, family, and individual well-being | II.   Non-traditional Economy Indicators |

Figure 6.2 Socio-economic indicators used in Communities and Diamonds reports

government; and sustainable development—and provides 22 measurements of socio-economic areas (see Figure 6.2). Different government agencies and departments provide data, most of which comes from surveys (e.g., NWT Community Survey, Community Impact Survey, the Census). With new mines opening, CAD Reports have become more detailed and CAD monitoring and follow-up frameworks provide detailed analysis of socio-economic factors within communities affected by mining. CADs are often inconclusive regarding attribution of socio-economic changes to mining activities; only a few monitored variables are considered impacted by resource extraction. Such uncertainties diminish the ability of CADs to detect and measure impacts effectively and indicators about health, demographics, and migration should be included.

SEIA in Greenland

Strategic environmental impact assessment became binding for EU member countries in 2004 and is a process to assess possible environmental effects stemming from policy, policy planning, action programs, and other strategic documents and regulations (SEA and Integration of the Environment into Strategic Decision-Making, Final Report, May 2001). In 2010, Greenland agreed to comply with EU protocol on SEA; the Faroe Islands remain undecided. Although compliant jurisdictions adhere to the protocol, details may be modified by country. The Bureau of Minerals and Petroleum (BMP) oversees these modifications and 2009 guidelines describe the Social Impact Assessments (Government of Greenland, 2009).

Mining and energy extractive development in Greenland are guided by four essential factors: (1) recruiting Greenlandic labor; (2) engaging Greenlandic enterprises; (3) focusing on knowledge transfer (e.g., education) to ensure long-term local capacity building within mining and mining support industries; and (4) preserving socio-cultural values and traditions. The main objectives of the SIA process are to engage all relevant stakeholders in consultations and public hearings; provide description and analysis of the social pre-project baseline, mitigation, and future monitoring; provide assessment based on data, identifying positive and negative social impacts at local and national levels; optimize positive impacts and mitigate negative impacts throughout a project lifetime; and develop a Benefit and Impact Plan to implement the Impact Benefit Agreement.

An SEA report includes eight elements: (1) description of social baseline conditions including anticipated changes before project commencement, using quantitative parameters where possible; (2) presentation and analysis of alternatives and comparison of feasible alternatives to the proposed project design, including how the community and Greenland would develop without the project; (3) outline of potential impacts by presenting predicted and assessed possible social impacts, using quantitative parameters where possible and describing relevant local, regional, or national development plans; (4) presentation of development oppor-tunities and mitigation of negative impacts; (5) development of a Benefit and Impact Plan with proposals for maximizing development opportunities and miti-gating negative impacts; (6) proposal of a Monitoring Plan regarding the effects of the implemented Benefit and Impact Plan; (7) proposal of an Evaluation Plan to evaluate monitoring results and discussion of adjustments to the Benefit and Impact Plan; and (8) description of plans for public participation during the SEIA process and project lifetime.

Case study: aluminum smelter in Maniitsoq

Discussion about a possible aluminum smelter began in 2006 when Alcoa contacted Greenlandic authorities. Alcoa wished to initiate preliminary surveys to assess the potential for a smelter in central West Greenland between the towns of Sisimiut and Nuuk (Hansen et al., 2009). A Strategic Environmental Assessment (SEA) began in April 2007. An SEA was recommended, which was published in parts between 2007–2009. The SEA report covered five domains: (1) *Nature and environment*: The analysis included physical and biological components of the environment, including the aquatic environment, water resources, waste, waste-water, air emissions, noise, and dust; (2) *Health*: Review of data from the National Institute of Public Health provided information about a wide cross-section of the population, including information about pregnancies, newborn babies, children aged 0–12 years, schoolchildren aged 15–17 years, and adults 18 years and older. Data included socio-demographic health, and living conditions and lifestyles, which made analysis of the volume and spread of disease and risk factors for chronic diseases possible; (3) *Culture*: Kangerlua is the largest and most resource-rich inland area in Greenland and is relatively easily accessible from the coast to inland through valleys, rivers, and lakes. Recognizing the rich heritage of the hinterland behind the planned smelter between Kangerlussuaq and Nuuk, a main focus was on archaeology and history in the SEA; (4) *Regional development and economy*: This domain describes the impact of aluminum smelter placement near three cities—Nuuk, Maniitsoq, and Sisimiut—and focuses on long-term regional and socio-economic consequences and on the construction and operational phases of the smelter and hydroelectric power plants; (5) *Cumulative impact assessment*: Cumulative analysis was conducted by examining existing and generated knowledge alongside a review of literature, reports and GIS data. Of note are the potential consequences of the project on population mobility.

This analysis resulted in two important outputs. First, an overview of the potential consequences of establishing an aluminum smelter was created. Second, scenarios the human and territorial consequences of smelter development were developed. The SEIA approach revealed gaps in more conventional EIA and SIA approaches and it became obvious that involvement of the public sphere early was crucial, and several rounds of public hearing and consultation were included. Analysis of public debate timing indicates that after a slow start the public debate flourished, especially after significant decision-making processes were completed. Indicators suggest that a democratic deficit exists regarding the timing of significant administrative and internal political decisions and public debate (Hansen et al., 2009).

SEIA in Northern Russia

In the Russian North, SEIA emerged in the 1990s. It is important to realize that studies of environmental and socio-economic impacts of resource development were conducted during the Soviet times, but remained largely unpublished, confidential, and thus unutilized (Sokolova and Pivneva, 2004, 2005, 2006, 2007). A formal process was established in 1995 as part of environmental impact assessment legislation (under the Federal Law "On environmental expertise") and specified in the "Regulation of assessment of impact of planned economic and other kinds of activity" ratified by the Order of the State Committee of Ecology of Russia 16.05.00 #372 (Goskomekologia, 2000). Legislation about Indigenous peoples and traditional nature-use territories gave further rights to these Russian citizens and the Federal Law "On the Guarantees of the Rights of the Indigenous Small-Numbered Peoples of the Russian Federation" (1999) introduced the term *ethnologic assessment (etnologicheskaya ekspertiza)*, which is a scientific examination of the impacts of environmental and socio-economic change on the development of an ethnic group. Article 8 of the law allows Indigenous people to participate in EIAs and ethnological assessments.

Many ethnological assessment methodologies were developed in the 2000s. For example, Tishkov and Stepanov (2004) suggested an ethnological monitoring system consisting of 46 standardized indicators. The most extensive SEIAs in the Russian North have taken place in Yamal-Nenets okrug (Bogoyavlenskiy et al., 2002) and Sakhalin oblast (Obzor ..., 2008; Murashko, 2006). Ethnological assessment, however, has been minimal in decision-making processes for regional development (Martynova and Novikova, 2012). Furthermore, according to the 2005 amendments to the Federal Law, which took effect in 2007, the scope of EIA was narrowed and resultingly, SEIAs have limited influence on decision-making unless given priority by a company or demanded by local residents.

A typical Russian SEIA has two parts: (1) research to identify communities in zones of direct and indirect impacts, and (2) monitoring of environmental and social-economic changes during project design and implementation. Because impacts are on both Indigenous and local communities and based on anthropological field studies, international and Russian experience, and legislation, SEIAs provide recommendations to mining companies and investors about strategies and cooperation with Indigenous and local residents. The ethnographic approach illuminates SEI on Indigenous communities but non-Indigenous residents are typically not included, except when impacted communities are majority non-Indigenous (i.e., in the case of Prigorodnoe on Sakhalin Island; see Bacheva, Kochladze and Dennis, 2006). Reliance on anthropological research and ethnographic methods has created this bias in research focus and design in SEIAs in the Russian North.

The qualitative and quantitative indicators used in SEIAs typically examine critical natural resources and their use by humans through hunting, fishing, and gathering; economics of the territory and development prospects (jobs, educational opportunities, services); population (history, settlement descriptions and sizes, ethnic and social structures, Indigenous societies, demographic characteristics); social conditions (employment, health care, wages) and built social environment; historical and cultural heritage (sacred places, archaeological sites); technogenic risks of mineral and hydrocarbon resource development; and social risks (prostitution, sexually transmitted diseases, cost of living, crime).

Where resource development occurs in Indigenous homelands, researchers examine the impacts of resource development on economic development, social and family relations, preservation of native language and culture, and reindeer herding and fishing when it is the basis for subsistence living for Indigenous peoples (Vasilkova, Martynova and Novikova, 2011). In territories of traditional nature-use by Indigenous peoples, the impacts on traditional land use are assessed. Usually, impacts are described as environmental ethics and not in economic terms. SEIA assesses possible effects on traditional land-based activities, identifying risks of oil extraction and transportation on traditional land use.

SEIA implementation in Russia relies on weak legislation and the lack of legal mechanisms to enforce recommendations is problematic. Sometimes, internal systemic corruption, where local experts conduct SEIAs, also contributes to incomplete or biased reporting. The quality and extent of EIAs and SEIAs depends on which shareholders participate in the projects. For example, the European Bank for Reconstruction and Development has an equity stake in the Irkutsk Oil Company since 2008 and, resultantly, they must conduct an SEIA in which social indicators, such as tax revenues, infrastructure development, job creation jobs, and support for various cultural events that include local and Indigenous residents, are quantified (Bourtsev and Cherdantsev, 2008). In other cases, specific variables address local stakeholder concerns. In the Verchnechonskneftegas SEIA, the representatives of Indigenous peoples requested an assessment of health and quality of life for local and Indigenous peoples that included indicators such as natural climate peculiarities; quality of water and soils; nutrition; radiation exposure, illness rate by age group; and infectious, parasitic, zoonotic, and social diseases.

SEIA indicators case study: Sakhalin-2

Offshore oil and gas fields dot the northern and eastern floor of the Sea of Okhotsk near Sakhalin Island. Sakhalin-2 includes development of the Piltun-Astokhskoye offshore oil and the Lunskoye offshore natural gas fields and much of the infrastructure required for oil and gas extraction and transportation is located on the island. In the case of the natural gas, transportation occurs onshore, to Russia's first Liquefied Natural Gas (LNG) plant. Currently operated by Sakhalin Energy Investment Company Ltd., corporate partners in this transnational operation have changed many times, but the most important change occurred in 2006 when the company went from being majority owned by transnational corporations to majority owned by Russia's national gas company, Gazprom (51% share; Bradshaw 2006). The current composition of transnational corporations with shares in Sakhalin-2 is Gazprom, Shell, Mitsui, and Diamond.

Before December 2006, Sakhalin-2 was Russia's only large-scale energy project operating without a Russian partner. In this era, Sakhalin-2 sought international financing, such as from the European Bank for Reconstruction and Development for project development, which meant that international standards and procedures for hydrocarbon development were followed, including SEIA development. Indeed, when managed by Shell Corporation prior to 2006, Sakhalin Energy prepared environmental, social, and health impact assessments for phase 2 of the project according to international best practice (Sakhalin Energy 2002). Environmental Resources Management, Ltd, a UK-based company, used local and non-local contracted experts to gather baseline information and identify key local concerns. Their Social Impact Assessment consulted over 5,000 residents and included 52 communities, focusing on 22 mostly rural communities where temporary construction and permanent facilities are sited or likely to be sited.

The SEIA developed for the Sakhalin-2 project has proven difficult for national and international development actors as the weight and roles of actors has shifted. When bolstered by international non-governmental organizations, such as GenderAction, Pacific Environment, Wild Salmon Center, and Friends of the Earth Japan, local environmental and cultural activist groups battled transnational development from the mid-1990s until 2006, ensuring that Sakhalin-2's SEIAs addressed social and environmental concerns for local and Indigenous peoples in multiples sites across the island that were earmarked for hydrocarbon development. Until 2006, both local and global environmental and cultural activists and the local and non-local (both Russian and foreign) ethnographers hired as consultants by the transnational hydrocarbon companies helped guarantee that understandings of the struggle over resources, both hydrocarbon and subsistence, were widely and continually publicized (Bradshaw 2005; Graybill 2013b). Since 2006 and the transfer of majority holding to Gazprom, international financing institutions will not finance this project (i.e., the European Bank for Reconstruction and Development; Williams 2007) and much of the contract work for SEIAs has gone to local (i.e., from Sakhalin) people. Previously, SEIAs were written for international lenders, ensuring that impacts would be addressed during or before development. Many activist groups were emboldened, requesting "cultural" (in the case of Indigenous peoples) or "gender" (see Bacheva, Kochladze and Dennis, 2006) impact assessments in protests against extraction projects (Graybill, 2009). This checks-and-balances approach, alongside local–global environmental and Indigenous activism, helped ensure that hydrocarbon development followed international standards.

Gazprom's majority control since 2006 has hushed discussions and knowledge of social and environmental conditions in the Sakhalin-2 project, leading to speculation that environmental greening is unlikely with Gazprom as the major operator (Bradshaw, 2006). Many SEIA components are now monitored by local experts, who are sometimes part of local networks and provide minimal services for maximum fees (Graybill, personal communication). Others provide very powerful, thoughtful data and expertise, making for an uneven landscape of SEIA knowledge and monitoring. Whether environmental and cultural activists will regain the power attained in the early 2000s is questionable and what the actual impacts on communities across Sakhalin have been remains unstudied. For example, Sakhalin Energy prior to 2006 had expressed interest in monitoring impacts on communities and the environment of hydrocarbon infrastructure (e.g., roads, pipelines, waste facilities), construction camps, Indigenous enclaves (especially in Val, Nogliki and Chir-Unvd) and in-migration of local people for jobs associated with hydrocarbon development. Since 2006, no reports have been produced illustrating attention to these concerns (Graybill, 2013a).

Summarizing the overview of the SEIA measurement frameworks in the Russian North, we conclude that a truncated legal base for SEIA and lack of implementation incentives and mechanisms make most indicator frameworks insufficient and ineffective. Simultaneously, ethnological assessments are well aligned with international practice and advance qualitative methodologies, which are often limited in other Arctic SEIAs. Other problems and limitations of Russia's SEIA process include the fact that public hearings are not widely publicized and conducted only after project initiation; geological exploration does not require impact assessments; SEIA are focused on Indigenous peoples (narrowly defined by the Russian legislation) and thus do not consider concerns of non-Indigenous residents who might perceive themselves as "native" and have strong attachment to place; lack of documentation about sacred places may leave them unassessed; territories traditionally used by Indigenous peoples are often located near settlements with predominantly non-Indigenous populations who may represent a numerical majority with different views on resource development; and lack of alternative development scenarios that could provide local communities with other possible development choices. Finally, SEIAs are often conducted by consultants (local and non-local) or by a division of extractive companies tasked with social development issues, thereby introducing bias into SEIA results and the lack of monitoring of (S)EIs during project implementation and after project termination remains problematic.

Milestones, limitations, and lessons from existing SEIA frameworks

Our survey of SEIA measurement frameworks demonstrates variability among SEIA processes among circumpolar locales, yet similarity of indicators and their selection processes. Domains and specific measures are often based on similar interests (population, health, economy, traditional culture) and data availability. However, representation of different domains is highly uneven and measures vary depending on national and regional data resources. This makes comparison, evaluation, validation, and transfer of SEIA measurement methodologies difficult. Creating a circumpolar impact indicators framework would require including common variables from SEIAs across the Arctic and could be based on ASI's six domains and seven indicators (Larsen, Schweitzer and Fondahl, 2010). Incorporation of SEIA in EIA is occurring across the Arctic, and most circumpolar jurisdictions have integrated SEIA and measurement frameworks into legal doctrine and actual EIAs. Although integration varies, legal and practical frameworks guide implementation of SEIA. Some jurisdictions require SEIAs as part of larger EIA frameworks. Many recommend or mandate the domains to be assessed, but only a few have developed specific criteria for assessment. Below, we focus on emerging trends and best practices that could aid SEIA development in Arctic countries.

Best practices

One emerging practice is collaboration among government, proponents, and local communities to create comprehensive, structured, and transparent frameworks and processes for SEIA. This includes inter-jurisdictional collaboration at national and international scales. Arctic regions belong to multiple jurisdictions (the EU, the US, Russia, Canada) and must follow externally imposed legal frameworks and practices, necessitating collaboration among Arctic regions. For example, external legal frameworks may unify SEIA practices (e.g., among EU members), but should be performed to address local conditions and incorporate suitable practices from other Arctic regions.

The shift from formalized assessments based on statistical data to locally focused, community-based approaches is considered as best practice. Other best practices include standardization of SEIA terms, development of clear methodological guidelines to ensure community-based research, valuation of hard-to-measure but community-relevant impacts and utilization of appropriate analytical tools, implementation of socio-economic benchmarks, and deployment of best available mitigation practices (MVEIRB, 2005).

Increasingly, these assessments include the concerns of Indigenous and local communities in SEIAs. Community engagement in Arctic SEIA processes has become a common practice and in addition to data collection using standard techniques, SEIA frameworks emphasize other types of information gathering, such as town hall meetings, focus groups, and interviews. Assessments currently span more domains and rely less on secondary data and aggregated statistics. Greenland's Aluminum Smelter, for example, used elaborate quantitative and qualitative methodologies. An emerging practice is to involve community members in gathering and analyzing information before the start of the project and after, for the purposes of follow-up and monitoring (Knopp et al., 2013; Petrov et al., 2013). This kind of community-based monitoring (CBM) is an important innovation that ensures community involvement in and relevance of the SEIA and, potentially, reduces costs of monitoring and assessment. CBM is likely to become a key tool in future SEIA practices.

Development of indicator systems reflecting local needs is gradually being incorporated in modern SEIA frameworks. These frameworks help guarantee sensitivity to socio-economic vulnerabilities and the concerns, demands, and desires of Arctic communities. Overarching systems, such as ASI, provide guidance regarding domains and indicators for Arctic regions; community-based indicators ensure adequacy and relevance.

Best practices clearly demonstrate that follow-up monitoring is critical for assessing resource activity impacts. While follow-up plans and long-term monitoring programs are becoming integral to SEIAs they are not universal and are sometimes limited to monitoring the correctness of the initial SEIA assumptions or conclusions rather than assessing actual resource development impacts. More detailed plans and mitigation strategies are needed to make follow-up meaningful. SEIA must also be linked to social impacts management plans, which implement adaptive management strategies (Franks et al., 2009).

Barriers to successful implementation of SEIA measurement systems and recommendations for research exist, certainly, but are decreasing as SEIAs become more standard, detailed, and known and utilized by communities, governments, and companies. Ultimately, the SEIA practices that follow the recommendations provided above and address the gaps outlined below will ensure that resource development benefits Northern communities more with less negative impacts.

Gap analysis

From early SEIA reviews (MVEIRB, n/d) and actual SEIA from different Arctic jurisdictions, we find 13 gaps in SEIA measurement systems:

Lack of theoretical frameworks. Common sense often guides selection of domains and measures, which can lead to imbalanced or biased choices of variables based on available data or particular interest. While regulatory documents, and thus SEIAs, sometimes specify what should be measured, clarity about the conceptual foundations used to determine the assessment parameters is often lacking. This is a concern for methodologies and data sets intended to be identifiable, replicable, and available for meaningful validation and follow-up. Fundamentally, SEIAs ask what to measure and how to measure it. Since both questions must be addressed scientifically, we advocate for continued development of interdisciplinary theoretical SEIA frameworks.

Fragmented, un-integrated measurement and monitoring systems. Existing systems provide un-integrated, fragmentary measurement and monitoring that do not reflect the complexities of human well-being. SEIAs are often comprehensive for specific locales, but lack a holistic picture of cumulative resource development impacts on human society and the environment. A comprehensive, nuanced understanding of SEI related to resource development is achieved via integrated analysis, which reveals the differential impacts on multiple societal groups (e.g., gender, age, ethnicity, occupation). Broadened participation in domain and variable selection and monitoring processes, inclusion of TEK, and public participation may help bridge this gap.

Lack of integration among indicators, monitoring and management. SEIA measurement practices are largely detached from resource management. Because SEIAs are not linked to local land use or resource management frameworks, indicators do not relate to actual use or management. This makes SEIAs less able to capture impacts related to use and management and less usable for decision-making and mitigation purposes.

Narrow definitions mean limited value. SEIAs are often narrowly focused on the direct impacts of resource development on communities. While understandable, this focus neglects broader activities, sites, and impacts (e.g., institutional change, policies, politics) of industrial activity. The narrow scope of SEIAs and focus on local indicators may insufficiently represent socio-economic concerns. While SEIA cannot resolve all policy matters, they may illuminate other fundamental issues (see Usher, 1993).

Limited data sets hinder predictive modeling. Data limitations constrain comprehensive, accurate SEIAs. Many Arctic regions are data poor; availability of socio-economic variables is limited to pre-existing indicators collected by statistical agencies, which are rarely ideal for measuring the resource development effects. Baseline data may also be limited in quantity and quality. Data collection for SEIAs may benefit assessment, but one-time data acquisition precludes using relevant indicators for future monitoring. Modeling and prediction of impacts based on inadequate data jeopardizes SEIA accuracy and thus validity. We recommend mandated data collection within business-community agreements and developing frameworks and local expertise for community self-monitoring.

Dependency on non-Arctic indicators. Globally, SEIA measurement systems have evolved considerably regarding the number and appropriateness of indicators. Inclusivity in the assessment process and collaborative planning for SEIA execution are important improvements for SEIA frameworks. However, knowledge of the indicators most effective for measuring human well-being and human development in the Arctic remains limited. As a result, most Arctic SEIA use indicators designed to assess other regions. Many traditional indicators (e.g., income, demography, employment) may not adequately portray socio-economic well-being in the Arctic because they do not consider traditional economy or different cultural, family, and community relationships. Development of northern-specific impact measurement frameworks must consider different and/or additional domains and indicators, and indicators must be adjusted to represent Arctic realities.

Lack of understanding of thresholds and impact significance. Insufficient understanding of socio-economic thresholds challenges the identification of the significance and threshold of an impact. Not unique to the Arctic, many environmental thresholds are specified in legislation, directives, and guidelines but socio-economic assessment is more complicated because understanding of impacts and thresholds is largely under-researched.

Quantitative tilt. Arctic SEIA measurement systems are largely quantitative, but qualitative data are also used because they describe heterogeneities in human societies (e.g., differences among stakeholders, their understandings of reality, discourses), which improves the overall understanding of impacts. While many SEIAs consult communities prior to and during assessment, a disconnect exists between qualitative and quantitative methods for indicator development. Research to integrate these two important aspects of SEIA is needed. Participatory actions as a part of SEIA can be also advantageous (Novikova, 2010).

Limited monitoring during and after a project. Modern SEIAs typically contain provisions for monitoring after SEIA completion and during project implementation. However, emphasis is placed on verifying the accuracy of impact prediction rather than on impact management and evaluation (Noble and Storey, 2005). Follow-up activities commonly are very limited (Galbraith, Bradshaw and Rutherford, 2007; Hunsberger, Gobson and Wismer, 2005). Follow-up monitoring of SEIA components will track the effect of resource development on Arctic communities, ensuring the receipt of benefits and mitigation of negative impacts.

Lack of community participation in measurement/monitoring/follow-up stages. Credibility and effectiveness are enhanced when the public participates in SEIA. Including local stakeholders creates trust between communities and industry. In some places, such as Russia, meaningful inclusiveness is the exception rather than standard practice. Even in countries that include public participation, local communities may not be effectively engaged in all SEIA stages. Typically, community involvement is high initially (i.e. preparation, submission) but is lower later (e.g., monitoring, follow-up). One impediment is a lack of CBM, which engages local stakeholders by empowering them with the collection, analysis, and sharing of meaningful information that is also valuable for SEIAs (Knopp et al., 2013; Petrov et al., 2013; Udofia, Noble and Poelzer, 2017). We recommend the development of Arctic-specific CBM frameworks to aid community participation in follow-up activities and build community capacity.

Developer-driven assessment. SEIA are often conducted and funded by development proponents, which may result in bias. In jurisdictions with co-management institutions or other similar arrangements, inherent bias is minimized. However, regions without inclusive review boards, transparency requirements, and

Table 6.3 Key SEIA subjects to be addressed

Specific subjects to be addressed	Priorities/strategies
Local workforce	Evaluation of local skilled work-force availability, involvement and development
Human mobility and its impacts	Evaulation of settlement consequences for other places of pulling skilled work force out of the labor markets
Gender	Evaluation of gendered potential out-migration, employment and educational opportunities
FIFO (fly-in/fly-out)	Evaluation of local impacts of major FIFO labor force from other places
Local/Indigenous cultures	Evaluation of impacts on local/Indigenous cuiltures
Education and human capital	Evaluation of impacts on educational attendance and attainment
Long-term economic consequences and behavior	Evaluation of long-term economic behavior (e.g., savings)

public oversight are at risk, which is mitigated by advancing co-management and oversight institutions (Papillon and Rodon, 2017).

Negotiated indicators. Indicator selection and implementation procedures vary across SEIAs, and usually result from negotiations among multiple stakeholders (e.g., the project proponent, government, and local communities). While bringing stakeholders together and allowing for community participation, creating negotiated measures may be inherently biased and result in "comfort monitoring" if indicators are used to measure impacts in certain ways or places that appease local communities or other actors but actually undermine effective monitoring (Noble and Birk, 2011). Establishing flexible but firm frameworks and standards for SEI monitoring ensures consistency and limits biases.

Lack of tools and techniques for addressing smaller projects. Many SEIAs are designed to serve large projects, making them unsuitable for smaller development projects (roads, small mines, etc.). Guidelines and standards for conducting SEIA are needed for small projects, especially those with smaller financial resources, which required cost-effective SEI indicators. One solution is to implement community-based or co-monitoring, where costs and responsibilities are shared among proponents, communities, and government. We also recommend a number of key subject areas of SEIA that need prioritized development in the near future (Table 6.3).

A circumpolar socio-economic impact assessment framework?

Arctic SEIA frameworks continue to evolve, reflecting changes in national legislation, globally accepted practices, and a turn to the regionalization of SEIA requirements. Yet few frameworks used in specific Arctic locales build on experiences from other regions in the Arctic or similarly remote resource frontiers, such as those in Australia. Indeed, scholars argue that the primary impediment for successful socio-economic monitoring and understanding SEIAs in a comparative circumpolar perspective is the lack of data comparability across different geographic scales and nations (Berman, 2011; Larsen, Schweitzer and Petrov, 2013). For these reasons, we argue that it is time to develop a circumpolar SEIA framework. One possible solution may be for the inter-governmental Arctic Council to work with statistical agencies to develop comparable data for monitoring socio-economic conditions and impacts. An important second step would be sharing SEIA data and methodologies in circumpolar collaboration. A partial framework for such collaboration is outlined in ASI II (Larsen, Schweitzer and Petrov, 2013). Finally, more collaborative efforts among local–international scales and among different actors (state, proponents, communities, Indigenous organizations) would result in a more comprehensive, Arctic-oriented, and community-focused SEIA process.

Acknowledgements

We are indebted to numerous colleagues who discussed socio-economic impact measurement with us in Nuuk and other venues. Specifically, we acknowledge Keith Storey, Lawrence Hamilton, Timothy Heleniak, Chris Southcott, Peter Schweitzer, Joan Nymand Larsen, and Gail Fondahl. This work was supported by ReSDA and NSF (PLR #1338850).

References

ADNR (Alaska Department of Natural Resources). 2008. North Slope Areawide Oil and Gas Lease Sale: Final Finding of the Director. July 15, 2008.
ADNR (Alaska Department of Natural Resources). 2011. North Slope Foothills Areawide Oil and Gas Lease Sales: Final Finding of the Director. May 26, 2011.
Armour, A. 1990. "Integrating Impact Assessment into the Planning Process." *Impact Assessment Bulletin* 8(1/2): 3–14.
Bacheva, F., Kochladze, M., and Dennis, S. 2006. Boom Time Blues: Big Oil's Gender Impacts in Azerbaijan, Georgia and Sakhalin. Report for CEE Bankwatch and Gender Action. http://genderaction.org/images/boomtimeblues.pdf.
Becker, H. A. 1997. *Social Impact Assessment: Method and Experience in Europe, North America and The Developing World.* London, UK: UCL Press.
Bekendtgørelse om vurdering af større anlægs virkning på miljøet (VVM). 1989. Bekendtgørelse nr. 446 af 23 juni 1989.
Berman, M. 2011. "Next Steps Toward an Arctic Human Dimensions Observing System." *Polar Geography* 34(1–2): 152–143.
Bjarnadottir, H. (ed.) 2001. *A Comparative Study of Nordic EIA Systems.* Stockholm: Nordregio.
Bogoyavlenskiy, D., Martyniva, E., Murashko, O., Khmeleva, E., Yankel, Y., and Yakovleva, O. 2002. *Opyt provedeniya etnologicheskoi ekspertizy.* Moscow: Raipon.
BOR (US Bureau of Reclamation) 2002. *Social Analysis Manual Volume I: Manager's Guide to Using Social Analysis; Volume II Social Analyst's Guide to Doing Social Analysis.* Resource Management and Planning Group. Technical Service Center, Denver Federal Center D-8580, Bldg. 67. Denver, CO.

Bourtsev, S. and Cherdantsev, A. (eds) 2008. Environmental and Social Impact Assessment of the Yarakta Oil and Gas Field Development. Irkutsk Oil Company.

Boverket (National Board of Housing, Building and Planning) 1994. The Planning and Building Act – The Law on Technical Qualities of Buildings and The Environmental Code with Ordinances of Relevance.

Bradshaw, M. J. 2005. "Lessons from the First Generation Sakhalin Oil and Gas Projects," for BP Exploration Ltd.

Bradshaw, M. J. 2006. "Internal Dynamics of the Russia Energy Sector: Lessons from Sakhalin." Aleksantri Institute, University of Helsinki, for the Finnish Ministry of Foreign Affairs.

Burdge, R. J. 1994. *A Community Guide to Social Impact Assessment*. Middleton, WI: Social Ecology Press.

Burdge, R. J. (ed.) 1998. *A Conceptual Approach to Social Impact Assessment: Collection of Writings by Rabel J. Burdge and Colleagues* (revised edition). Middleton, WI: Social Ecology Press.

Burdge, R. J. 1999. *A Community Guide to Social Impact Assessment* (revised edition). Middleton, WI: Social Ecology Press.

Burdge, R. J. 2004. *The Concepts, Process and Methods of Social Impact Assessment*. Middleton, WI: Social Ecology Press.

Burdge, R. J. and Vanclay, F. 1995. "Social Impact Assessment." *Environmental and Social Impact Assessment*, 31–66.

Bureau of Mining and Petroleum. Greenland Government. 2009. Guidelines for Social Impact Assessments for Mining Projects in Greenland.

CAD. 2013. Communities and Diamonds 2011 Annual Report of the Government of the Northwest Territories under the BHP Billiton, Diavik and De Beers Socio-economic Agreements. GNWT. Yellowknife.

CEAA 2012 Canadian Environmental Assessment Act, 2012 (S.C. 2012, c. 19, s. 52).

Convention on Environmental Impact Assessment in a Transboundary Context (Espoo, 1991).

Duinker, P. N. and Greig, L. A. 2006. "The Impotence of Cumulative Effects Assessment in Canada: Ailments and Ideas for Redeployment." *Environmental Management* 37(2): 153–161.

EIA Directive 97/11/EC.

Force, J. E. and Machlis, G. E. 1997. "The Human Ecosystem Part II: Social Indicators in Ecosystem Management." *Society and Natural Resources* 10: 369–383.

Franks D., Fidler C., Brereton D., Vanclay F., and Clark, P. 2009. Leading Practice Strategies for Addressing the Social Impacts of Resource Developments. Centre for Social Responsibility in Mining, Sustainable Minerals Institute, The University of Queensland. Briefing paper for the Department of Employment, Economic Development and Innovation, Queensland Government.

Freudenburg, W. R. 1986. "Social Impact Assessment." *Annual Review of Sociology* 12: 451–478.

Galbraith, L., Bradshaw, B., and Rutherford, M. B. 2007. "Towards a New Supraregulatory Approach to Environmental Assessment in Northern Canada." *Impact Assessment and Project Appraisal* 25(1): 27–41.

Galisteo Consulting. 2002. "Social, Cultural, Economic Impact Assessments: A Literature Review." Albuquerque, NM.

GNWT. 2003. NWT Social Insdicators. NWT Bureau of Statistics.

Goskomekologia. 2000. Provision on the Assessment of the Proposed Economic or Other Activity on Environment. Executive Order #372.

Governance of energy and climate adaptation in the Nordic periphery (K-Base). 2009. Nordic Council of Ministers Research Programme Report.

Graybill, J. K. 2009. "Places and Identities on Sakhalin Island: Situating the Emerging Movements for 'Sustainable Sakhalin'." In Agyeman, J. and Ogneva-Himmelberger, E. (eds). *Environmental Justice of the Former Soviet Union*. Boston, MA: MIT Press.

Graybill, J. K. 2013a. "Mapping an Emotional Topography of an Ecological Home-land: The Case of Sakhalin Island, Russia." *Emotion, Space and Society* (Available online first).

Graybill, J. K. 2013b. "Environmental Politics on Russia's Pacific Edge: Reactions to Energy Development in the Russian Sea of Okhotsk." *Pacific Geographies* 40: 11–16.

Hansen, K. G., Lund Sørensen, F. and Jeppson, S. R. 2009. "Decision Processes, Communication and Democracy: The Aluminium Smelter Project in Greenland." In Hukkinen, J., Hansen, K. G., Langlais, R., and Rasmussen, R. O. (eds). *Knowledge-based Tools for Sustainable Governance of Energy and Climate Adaptation in the Nordic periphery*. Nordic Council of Ministers Research Programme Report.

Hansen, K. G. and Rasmussen, R. O. 2013. "New Economic Activities and Urbanisation: Individual Reasons for Moving and for Staying – Case Greenland." In Hansen, K. G., Rasmussen, R. O., and Weber, R. (eds). *Proceedings from the First International Conference on Urbanisation in the Arctic*. Conference August 28–30, 2012, Ilimmarfik, Nuuk, Greenland. Nordregio Working Paper.

Hunsberger, C., Gibson, R., and Wismer, S. 2005. "Citizen Involvement in Sustainability-centred Environmental Assessment Follow-up." *Environmental Impact Assessment Review* 25(6): 609–627.

Huskey, L. 1979. Statewide Impacts of OCS Petroleum Development in Alaska. Special Report Number 1, Alaska OCS Socioeconomic Studies Program, Anchorage: U.S. Bureau of Land Management, Alaska Outer Continental Shelf Office.

Icelandic Ministry of the Environment and Natural Resources. 1993. The Act on Environmental Impact Studies.

The Interorganizational Committee on Principles and Guidelines for Social Impact Assessment. 2003. "Principles and Guidelines for Social Impact Assessment." *The USA Impact Assessment and Project Appraisal* 21(3): 231–250.

The Interorganizational for Social Impact Assessment. 1994. "Guidelines and Principles for Social Impact Assessment, US Department of Commerce NOAA Tech Memo NMFS-F/SPO-16." *Impact Assessment* 12(2): 107–152.

Juslen J. 1995. "Social Impact Assessment: A Look at Finnish Experiences." *Proj Appraisal* 10(3): 163–170.

Knopp, J., Pokiak, F., Gillman, V., Porta, L., and Amos, V. 2013. Inuvialuit Settlement Region Community-based Monitoring Program (ISR-CBMP): Community-driven Monitoring of Locally Important Natural Resources. White Paper for the Arctic Observing Summit (AOS). April–May. Vancouver, CA.

Kruse, J., Lowe, M., Haley, S., Fay, G., Hamilton, L., and Berman, M. 2011. "Arctic Observing Network Social Indicators Project: Overview." *Polar Geography* 34(1–2): 1–8.

Kusel, J. 2001. "Assessing Well-being in Forest Dependent Communities." *Journal of Sustainable Forestry* 13(1–2): 359–384.

Larsen J., Schweitzer, P., and Fondahl, G. (eds). 2010. *Arctic Social Indicators I. Steffansson Arctic Institute*. Akureyri.

Larsen J., Schweitzer, P., and Petrov, A. (eds). 2015. *Arctic Social Indicators II*. Steffansson Arctic Institute. Akureyri.

Mackenzie Valley Environmental Impact Review Board (MVEIRB). 2005. Issues and Recommendations for Social and Economic Impact Assessment in the Mackenzie Valley. Delivering on the Mackenzie Valley Review Board's Mandate to Assess the Socio-Economic Impacts of Proposed Developments.

Martynova, E. and Novikova, N. 2012. *Tazobskiye nentsy v usloviyakh neftegazovogo osvoyeniya.* Moscow: Miklukho-Maklay Institute of Anthropology.

Mitchell, R. E. and Parkins, J. R. 2011. "The Challenge of Developing Social Indicators for Cumulative Effects Assessment and Land Use Planning." *Ecology and Society* 16(2): 29.

Murashko, O. 2006. *Etnologicheskaya ekspertiza v Rossii I mezhdunarodnya standarty otsenki vozdeistviya proektov na korennye narody.* Moscow.

MVEIRB, Mackenzie Valley Environmental Impact Review Board. 2007. *Socio-economic Impact Assessment Guidelines.* 2nd edition. Yellowknife.

MVRMA. 1998. Mackenzie Valley Resource Management Act (S.C. 1998, c. 25).

National Environmental Policy Act. 1970. Public Law 91–190:852–859.42, U.S.C and as amended Public Law 94–52 and 94–83 42 U.S.C., pages 4321–4347.

Noble, B. F. and Birk, J. 2011. "Comfort Monitoring? Environmental Assessment Follow-up Under Community-industry Negotiated Environmental Agreements." *Environmental Impact Assessment Review* 31(1): 17–34.

Noble B. F. and Fidler, C. 2011. "Advancing Indigenous Community – Corporate Agreements: Lessons from Practice in the Canadian Mining Sector." *Oil, Gas, & Energy Law* 9(4): 1–30.

Noble, B. F. and Storey, K. 2005. "Toward Increasing the Utility of Follow-up in Canadian EIA." *Environmental Impact Assessment Review* 25(2): 163–180.

Norwegian Ministry for the Environment. 1990. *Environmental Impact Assessment in Norway: Provisions in the Planning and Building Act.*

Novikova, N. 2010. Etnologicheskaya ekspertiza: ne perekrestke istorii, etnologii I yuridicheskoi antropologii. In *Integratsiya arkheologicheskikh i etnograficheskikh issledovaniy.* Kazan: Shardzhani Institute of History, pp. 76–80. Obzor documentatsii proekta Sakhalin-2 kasauscheisya korennykh malochislannykh narody. 2008. *Ludi Severa. Prava, resursy, expertiza* Moscow, pp. 288–311.

Papillon, M. and Rodon, T. 2017. "Proponent-indigenous Agreements and the Implementation of the Right to Free, Prior, and Informed Consent in Canada." *Environmental Impact Assessment Review* 62: 216–224.

Peat, M., Mitchel & Co., URSA, CCC/HOK, and Dames & Moore. 1977. Alaska OCS Socioeconomic Studies Program Task Report: Literature Survey. Technical Report 2, Alaska OCS Socioeconomic Studies Program, Anchorage: U.S. Bureau of Land Management, Alaska Outer Continental Shelf Office.

Petrov, A., Southcott, C., Simpson, B., Routh, S., and Cavin, P. 2013. Developing Inuvialuit Baseline Indicators System for (Self) Monitoring Community Well Being and Impacts of Resource Development. White Paper for the Arctic Observing Summit (AOS). April–May. Vancouver, CA.

Rasmussen, R. O. 2010. *Mobilitet i Grønland. Sammenfattende analyse. (Mobility in Greenland. The Comprehensive Analysis).* Nordregio, Stockholm. www.smv.gl/DK/mobilitet/rapporter/samlet_mobilitet_v04_feb2010.pdf.

Report of the United Nations Conference on the Human Environment, United Nations Environment Programme Stockholm 1972.

Sakhalin Energy. 2002. Overview of Environmental, Social and Health Impact Assessment. Sakhalin Energy Investment Company.

Schweitzer, P., Stammler, F., Ebstein, C., Ivanova, I., and Litvina, I. 2013. ReSDA Gap Analysis Report #2A: Social Impacts of Non-Renewable Resource Development on Indigenous communities in Alaska, Greenland and Russia. Whitehorse.

SEA and Integration of the Environment into Strategic Decision-Making. Final Report, May 2001.

Slootweg, R., Vanclay, F., and van Schooten, M. 2001. "Function Evaluation as a Framework for the Integration of Social and Environmental Impact Assessment." *Impact Assess Proj Appraisal* 19(1): 19–28.

Sokolova, Z. and Pivneva, E. 2004. Этнологическая экспертиза: Народы Севера России. 1956–1958 годы. – М., 2004. – 370 с.

Sokolova, Z. and Pivneva, E. 2005. Этнологическая экспертиза: Народы Севера России. 1959–1962 годы – М.: Ин-т этнологии и антропологии РАН, 2005. – 411 с. .

Sokolova, Z. and Pivneva, E. 2006. Этнологическая экспертиза: Народы Севера России. 1963–1980 годы – М.: Ин-т этнологии и антропологии РАН, 2006. – 380 с.

Sokolova Z. and Pivneva, E. 2007. Этнологическая экспертиза: Народы Севера России. 1985–1994 годы– М.: Ин-т этнологии и антропологии РАН, 2007. 316 с.

Stepanov, V. 2006. Nauchnaya exspertiza i zakonodatelstvo o natsionalno-kulturnom razvitii v Rossii. In *Etnologiya obschestvu: Prikladnye issledovaniya v etnologii.* Moscow: Orgservis, pp. 79–109.

Taylor, C. N., Bryan, C. H., and Goodrich, C. G. 1995. *Social Assessment: Theory, Process and Techniques.* 2nd edition. Christchurch: Taylor Baines & Associates.

Taylor, N. C., Hobson, B., and Hobson, C. G. 2004. *Social Assessment: Theory, Process and Techniques.* Middleton, WI: Social Ecology Press.

Tetra Tech, Inc. 2009. *Red Dog Mine Extension – Aqqaluk Project: Supplemental Environmental Impact Statement, Volume 1.* Seattle, WA: Environmental Protection Agency.

Therivel, R. and Ross, B. 2007. "Cumulative Effects Assessment: Does Scale Matter?" *Environmental Impact Assessment Review* 27(5): 365–385.

Tishkov, V. and Stepanov, V. 2004. *Izmereniya konflicta: Metodikai rezultaty etnokon fessionalnogo monitoring v seti EAWARN v 2003 gody.* Moscow: Institute of Ethnology and Anthropology.

Udotha, A., Noble, B., and Poelzer, G. 2017. "Meaningful and Efficient? Enduring Challenges to Aboriginal Participation in Environmental Assessment." *Environmental Impact Assessment Review* 65: 164–174.

U.S. Council on Environmental Quality. 1978. *Reguluationos for Implementing the Procecural Provisions of the National Environmental Quality Act (40 CFR 1500–1508).* Washington, DC: U.S. Council on Environmental Quality.

U.S. Forest Service. 1982. "Guidelines for Economic and Social Analysis of Programs, Resource Plans, and Projects: Final Policy." *Federal Register* 47(80): 17940–17954.

Vanclay, F. 1999. "Social Impact Assessment." In Petts, J. (ed). *Handbook of Environmental Impact Assessment, Volume 1.* Oxford: Blackwell, pp. 301–326.

Vanclay, F. 2002. "Conceptualising Social Impacts." *Environmental Impact Assessment Review* 22(3): 183–211.

Vasilkova, T., Evai, A., Martynova, E., and Novikova, N. 2011. *Korennye malochislannye narody n promyshlennoye razvitie Arktiki.* Moscow: Shadrinsk Printing House.

Williams, A. 2007. EBRD No Longer Considers Current Financing Package for Sakhalin II. www.ebrd.com, accessed January 2011.

7 Resource development and well-being in Northern Canada

Brenda Parlee

Many Arctic nations, including Canada, are highly dependent on resource extraction as a means of development; many mining, petroleum and hydro projects are promoted in the north as a pathway to improved well-being. However, over the past decades, research in the Canadian Arctic, Alaska, Greenland, Scandinavia and Russia tell a different story. While there are some positive case studies, the net effect of resource extraction has largely been adverse in the circumpolar north (Abele et al., 2009; Duhaime and Caron, 2006), a story which has parallels in Africa, Australia and Oceania, Latin America and Asia (Bebbington, Burneo and Warnaars, 2008). What is well-being? What are the issues behind these contradictory narratives?

Well-being has many different meanings within different disciplines and across socio-cultural contexts. Emerging from the disciplines of psychology and sociology, comprehensive studies on patterns of well-being have been carried out in many parts of the world but have been limited in the Arctic until recent years (Diener et al., 1999; Larson and Fondahl, 2015). As a concept in community planning and development, it is strongly associated with concepts such as quality of life and life satisfaction (Diener and Suh 2000). For the purposes of this chapter, it is considered also as the set of tools needed to deal with the stresses and challenges of everyday life (Giri, 2000; Diener et al., 1999).

Well-being, as measured by conventional social and health indicators, varies across the circumpolar Arctic. By most conventional measures Nordic populations seem much better off when compared to those in Canada, Alaska, Greenland and Russia. The outcomes are different, however, when using such subjective measures as "perceived life satisfaction". Data from the SLiCA (Survey of Living Conditions in the Arctic) reveals Canada ahead of other circumpolar nations; Russian communities still fall far behind all other nations (Poppel, 2015: 66). There are also important differences within circumpolar nations; gaps between the well-being of Indigenous and non-Indigenous peoples are among the highest – particularly in the Canadian Arctic. Based on 2001 and 2007 data, nearly 50% of First Nations communities reportedly occupy the lower half of the index range while less than 5% of other Canadian communities fell within this range (Indian and Northern Affairs Canada, 2004; Indian and Northern Affairs Canada, 2010).

Caution is needed however, in the interpretation and use of such well-being data and comparative analyses. Indigenous scholars have been among those critical of such socio-economic data and the associated data collection processes. Although methods vary, most large-scale health and social data collections and reporting efforts are perceived with scepticism given histories of colonial surveillance (Kukutai and Walter, 2015; Andersen, 2013; O'Neil, 1998). As noted in Chapter 6 of this volume, critiques of statistics and how they are collected, interpreted and utilized have spurred many academics and governments to rethink how and why they collect such data and its value in decision-making, particularly in relation to Indigenous peoples. However, even with improved intentions and reflectivity in the data collection process, there are limitations to the kind of "data" that can be feasibly collected at national and multi-national scales. Conventional kinds of data and data collection may not be done with the aim of promoting development or masking other more complex quality of life issues, but because no other data exists.

This chapter discusses some core themes of data related to well-being and resource development with the aim of summarizing the knowledge that exists and illuminating key gaps in the literature.

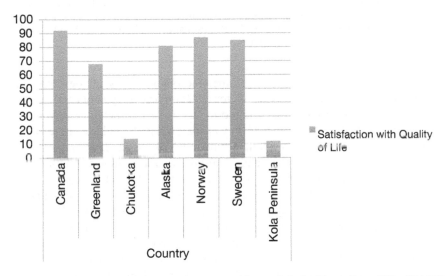

Figure 7.1 Percentage of those very or somewhat satisfied with quality of life (SLiCA) Reported in www.arcticlivingconditions.org and the SLiCA database (adapted from Poppel 2015)

Resource development: the challenge to well-being

Resource development in the circumpolar north represents a growing percentage of national GDP (Haley et al., 2011; McDonald, Glomsrød, and Mäenpää, 2006). By some accounts the economic contribution of resource extraction in Canada ranks highest among all circumpolar nations next to Arctic Russia (Figure 7.2).

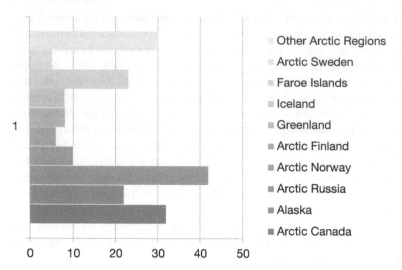

Figure 7.2 Dependence on Primary Resource Sector by Arctic Nations (Percentage of National GDP of Primary Resource Sector – Non Renewable Resource Extraction and Renewable Resources) (adapted from Glomsrød 2009) https:// oaarchive.arctic-council.org/bitstream/handle/11374/840/ACSAO-NO03_ 7_2_ECONOR.pdf?sequence=1

As noted in Chapter 8 of this volume, the opportunities for development associated with the primary resource sector are complicated by boom-bust cycles triggered by oscillations in global demand for raw materials including petroleum, metals, diamonds and other resources. In Canada, such highs and lows of resource development are felt across the north. While much of the economic reporting is positive, local communities consistently question whether current and anticipated resource development activity in the Canadian Arctic will result in long-term and sustained benefits for northern peoples. Classical sociological theory would suggest people are better off when change is slow and of limited scope. However, cumulative effects theorists warn of the costs of many incremental effects or "death by a thousand cuts", particularly for Indigenous peoples (Walker et al., 2011).

Lack of longitudinal data and limited comparative analysis makes it difficult to theorize in detail about patterns and trends in the relationship between well-being and resource development; many kinds of effects can neither be predicted due to the dynamic and complex nature of social and ecological systems (i.e., communities); impacts are highly varied and local capabilities to deal with them are also complex. As a result, the conceptual and analytical frameworks are highly varied.

In the absence of good data, simplistic, dichotomous and sometimes contradictory theories about resource development in the Arctic are often put forward and popularized. For example, beginning in the Berger years in north western Canada (1970s), a pipeline was proposed as a means of developing the northern *frontier* whilst Indigenous peoples fought to protect their *homelands* (Berger, 1977;

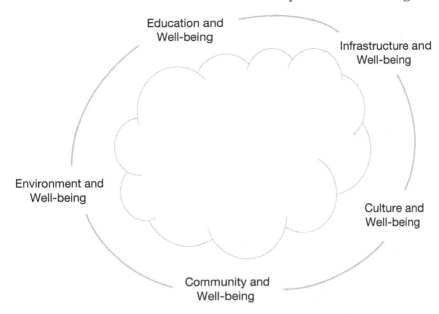

Education and Well-being

Infrastructure and Well-being

Environment and Well-being

Culture and Well-being

Community and Well-being

Figure 7.3 Dimensions of well-being: framework for thinking about issues of well-being and resources development

Usher, 1993). Many northerners speak about their choice to follow a modern *or* a traditional way of life (Ørvik, 1976; Fogel-Chance, 1993). The choice to participate in the local rather than a global economy is another binary perspective found in the literature on northern resource development.

The framing of northern peoples as a "vulnerable" population is particularly problematic. It is a label which implies those with power and capacity are located in the south (Ford and Smit, 2004; Cameron, 2012). "the label 'vulnerable' can shape how northern indigenous peoples come to see themselves as they construct their own identities and that identifying themselves as 'vulnerable' may ultimately hinder their efforts to gain greater autonomy over their own affairs" (Haalboom and Natcher, 2012: 324).

The Arctic is large, resource development projects are complex and communities are diverse. How mining, petroleum extraction and other related development activities feature in the narratives and efforts to achieve well-being need to be considered within the social-cultural and ecological context from which it emerges – every community has their own ways of defining well-being and their own particular lens through which they interpret both the opportunities and threats posed. Nonetheless, there are some broad themes and issues which are cross-cutting in the literature and appear relevant in consideration of the opportunities and threats of resource development,

In an effort to unpack some of the simplistic generalizations about the impact of resource development on community well-being, this chapter explores five aspects or domains of well-being discussed in the literature. Guided by theories on capabilities and their importance to health, this chapter examines how culture and community, education, financial capital, the environment, infrastructure and governance are important tools for achieving well-being.

Culture and well-being

Culture is a key determinant of well-being in many parts of the circumpolar north, particularly in Northern Canada (Nuttall, 1998; Bals et al., 2010; Figueroa, 2011). This focus on culture as a theme in the development process in Canada can be attributed to colonial histories aimed at cultural assimilation (O'Neil, 1986). As evidenced in the text of land claim agreements, land use plans and the terms of reference of environmental assessment processes, northern communities and governments are preoccupied with ensuring that development does not repeat the mistakes of the past but rather nurtures the unique way of life of northern Indigenous peoples. This effort to nurture cultural difference is a departure from earlier periods of Canadian governance which assumed development was only achievable through cultural assimilation; many kinds of social policies of the Canadian government for example were based on the assumption that cultural differences were to blame for lack of development (Macklem, 2001).

Cultural continuity is increasingly recognized as protective of well-being as indicated by research on alcohol addiction and Aboriginal youth suicide (Chandler and Lalonde, 1998; Kirmayer et al., 2011; Currie, Schopflocher and Wild, 2011). The notion and evidence that culture is "protective" is a departure from biomedical interpretations of well-being and health from the past which situated "culture" as the cause of poor health (Waldram, 2000). The shift in viewing culture as the problem and more to its view as the solution is attributed to a variety of socio-political shifts as well as more pluralistic approaches to the study and treatment of illness. As a result, cultural content is now viewed as important in many aspects of health care and social programs, education and governance, including those related to resource development.

A wide range of surveys on health and well-being from the *Inuit Health Survey* to the *Aboriginal Peoples Survey* have attempted to measure culture using a variety of indicators. This can range from the tracking of harvest practices, fluency in an Aboriginal language or participation in ceremonies or cultural activities. While such efforts are aimed at highlighting the protective significance of culture against such social pathologies as trauma, drug and alcohol addiction and suicide (Chandler and Lalonde, 1998), the essentialization of culture into measureable bits or "things" is seen as overly simplistic (Brady, 1995) and in some cases part of a colonial and neo-colonial process of Aboriginal surveillance (O'Neil et al., 1998).

Cultural knowledge, norms, practices and beliefs, or Traditional Knowledge, may be equally protective of well-being. As Chapter 13 in this volume discusses, Traditional Ecological Knowledge (TK) is a cumulative body of knowledge developed over many generations (Berkes, 2008). In many parts of the north, including Nunavut and settled land claim areas of the Northwest Territories,

Traditional Knowledge is to be considered equally with science in decision-making about lands and resources.

The recognition of "culture" and support of cultural continuity in the governance of resource development is, however, challenging. Some sociologists point out there is lack of clarity about what kinds of practices, behaviours or relations might be considered assimilative or neo-colonial in nature and what might be considered culturally sensitive; indeed the line between cultural respect and cultural appropriation can sometimes be blurry. Beyond recognition and use of material culture (e.g., wearing of caribou hide moccasins at the workplace) and use of symbolic culture from time to time (e.g., use of Indigenous language in the naming of a mine site), more in-depth research is needed on the successes and failures of cultural sensitivity policies and more broadly how Indigenous cultural practices and identities are influenced by resource development activity and their governance (Missens, Dana and Anderson, 2007).

Education and well-being

A large body of research has focused on the role of education as a tool for achieving well-being with a lesser focus of the intersection between education, well-being and resource development. In addition to enabling individuals and communities to *capitalize* on resources and opportunities in ways that are sustainable and beneficial, knowledge and the social processes and institutions that produce it, are also interconnected with the socio-cultural practices, values and norms which underlie well-being.

In the circumpolar north, as elsewhere, education has many faces and is often dualistically conceptualized as both formal and informal; formal education is defined here as the form of education delivered by the state (governments) and the latter rests in the domain of the community and household (MacGregor, 2011). Formal educational achievement rates in the circumpolar north are highly varied. Norway has the highest post-secondary education rates while Canada has the lowest of all circumpolar nations (Hirshberg and Petrov, 2015).

Explanations for these statistics from Canada's north are varied. The colonial histories within Canada including policies aimed at cultural assimilation are among the most well supported and accepted theories. The traumas associated with the residential school system, for example, are among the reasons why the education system today continues to be perceived as a threat to well-being rather than an opportunity (TRC, 2015).

There are also consistent concerns about the lack of fit between Euro-Canadian systems of education and those of northern Indigenous peoples (Berger, Epp and Møller, 2006), a problem theorized in social psychology as cultural dislocation or discontinuity. Efforts to deliver formal education in a more culturally meaningful way are consequently being spearheaded in the Canadian north and other circumpolar nations (Hirvonen, 2004). "One of the greatest challenges facing communities, educators and researchers in the Arctic is that of developing genuine

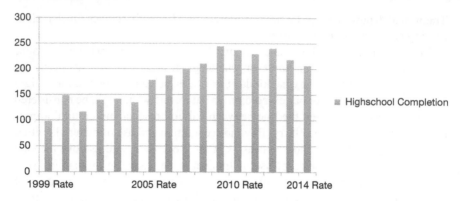

Figure 7.4 Secondary school graduation rates in Nunavut (1999–2014). Based on data from
Nunavut Bureau of Statistics (accessed 2015 November via: stats.gov.nu.ca)

Inuit, Dene and other approaches to education, not just sprinkling cultural materials
into approaches designed for southern systems" (Vick-Westgate, 2002: 17).

Consequently there is a growing body of work in education policy and elsewhere
focused on how to build more holistic models of formal education and learning
that reflect traditional ways of knowing, as well as skills for survival in a globalized
Arctic economy (Battiste, 1998).

Comparative analysis of educational attainment rates across the circumpolar
north is thus only marginally useful given the different kinds of socio-cultural
histories. A more helpful approach is to explore changes within each region. In
Nunavut, for example, educational attainment rates between 1999–2014 tell a more
hopeful story about the progress in addressing some of the education issues noted
above. Between 1999–2010 for example, high school completion rates almost
doubled. It is not, however, clear what kinds of strategies to improve these educational outcomes were most meaningful and successful within the territory.

To encourage greater participation and success in formal education, industry
is playing an increasingly more influential role in the dynamics of education and
training through the offering of adult education programs that are directly linked
to employment opportunities. While many of these programs are equally or more
successful than government initiatives, there are concerns that over-emphasis on
resource development employment versus those jobs of value in communities or
other sectors leading to a brain drain in communities (Hodgkins 2008).

Although creating short-term benefits, it may also reshape the kinds of
educational opportunities and the short- and medium-term characteristics of the
labour force.

Infrastructure – material well-being

Resource development has the potential to both improve and adversely affect
material well-being through changes in infrastructure (Hajkowicz, Heyenga and
Moffat, 2011). It is evidenced in research in many parts of the globe that "poor

infrastructure within a nation constrains opportunities for stable relationships, personal expressiveness, and productivity" (Ryan and Deci, 2001: 153). Communities in some parts of the world have benefited from the investment in infrastructure made by mining companies (e.g., recreation facilities). There are case studies in various regions where resource companies have resolved infrastructure problems through various kinds of private-public partnership arrangements (Wilson, 2004). Other communities have been negatively affected by the draw on infrastructure of large-scale industrial activity when corporations access but do not compensate for draws on educational and health facilities, water and sewage infrastructure and housing (MMDS, 2002). It is the absence of such basic infrastructure which often leads to poor health and well-being outcomes (e.g. tuberculosis) that are not dissimilar from the developing world (Christensen, 2012; Tester, 2009). At the same time, evidence suggests that improvements in infrastructure, particularly in areas suffering from significant infrastructure short-falls, substantively contributes to community well-being.

Housing is among the most critical infrastructure issues in Northern Canada (Christensen 2012). The problem is seen as a growing crisis for two reasons. First, federal investment in housing which began in the 1960–1970s created needs which did not previously exist. The declining condition of the housing projects constructed over twenty to thirty years ago, limited investment in housing maintenance and new homes, coupled with a growing population have created a gap between the number of houses available and the number of families in need. In Nunavut it is estimated that half of family homes are "overcrowded" with consequent statistics of social illness, particularly among children and youth becoming issues of national and international attention. Previous research on homelessness suggests housing insecurity is likely to be compounded by boom-bust cycles of resource development and uneven development across the territories with vulnerable populations (e.g. homeless, single mothers) facing the greatest dis-benefits.

Public-private partnerships, in which governments work collaboratively with industry to resolve infrastructure short-falls during boom periods in areas affected by resource development is a powerful model in other parts of the world including Australia (Grieve and McKenzie, 2010).

However, when development projects (i.e., mines) close and communities face periods of bust, there are other kinds of infrastructure challenges. Historically, mining towns that popped up during period of construction became "ghost-towns", creating various kinds of infrastructure maintenance and decommissioning challenges. In 1998, for example, all 1,000 residents deserted the town of Pyramiden, Svalbard (Norway) after the Russian coal company operating there for 50 years closed its operation (Andreassen, Bjerk and Olsen, 2010: 11). Although some ghost-towns have become viable tourist economies, many abandoned mining towns have created public liabilities owing to the toxic tailings ponds and mining shafts which require constant maintenance and monitoring. In Canada, the issue is a preoccupation of the federal government as well as affected communities: "there are literally 1000s of contaminated sites which are under federal jurisdiction and many more which we describe as orphan sites where there is an abandoned mine and there is no possibility of finding an organization who will pick up the clean-up cost" (Anderson, 1999).

As Keeling et al. note in Chapter 10 of this volume the issues associated with abandoned or decommissioned mines are complex, multi-layered and not evenly experienced (see also Keeling and Sandlos, 2009; Sandlos and Keeling, 2012; Keeling and Sandlos, 2015). Indigenous communities may be most affected. To avoid the problems associated with company towns, fly-in / fly-out employment scenarios became common practices in Northern Canada with the aim of increasing opportunities for Aboriginal employment (Storey, 2010).

Financial capital and well-being

Income from employment is often a selling point and focus during socio-economic impact assessments of large-scale resource development projects; indeed employment in the mining, petroleum and related primary service sectors (e.g., construction) can be overemphasized as the magic bullet for addressing circumstances of poverty in Northern Canada (Simeone, 2008). But does increased income improve well-being? Chapter 5 has already indicated that in Northern Canada this relationship is not clear.

There is a broad body of literature focusing on the relationship between income and well-being (Diener and Oishi, 2000). It is assumed that as income increases, poverty will decrease with resulting improvements in well-being. This is not true across all income levels and may also not be the case in Northern Canada where many other factors of culture, society and the land figure into peoples sense of a "good life". Based on research in other countries, the relationship between well-being and growth in income are strongest at lower income levels (i.e. people whose income rises above the basic needs level are more likely to see steeper improvements in well-being than those at higher income levels). Significant improvements in health (e.g. life expectancy) have been made in the last 50 years in many countries but are not accompanied by similar improvements in well-being. "The proportion of people in developed societies who are happy or satisfied with their lives has remained stable over the past decades even though they have become, on average much richer" (Eckersley, 2001: 77). In North America, it is often stated that people are more interested in their relative income than the absolute income. In other words, we care more that we have more money than our neighbours than about the amount of money itself. In Northern Canada, however, where incomes are low in comparison to the high costs of living, absolute income (making ends meet) likely has more significance.

However, equally as important to well-being may be the social networks, extent of social supports and feelings of belonging to a community. Individualistic traits and lack of social connectedness are conversely associated with the absence of well-being (Eckersley, 2007).

With that in mind, a key misconception of resource development policy in the north may be the need for full-time employment for all. Seasonal or flexible part-time employment may be the most desirable for many families (Abele, 2009).

Full-time employment has the potential to increase food insecurity as households have less time to procure country/traditional foods and become increasingly dependent on less nutritious and more costly food alternatives from the store (Todd, 2010). Research on the household and mixed economy suggests traditional economies may be protective, in the context of boom-bust effects (Usher, Duhaime and Searles, 2003). As housing and food prices increase during a boom or wages to meet those needs disappear in an economic collapse, those who have skills and knowledge for living on land (e.g. procuring traditional/country food) are considered better off than those who are entirely dependent on the wage economy. Strong social networks and a strong social-economy may also mitigate some of community-wide effects that may result during a "bust" period of development.

Revenues from resource development can contribute to well-being in other ways. Although employment creates opportunities to improve individual circumstances, business contracts and enterprises, taxation, royalties and negotiated agreements (triggered by land and resource rights) can enable Indigenous communities to capture financial benefits at community and regional scales (see Chapter by Bradshaw et al.). As discussed elsewhere in this volume, there are many questions about how revenues from some arrangements translate into benefits for communities and contribute to well-being more generally. Much may depend on the distribution mechanism; in some cases revenues are allocated out on a per capita basis to individual beneficiaries while in other cases moneys are retained by local and regional governments in order to invest or support community projects and programs. Trust funds are a useful mechanism for communities seeking to invest in order to create long-term benefits (Duncan, 2003: 320). But there are a variety of factors which affect whether or not communities can negotiate "good" contracts and agreements. In addition to the knowledge and expertise of individual negotiators, inequities in power relations between Indigenous peoples, governments and corporations affect the ability of Indigenous communities to effectively negotiate in ways that meaningfully contribute to their communities (Snipp, 1986).

The environment and well-being

While there are many factors that affect the well-being of northern communities, changes in the health of the environment are among the most broad-ranging and poorly understood. The relationship between the well-being of Indigenous peoples and the health of the environment has been a key focus of concern due to the strong social, economic cultural and health dimensions of this relationship: "if the land is not healthy, how can we be?" poignantly highlights this interconnection (Adelson, 1998). Owing to this multi-dimensional relationship to the environment, some scholars suggest that Indigenous people bear the greatest burden of environmental "bads" of resource development when compared to other populations (Clay, 1994; Gedicks, 1994). When something happens to the land, water and wildlife, there are reverberating effects for community well-being. People worry about what will happen to the land and their children in the future. As in other communities that depend significantly on the land and resources for their livelihood, these unnatural changes are the cause of significant anxiety (Usher, 1991; Bielawski and Lutsel K'e, 1992).

Chapter 10 in this volume discuss the environmental legacies that have resulted from extractive resource development in the North. Chemical contamination, loss and disturbance of valued environments and resources are among the most prevalent issues in the literature. This is true across the circumpolar north, particularly in the Russian north. With the expansion of resource development activity, there are growing concerns about both the local and circumpolar risks posed by resource development and the implications for various aspects of human security including the food security of northern Indigenous peoples.

Much concern around the contamination of particular environments and resources is attributable to the importance of these environments and resources for subsistence. Although the dependence on renewable resources in the Arctic as traditional/country food varies across the north (AMAP, 2007; Stefánsdóttir, 2014), resource development is considered a consistent threat to wildlife, water, fisheries and biodiversity valued by Indigenous peoples. Case studies can be found in almost every part of Northern Canada, Alaska, Greenland, Norway and Russia. While many impacts are localized, other kinds of resource development impacts transcend political and geographic borders. As evidenced by diamond mining activity in the Northwest Territories, development in one part of the migratory caribou range, for example, has implications on communities in other parts of the range (Kendrick and Lyver, 2005; Parlee and Manseau, 2005; Gunn et al., 2011). Shipping of fuel and raw materials from various ports of extraction to markets elsewhere is another critical pathway. Although individual nation states make policies concerning the well-being of their own citizens, there are no clear mechanisms for addressing the transboundary implications of development across nation states in the Arctic.

The research on subsistence economies is detailed in Natcher's chapter in this volume. The quantification of some aspect of land and resource use is often used as the go-to reference for understanding the significance of the environment or resource to a particular community or cultural group with health and well-being interpretable through various aspects of harvest practice, yield, consumption or other material outcome (e.g., caribou hide moccasins).

The tracking of harvest yields is indeed common practice across many parts of Northern Canada and Alaska. "Native harvest studies" date back to the James Bay and Northern Quebec Agreement (1976) and have been more recently included as part of the settlement of land claims in the region of Nunatsiavut (Berkes, 1983; Usher and Wenzel, 1987; Usher, 2002). These studies indicate that subsistence harvesting as well as other aspects of land use are declining across the north. Greater engagement or harvesters into the wage economy is one of the key drivers behind declining participation in hunting, trapping and fishing in many regions, a trend precipitated and compounded by the rapid pace and scale of socio-cultural change across the north. A critical concern with declining harvest of species such as caribou, whale, fish and moose are the associated implications for diet and health. Shifts away from traditional/country food consumption has been associated with an increase in lifestyle-related diseases such as Type II diabetes, chronic heart disease and some cancers (Egeland et al., 2011).

There are many other dimensions of human-environment relations in the Arctic that go beyond harvesting and food consumption. Research on place and place names indicate the depth of meaning around human-environment relations. While resource development and place-attachment has been a theoretical lens often used elsewhere, it has not been well used in the Arctic as a means of understanding differences in the differential effects of resource development activity on different populations.

The reification of "environmental change" as a new problem caused by development or climatic effects is problematic, however, given that ecological variability has been characteristic of Arctic life. As such some aspects of eco-logical variability (cycles, patterns etc.) are accepted and entrenched in livelihood practices.

> Arctic hunters and herders have always lived with and adapted to shifts and changes in the size, distribution, range and availability of animal populations. They have dealt with flux and change by developing significant flexibility in resource procurement techniques and in social organization.
>
> (Nuttall et al. 2005: 11–27)

Differentiating between known kinds of variation and new kinds of patterns and extremes is key however, is much overlooked in the climate change literature (Duerden 2004). Elders in Lutsel K'e distinguish between natural change and development effects – edo and change that is perceived as unnatural – edo aja – which translates directly as "something has happened to it" (Parlee et al. 2005). The ways in which the environment variability and change feature positively (as well as adversely) in narratives and experiences of well-being needs further exploration.

Community and well-being

"Community" is also a fundamental contributor to well-being. Although it has many definitions and moorings, social capital is among the conceptual frame-works useful in this chapter. Putnam defined social capital as "features of social organization, such as networks, norms and social trust, that facilitate coordination and cooperation for mutual benefit" (Putnam, 1995: 67). It differs from other forms of capital discussed above in its relative intangibility, "for it exists in the relations among people" (Coleman, 1988: S101). Numerous scholars have operationalized the concept, demonstrating that without adequate forms of social capital, individuals and communities may be unable to develop other forms of capital (e.g., human capital, effective) and may be limited in their ability to achieve critical goals such as cultural sustainability, political efficacy and economic development (Portes, 1998).

Its academic definitions are only partially useful but it also has currency within Indigenous communities. Mignone and O'Neil suggest that social capital in the First Nation context can be measured by the degree that its resources are socially invested; that it presents a culture of trust, norms of reciprocity, collective action and participation; and that it possesses inclusive, flexible and diverse networks (Mignone and O'Neil, 2005). Social capital of a bonding nature is closely reflected in such indicators as intergeneration knowledge sharing (elders sharing knowledge with youth), family cohesion (parents supporting youth), volunteerism, civic participation (participation in public meetings), social interaction and communication, demonstration of traditional values (respect for the land) and participation in cultural events such as caribou hunting and spiritual gatherings (Parlee, O'Neil and Lutsel K'e, 2007).

Many aspects of such social capital are grounded in the cultural traditions and subsistence economies of Indigenous communities (Usher et al., 2003). But as Duhaime and others have noted, subsistence does not simply involve hunting, fishing, and other food gathering activities: "it is a powerful ideology that extends into other areas of life including raising of children, and the treatment of elders. It also contributes to the structure of social relations, community leadership and moral authority" (Duhaime et al., 2004).

The erosion of community has emerged as a key concern within the context of resource development. Finding ways to maintain or build upon the social capital associated with subsistence economies in emerging resource development economies will create opportunities for real economic opportunity and benefit. The lack of continuity in social norms, or the disruption of social systems, may greatly limit the continuity of identity and community known to be protective of many aspects of community health and well-being (Chandler and Lalonde, 2008).

Different aspects of social capital may be critical ingredients to achieving well-being in the context of resource development (Light and Dana, 2013; Casey and Christ, 2005); these can be classified as bonding and bridging social capital. The former refers to the value assigned to social networks between homogeneous groups of people and the latter to that of social networks between socially heterogeneous groups. Those communities with strong social capital of a bonding nature may have a greater capacity to offset socio-economic inequities associated with the resource curse. Furthermore, those communities able to build social capital outside of their own communities to other locales may be less vulnerable to external pressures from government or industry and may be able to take advantage of new opportunities and innovations (Woolcock, 2001: 12).

The implied causal chain starts with membership in civic and social organizations, creating generalized bonds of trust within a community. These in turn serve to lower economic risk and reduce transaction costs (by increasing the "social costs" of malfeasance and free riding), which facilitates the dissemination of organizational and technical knowledge, enhancing both economic and governmental efficiency and, finally, enhancing community prosperity (Casey and Christ, 2005: 828).

However, this pattern of social capital and development is not similar in all cases. Miguel, Gertler and Levine studied social capital indicators across 274 Indonesian states and found no correlation with economic development outcomes from industrial activity (Miguel, Gertler and Levine, 2005). More research in the circumpolar north about the role of social capital in economic development is needed.

Governance and well-being

Governance of resource development is an overarching theme in the literature on well-being. Various literatures including those from political science, psychology and sociology suggest that the more individuals, communities and governments perceive themselves to be in control over their lands and resources and experience efficacy in their efforts to influence resource development outcomes, the higher the level of subjective well-being.

Traditional social norms, "rules" or Indigenous legal orders have long been recognized as key to the stability and security of Indigenous communities including those in the Arctic. Within the context for formal state and government arrangements, these social norms have been superceded in attention by institutions and processes of government that mirror state bureaucracies. The extent to which land claim settlements, and the institutions that have been created with their implementation, contribute to the well-being of Arctic communities is not well explored in the literature. Indeed there are concerns that the bureaucracies that have been created in some regions have done more to hinder rather than sustain the traditional ways of life of northern Indigenous peoples (Nadasdy, 1999). The extent to which such bureaucracies have created administrative control but not real power in decisions over issues that affect well-being (e.g., health, education and land) and resource management warrants further consideration.

Informal institutions including norms and rules that guide everyday social relations are an equally or more important aspect of northern governance as formal systems of government. Such norms play an important role in defining what matters and what does not, guiding social and social-ecological behaviours and decisions.

Human security is used globally to speak to human development; it is generally defined as "safety from such chronic threats as hunger, disease, repressing and hurtful disruptions in the patterns of daily life" (Axworthy, 2001). The evolution of the term "security" from one of militaristic meaning to one in vogue in sociology, geography and health discourse has resulted in a broader conceptualization of the socio-economic and cultural determinants of well-being. Human security in the Arctic is concerned with the intersection of power and governance and the social, economic political and environmental factors that contribute to the well-being of Arctic peoples (Daveluy, Lévesque and Ferguson, 2011). It has become a popular concept in Arctic policy discourse in Canada as federal and territorial governments seek to maintain sovereignty (i.e. military security) amidst mounting global interest in the mineral, oil/gas and hydro-electric potential and the sudden access to those resources that has resulted with melting sea ice in the north-west passage of Nunavut (Borgerson, 2008). Among the concepts of greatest currency is that of food security (Huish, 2008). The World Health Organization defines food security as existing "when all people at all times have access to sufficient, safe, nutritious food to maintain a healthy and active life". The concept generally considers both physical and economic factors that limit or facilitate access as well as cultural attributes that determine food preference. In Canada, 21% of Aboriginal households are at risk for being "food insecure", however, in Northern Canada, the statistics are twice as high in some communities in Nunavut (Egeland et al., 2011). Food insecurity is compounded in the north by high levels of unemployment and higher than average cost of food.

Conclusion

Sociologists, geographers, anthropologists, historians, economists and political scientists are in a unique position to learn from northerners and collaboratively address commonly identified issues. Unlike the silos of socio-economic, health and biophysical study and regulation that have long characterized the governance of natural resources and development in the north, well-being can make us think about local experiences of sustainability in a more holistic and integrated fashion. Rather than reflecting old social science paradigms, well-being can also account for alternative ways of knowing including local traditional knowledge of Indigenous peoples.

   Well-being is a useful concept in that it tends to transcend cultural, disciplinary and geographic boundaries. It is a useful concept for this chapter in that equivocal concepts can be found in many cultures and languages in the circumpolar north. Well-being, although statistically measureable by some accounts, is also normative. Values, beliefs and social norms of individual social groups and the society at large affect the ways in which individuals perceive and respond to the benefits and dis-benefits of resource development. As a consequence, well-being has many meanings which can lead to ambiguous dialogue debate. This chapter suggests a need to dig down below the jargon of well-being, create a taxonomy of the various determinants of well-being that often come into play in discussions about well-being in northern communities. In so doing communities will be better able to determine under what conditions extractive resource development can be useful for their long-term sustainability.

References

Abele, F., Courchene, T., Seidle, L. and St.-Hilaire, F., eds (2009). *Northern Development: Past, Present and Future. Northern Exposure: People, Powers and Prospects in Canada's North.* Montreal: The Institute on Public Policy Research.
ACIA (Arctic Climate Change Impact Assessment) (2004). *Impacts of a Warming Arctic–Arctic Climate Impact Assessment.* Cambridge, UK: Cambridge University Press.

Adelson, N. (1998). Health Beliefs and the Politics of Cree Well-Being. *Health: An Interdisciplinary Journal for the Social Study of Health, Illness and Medicine* 2(1): 5–22.

Andersen, C. (2013). Underdeveloped Identities: The Misrecognition of Aboriginality in the Canadian Census. *Economy and Society* 42(4): 626–650.

Andersen, T. and Poppel, B. (2002). Living Conditions in the Arctic. *Social Indicators Research* 58(1): 191–216.

Anderson, D. (1999). House Address, reply to the Throne Speech, Thursday, October 14, 1999, 36th Parliament, 2nd Session. Edited *Hansard* 3: 1710–1740.

Anderson, R. B., Dana, L. P. and Dana, T. E. (2006). Indigenous Land Rights, Entrepreneurship, and Economic Development in Canada: "Opting-in" to the Global Economy. *Journal of World Business* 41(1): 45–55.

Andreassen, E., Bjerck, H. B. and Olsen, B. (2010). *Persistent Memories: Pyramiden-a Soviet Mining Town in the High Arctic*. Central Norway: Tapir Academic Press.

Arctic Council (1997). *Arctic Pollution Issues: A State of the Arctic Environment Report*. Arctic Monitoring and Assessment Program (AMAP).

Arctic Human Development Report (2004). Stefansson Arctic Institute. Retrieved May 25, 2006, from www.svs.is/AHDR/AHDR%20chapters/AHDR_first%2012pages.pdf.

Axworthy, L. (2001). Human Security and Global Governance: Putting People First. *Global Governance* 7: 19.

Aylward, M. L. (2007). Discourses of Cultural Relevance in Nunavut Schooling. *Journal of Research in Rural Education* 22(7): 1–9.

Ballard, C. and Banks, G. (2003). Resource Wars: The Anthropology of Mining. *Annual Review of Anthropology*, 287–313.

Bals, M., Turi, A. L., Skre, I. and Kvernmo, S. (2010). Internalization Symptoms, Perceived Discrimination, and Ethnic Identity in Indigenous Sami and Non-Sami Youth in Arctic Norway. *Ethnicity & Health*, 15(2): 165–179.

Barnhardt, C. (2001). A History of Schooling for Alaska Native People. *Journal of American Indian Education* 40(1): 1–30.

Battiste, M. (1998). Enabling the Autumn Seed: Toward a Decolonizing Approach to Aboriginal Knowledge, Language and Education. *Canadian Journal of Native Education* 22(1): 16–27.

Bebbington, A., Hinojosa, L., Bebbington, D. H., Burneo, M. L. and Warnaars, X. (2008). Contention and Ambiguity: Mining and the Possibilities of Development. *Development and Change* 39(6): 887–914.

Belanger, Y. and Newhouse, D. (2004). Emerging from the Shadows: The Pursuit of Aboriginal Self-Government to Promote Aboriginal Well-being. *The Canadian Journal of Native Studies* 14(1): 129–222.

Berger, P. (2008). Inuit Visions for Schooling in One Nunavut Community, PhD Thesis, Lakehead University, Thunder Bay. Berger Commission Community Transcripts. (March 13, 1967).

Berger, P., Epp, J. R. and Møller, H. (2006). The Predictable Influences of Culture Clash, Current Practice, and Colonialism on Punctuality, Attendance, and Achievement in Nunavut Schools. *Canadian Journal of Native Education* 29(2): 182.

Berger, T. R. (1977). *Northern Frontier, Northern Homeland*. Ottawa: Government of Canada.

Berkes, F. (2017). *Sacred Ecology* (4th Editon). New York: Routledge.

Bielawski, E. and Lutsel K'e Dene First Nation (1992). *The Desecration of Nanula Tue: Impact of the Talston Hydro Electric Project on the Dene Soline*. Ottawa: Royal Commission on Aboriginal Peoples.

Bjerregaard, P., Kue Young, T., Dewailly, E. and Ebbesson, S. (2004) Review Article: Indigenous Health in the Arctic: An Overview of the Circumpolar Inuit Population. *Scandinavian Journal of Public Health* 32(5): 390–395.

Bosch, O. J. H., King, C. A., Herbohn, J. L., Russell, I. W. and Smith, C. S. (2007). Getting the Big Picture in Natural Resource Management – Systems Thinking as "Method" for Scientists, Policy Makers and Other Stakeholders. *Systems Research and Behavioral Science* 24(2): 217–232.

Boyer, Y. (2006). Self Determination as a Social Determinant of Health. Discussion Document for the Aboriginal Working Group of the Canadian Reference Group Reporting to the WHO Commission on Social Determinants of Health. Hosted by the National Collaborating Centre for Aboriginal Health and funded by the First Nations and Inuit Health Branch of Health Canada. Vancouver: June 29.

Brady, M. (1995). Culture in Treatment, Culture as Treatment. A Critical Appraisal of Developments in Addictions Programs for Indigenous North Americans and Australians. *Social Science & Medicine* 41(11), 1487–1498.

Cameron, E. S. (2012). Securing Indigenous Politics: A Critique of the Vulnerability and Adaptation Approach to the Human Dimensions of Climate Change in the Canadian Arctic. *Global Environmental Change* 22(1): 103–114.

Casey, T. and Christ, K. (2005). Social Capital and Economic Performance in the American States. *Social Science Quarterly* 86(4): 826–845.

Chance, N. A. and Andreeva, E. (1995). Sustainability, Equity, and Natural Resource Development in Northwest Siberia and Arctic Alaska. *Human Ecology* 25(2): 217–240.

Chandler, M. and Lalonde, C. (2008). Cultural Continuity as a Protective Factor Against Suicide in First Nations Youth. *Horizons* 10(1): 68–72.

Chapin, F. S. (2005) *Polar Systems. Ecosystem and Human Well-being: Current, State and Trends. Millennium Ecosystem Assessment*. Washington, DC: MEA.

Christensen, J. (2011). Homeless in a Homeland: Housing (In)security and Homelessness in Inuvik and Yellowknife, Northwest Territories, Canada, PhD Thesis, McGill University, Montreal.

Christensen, J. (2012). "They Want a Different Life": Rural Northern Settlement Dynamics and Pathways to Homelessness in Yellowknife and Inuvik, Northwest Territories. *The Canadian Geographer/Le Géographe Canadien* 56(4): 419–438.

Clay, J. (1994). *Who Pays the Price?: The Sociocultural Context of Environmental Crisis*. Washington, DC: Island Press.

Coryse, C. and Scott, K. (2006). The Determinants of Employment among Aboriginal Peoples. Ottawa, ON: Paper submitted to the Policy and Research Coordination Directorate, Strategic Policy and Planning Branch, Human Resources and Social Development Canada.

Currie, C. L., Schopflocher, D. P. and Wild, T. C. (2011). Prevalence and Correlates of 12-month Prescription Drug Misuse in Alberta. *The Canadian Journal of Psychiatry* 56(1), 27–34.

Daveluy, M., Lévesque, F., Ferguson, J. and International School for Studies of Arctic Societies (2011). *Humanizing Security in the Arctic*. Edmonton: CCI Press.

De Beers Canada. De Beers Canada Partners with NWT Literacy Council. Retreived from: http://debeerscanada.com/files_3/pdf_documents/2011-03-22_Media-Release_De-Beers-announces-partnership-with-NWT-Literacy-Council.pdf.

De Beers Snap Lake Mine Socio-Economic Report 2009. Retrieved April 8, 2011, from: www.debeersgroup.com/ImageVault/Images/id_2090/scope_0/ImageVaultHandler.aspx.

Diener, E. (1996). Subjective Well-being in Cross-cultural Perspective. In: Grad, H., Blanco, A. and Georgas, J. (eds), *Key Issues in Cross-cultural Psychology*. Liese: Swets and Zeitlinger.

Diener, E. and Oishi, S. (2000). Money and Happiness: Income and Subjective Well-being Across Nations. In: Diener, E. and Suh, E. M. (eds), *Culture and Subjective Well-being*. Cambridge, MA: MIT Press, pp. 185–218.

Diener, E. and Suh, E. M. (2000). *Culture and Subjective Well-being*. Cambridge, MA: MIT Press.

Duerden, F. (2004). Translating Climate Change Impacts at the Community Level. *Arctic* 57(2): 204.

Duhaime, G., Lemelin, A., Didyk, V., Goldsmith, O., Winther, G., Caron, A. . . . and Godmaire, A. (2004) *Arctic Economies. Arctic Human Development Report*. Stephanson Arctic Institute, Akureyri, Iceland, pp. 69–84.

Duncan, R. (2003). Agricultural and Resource Economics and Economic Development in Aboriginal Communities. *Australian Journal of Agricultural and Resource Economics* 47(3), 307–324.

Eckersley, R. (2006). Is Modern Western Culture a Health Hazard? *International Journal of Epidemiology* 35: 252–258.

Eckersley, R. M. (2007). Culture, Spirituality, Religion and Health: Looking at the Big Picture. *Medical Journal of Australia* 186: S54–56.

Eckersley, R., Dixon, J. and Douglas, R. (2001). *The Social Origins of Health and Well-Being*. Cambridge, UK: Cambridge University Press.

Eddy, J. A. (1974). Astronomical Alignment of the Big Horn Medicine Wheel. *Science* 184(4141): 1035–1043.

Egeland, G. M., Johnson-Down, L., Cao, Z. R., Sheikh, N. and Weiler, H. (2011). Food Insecurity and Nutrition Transition Combine to Affect Nutrient Intakes in Canadian Arctic Communities. *Journal of Nutrition* 141(9): 1746–1753.

Elias, P. D. (1991). *Development of Aboriginal People's Communities*. North York: Captus Press.

Espiritu, A. (1997). Aboriginal Nations: Natives in Northwest Siberia and Northern Alberta. In: Alden Smith, E. and McCarter, J. (eds), *Contested Arctic: Indigenous Peoples, Industrial States, and the Circumpolar Environment*. Seattle, WA: University of Washington Press, pp. 41–67.

Figueroa, R. M. (2011). Indigenous Peoples and Cultural Losses. *The Oxford Handbook of Climate Change and Society*. Oxford: Oxford University Press, pp. 232–250.

First Nations and Inuit Regional Health Survey National Steering Committee. (2004). *First Nations and Inuit regional health surveys: A Synthesis of the National and Regional Reports*. Ottawa: National Aboriginal Health Organization.

Flanagan, T. (2008). *First Nations? Second Thoughts*. Montreal and Ontario: McGill-Queen's Press.

Fogel-Chance, N. (1993). Living in Both Worlds: "Modernity" and "Tradition" among North Slope Iñupiaq Women in Anchorage. *Arctic Anthropology*: 30(1): 94–108.

Fonda, M. and Anderson, E. (2009). Diamonds in Canada's North: A Lesson in Measuring Socio-Economic Impacts on Well-being. *Canadian Issues*: 107.

Fondahl, Gail. (1997). Environmental Degradation and Indigenous Land Claims in Russia's North. In: Alden Smith, E. and McCarter, J. (eds), *Contested Arctic: Indigenous Peoples, Industrial States, and the Circumpolar Environment*. Seattle, WA: University of Washington Press, pp. 68–87.

Ford, J. D. and Smit, B. (2004). A Framework for Assessing the Vulnerability of Communities in the Canadian Arctic to Risks Associated with Climate Change. *Arctic* 57(4): 389–400.

Freilich, M. (1967). Ecology and Culture: Environmental Determinism and the Ecological Approach in Anthropology. *Anthropology Quarterly* 40(1): 26–43.

Galabuzi, G. (2004). Social Exclusion. In: Raphael, D. (ed.), *Social Determinants of Health: Canadian Perspectives.* Toronto: Canadian Scholars' Press Inc., pp. 235–252.

Gedicks, A. (1994). *The New Resource Wars: Native and Environmental Struggles Against Multinational Corporations* (Vol. 210). Montreal: Black Rose Books Ltd.

Gibson, G. and Klinck, J. (2005). Canada's Resilient North: The Impact of Mining on Aboriginal Communities. *Pimatisiwin*, 3(1): 116–139.

Giri, A. K. (2000). Rethinking Human Well_being: A Dialogue with Amartya Sen. *Journal of International Development: The Journal of the Development Studies Association* 12(7), 1003–1018.

Government of Northwest Territories, Department of Education, Culture and Employment. "Dene Kede Education: A Dene Perspective." Retrieved April 7, 2011, from: www.ece.gov.nt.ca/divisions/kindergarten_g12/curriculum/curriculum_ Services/Dene_ Keda_Resource_Manual/D_K_ Manual_Main.htm.

Gracey, M. and King, M. (2009). Indigenous Health Part 1: Determinants and Disease Patterns. *Lancet* 374(9683): 65–75.

Great Bear Lake Working Group. 2006. "The Water Heart": A Management Plan for Great Bear Lake and its Watershed. Directed by the Great Bear Lake Working Group and facilitated and drafted by Tom Nesbitt (May 31, 2005, with Caveat of February 7, 2006).

Grieve, S. and Haslam-Mckenzie, F. (2010) Local Housing Strategies: Responding to the Affordability Crisis. In: Alexander, I., Greive, S. and Hedgecock, D. (eds), *Planning Perspectives from Western Australia: A Reader in Theory and Practice.* North Fremantle: Fremantle Press, pp. 66, 84.

Gunn, A., Johnson, C. J., Nishi, J. S., Daniel, C. J., Russell, D. E., Carlson, M. and Adamczewski, J. Z. (2011). Understanding the Cumulative Effects of Human Activities on Barren-Ground Caribou. *Cumulative Effects in Wildlife Management: Impact Mitigation.* Boca Raton: CRC Press, pp. 113–134.

Haalboom, B. and Natcher, D. C. (2012) The Power and Peril of "Vulnerability": Lending a Cautious Eye to Community Labels in Climate Change Research. *Arctic* 65(3): 1.

Hajkowicz, S. A., Heyenga, S. and Moffat, K. (2011). The Relationship between Mining and Socio-Economic Well Being in Australia's Regions. *Resources Policy* 36(1), 30–38.

Hicks J. (2009). Toward More Effective, Evidence-Based Suicide Prevention in Nunavut. In: Abele, F., Courchene, T. J., Seidle, L. and St-Hilaire, F. (eds), *Northern Exposure: Peoples, Powers and Prospects in Canada's North.* Montreal: IRPP, pp. 467–498.

Hirshberg, D. and Petrov, A. (2015). Education and Human Capital. In: Larsen, J.-N. and Fondahl, G. (eds), *Arctic Human Development Report: Regional Processes and Global Linkages.* Copenhagen: Nordic Council of Ministers, pp. 347–395.

Hirvonen, V. (2004). *Sámi Culture and the School. Reflections by Sámi Teachers and the Realization of the Sámi School.* Norsk forskningsråd, Samisk høgskole og Čálliid-Lágadus.

Hodgkins, A. (2008). Marketing Adult Education for Mega-Projects in the Northwest Territories. *Journal of Contemporary Issues in Education* 3(2): 46–58.

Huish, R. (2008). Human Security and Food Security in Geographical Study: Pragmatic Concepts or Elusive Theory? *Geography Compass* 2(5), 1386–1403.

Indian and Northern Affairs Canada (2004). Measuring First Nations Well-being: First Nations Community Well-Being Index. In: *Statistics Canada Strategic Research and Analysis*. Ottawa: Government of Canada.

Indian and Northern Affairs Canada (2010a). Regional Variation and Disparities of First Nations and Inuit Well-Being. First Nation and Inuit CommunityWell-Being: Describing Historical Trends (1981–2006).

Indian and Northern Affairs Canada (2010b). Regional Variation and Disparities of First Nations and Inuit Well-Being – First Nation and Inuit Community Well-Being: Describing Historical Trends (1981–2006). In: *Indian and Northern Affairs Canada*. Ottawa: Government of Canada.

Inuit Tapiriit Kanatami (ITK) (2004). Backgrounder on Inuit and Housing: For Discussion at Housing Sectoral Meeting, November 24th and 25th 2004 in Ottawa. Ottawa: Inuit Tapiriit Kanatami.

Hicks, J. (2004). On the Application of Theories of "Internal Colonialism" to Inuit Societies. Paper Presented at the Annual Conference of the Canadian Political Science Association. Winnipeg (June 5, 2004).

Kaltenborn, B. (1998). Effects of Sense of Place on Responses to Environmental Impacts: A Study among Residents in Svalbard in the Norwegian High Arctic. *Applied Geography* 18(2): 169–189.

Kaul, I. (2002). *Challenges of Human Development in the Arctic*. New York: United Nations Development Program.

Keeling, A. and Sandlos, J. (2009). Environmental Justice Goes Underground? Historical Notes from Canada's Northern Mining Frontier. *Environmental Justice* 2(3): 117–125.

Keeling, A. and Sandlos, J. (2015). *Mining and Communities in Northern Canada: History, Politics, and Memory*. Calgary: University of Calgary Press.

Kendall, J. (2001). Circles of Disadvantage: Aboriginal Poverty and Underdevelopment in Canada. *American Review of Canadian Studies* 31(1–2): 43–59.

Kendrick, A. and Lyver, P. O. B. (2005). Denésqliné (Chipewyan) Knowledge of Barren-Ground Caribou (Rangifer tarandus groenlandicus) Movements. *Arctic* 58(2): 175 191.

King, M., Smith, A. and Gracey, M. (2009). Indigenous Health Part 11: The Underlying Causes of the Health Gap. *Lancet* 374: 76–85.

Kirmayer, L. J., Dandeneau, S., Marshall, E., Phillips, M. K. and Williamson, K. J. (2011). Rethinking Resilience from Indigenous Perspectives. *Canadian Journal of Psychiatry* 56(2): 84–91.

Korsmo, F. 1996. Claiming Territory: The Saami Assemblies as Ethnopolitical Institutions. *Polar Geography* 20(3): 163–179.

Kral, M., Idlout, L., Minore, J., Dyck, R. and Kirmayer, L. (2011). Unikkaartuit: Meanings of Well-being, Unhappiness, Health, and Community Change among Inuit in Nunavut, Canada. *American Journal of Community Psychology* 48(3): 426–438.

Kruse, J., Poppel, B., Abryutina, L., Duhaime, G., Martin, S., Poppel, M. . . . and Hanna, V. (2009). Survey of Living Conditions in the Arctic (SLiCA). In Møller, V., Huschka, D. and Michalos, A. C. (eds), *Barometers of Quality of Life Around the Globe: How Are We Doing?* Dordrecht: Springer Netherlands, pp. 107–134.

Kuyek, J. and Coumans, C. (2003). *No Rock Unturned: Understanding the Economies of Mining Dependent Communities*. Ottawa: Mining Watch.

Kukutai, T. and Walter, M. (2015). Recognition and Indigenizing Official Statistics: Reflections from Aotearoa New Zealand and Australia. *Statistical Journal of the IAOS* 31(2): 317–326.

Landau, J. (2007). Enhancing Resilience: Families and Communities as Agents for Change. *Family Process* 46(3): 351–365.

Larsen, J. N. and Fondahl, G. (eds) (2015). *Arctic Human Development Report: Regional Processes and Global Linkages*. Denmark: Nordic Council of Ministers.

Light, I. and Dana, L. P. (2013). Boundaries of Social Capital in Entrepreneurship. *Entrepreneurship Theory and Practice* 37(3), 603–624.

Macklem, P. (2001). *Indigenous Difference and the Constitution of Canada*. Toronto: University of Toronto Press.

MacPherson, J. (1991). *Dreams & Visions: Education in the Northwest Territories from Early Days to 1984*. Yellowknife: Department of Education.

McAllister, M. L., Fitzpatrick, P. and Fonseca, A. (2014). Unstable Shafts and Shaky Pillars: Institutional Capacity and Sustainable Mineral Policy in Canada. *Environmental Politics* 23(1): 77–96.

McDonald, H., Glomsrød, S. and Mäenpää, I. (2006). *Arctic Economy within the Arctic Nations. The Economy of the North*. Oslo: Statistics Norway.

McGregor, H. E. (2011). *Inuit Education and Schools in the Eastern Arctic*. Vancouver: UBC Press.

McHardy, M. (2001). *First Nations Community Well-Being in Canada: The Community Well-Being Index (CWB)*. Ottawa: Strategic Research and Analysis Directorate Indian and Northern Affairs Canada.

Merry, S. E. (2001). Changing Rights, Changing Culture. In: Cowan, J., Dembour, M-B. and Wilson, R. (eds), *Culture and Rights*. New York: Cambridge University Press.

Mignone, J. (2003). *Measuring Social Capital: A Guide for First Nations Communities*. Ottawa: Canadian Institute for Health Information.

Mignone, J. and O'Neil, J. (2005). Social Capital and Youth Suicide Risk Factors in First Nations Communities. *Canadian Journal of Public Health* 96(1): S51–4.

Miguel, E., Gertler, P. and Levine, D. I. (2005). Does Social Capital Promote Industrialization? Evidence from a Rapid Industrializer. *Review of Economics and Statistics* 87(4): 754–762.

Missens, R., Paul Dana, L. and Anderson, R. (2007). Aboriginal Partnerships in Canada: Focus on the Diavik Diamond Mine. *Journal of Enterprising Communities: People and Places in the Global Economy* 1(1), 54–76.

MMSD. (2002). Facing the Future. The Report of the Mining, Minerals and Sustainable Development (MMSD) Australia Project, Earthscan Publications, London and Sterling.

Northern News Service. Retrieved from: www.nnsl.com/frames/newspapers/2008–06/jun2_08nellie.html.

Norwegian, H. (2000). Deh Cho First Nations Call for Tar Sands Moratorium. (Press Release from Deh Cho First Nations).

Nuttall, M. (1998). *Protecting the Arctic: Indigenous Peoples and Cultural Survival*. London: Routledge.

Nuttall, M. (2000). Indigenous Peoples, Self-determination and the Arctic Environment. In: Nuttall, M. and Callaghan, T. V. (eds), *The Arctic: Environment, People, Policy*. Amsterdam: Harwood Academic Publishers, pp. 377–409.

Nuttall, M., Symon, C., Arris, L. and Hill, B. (2005) Hunting, Herding, Fishing, and Gathering: Indigenous Peoples and Renewable Resource Use in the Arctic. In: *Arctic Climate Impact Assessment*. New York: Cambridge University Press.

O'Faircheallaigh, C. and Ali, S. (eds) (2017). *Earth Matters: Indigenous Peoples, the Extractive Industries and Corporate Social Responsibility*. London: Routledge.

O'Neil, J. D. (1986). *Colonial Stress in the Canadian Arctic: An Ethnography of Young Adults Changing.* Dordrecht: Springer.

O'Neil, J., Reading, J. and Leader, A. (1998). Changing the Relations of Surveillance: The Development of a Discourse of Resistance in Aboriginal Epidemiology. *Human Organization* 57(2): 230–237.

Ørvik, N. (1976). Northern Development: Modernization with Equality in Greenland. *Arctic* 29(2): 67–75.

Parlee, B. and Furgal, C. (2012). Well-being and Environmental Change in the Arctic. *Climatic Change* 115(1): 13–34.

Parlee, B. and Lutsel K'e Dene First Nation. (2005). Understanding and Communicating about Ecological Change. *Breaking Ice: Renewable Resource and Ocean Management in the Canadian North* 1: 165.

Parlee, B., O'Neil, J. D. and Lutsel K'e Dene First Nation (2007). The Dene Way of Life: Perspectives on Health from the Canadian North. *Journal of Canadian Studies* 41(3): 112–133.

Poppel, B. (ed) (2015). *SLiCA: Arctic Living Conditions: Living Conditions and Quality of Life among Inuit, Saami and Indigenous Peoples of Chukotka and the Kola Peninsula.* Nordic Council of Ministers.

Portes, A. (1998). Social Capital: Its Origins and Applications in Modern Sociology. *Annual Review of Sociology* 24: 1–24.

Putnam, R. D. (1995). Bowling Alone: America's Declining Social Capital. *Journal of Democracy* 6(1): 65–78.

Rasmussen, D. (2011). Forty Years of Struggle and Still no Right to Inuit Education in Nunavut. *Interchange* 42(2): 137–155.

Rasmussen, K. (1927). *Across Arctic America.* London: Putnam.

Richmond, C., Ross, N. A. and Egeland, G. M. (2007). Societal Resources and Thriving Health: A New Approach for Understanding Indigenous Canadian Health. *American Journal of Public Health* 97(10): 1827–1833. doi:10.2105/AJPH.2006.096917.

Rolfe, J., Miles, B., Lockie, S. and Ivanova, G. (2007). Lessons from the Social and Economic Impacts of the Mining Boom in the Bowen Basin 2004–2006. *Australasian Journal of Regional Studies* 13(2): 134–153.

Rønning, W. and Wiborg, A. (2008). *Education For All in the Arctic? A Survey of Available Information and Research* (No. 1008/08). Working Paper.

Ryan, J. (1995). Doing Things the Right Way. *Arctic Institute of North America.* Calgary: University of Calgary Press.

Ryan, R. M. and Deci, E. L. (2001). On Happiness and Human Potentials: A Review of Research on Hedonic and Eudaimonic Well-being. *Annual Review of Psychology* 52(1): 141–166.

Salokangas, R. and Parlee, B. (2009). The Influence of Family History on Learning Opportunities of Inuvialuit Youth. *Inuit Studies* 33(1–2): 191–207.

Sandlos, J. and Keeling, A. (2012). Claiming the New North: Development and Colonialism at the Pine Point Mine, Northwest Territories, Canada. *Environment and History* 18(1): 5–34.

Satzewich, V. (1991) Aboriginal Peoples in Canada: A Critique of the Chicago School and Internal Colonial Models. *Innovation: The European Journal of Social Science Research,* 4(2): 283–302.

Schlag, M. and Fast, H. 2005. Marine Stewardship and Canada's Ocean Agenda in the Western Arctic Canada: The Role for Youth. In: Berkes, F., Huebert, R., Fast, H., Manseau, M., and Diduck, A. (eds), *Breaking Ice: Renewable Resource and Ocean Management in the Canadian North.* Calgary: University of Calgary Press.

Senécal, S. and O'Sullivan, E. (2006). The Well-Being of Inuit Communities in Canada: A Paper Prepared for the Strategic Research and Analysis Directorate, Indian and Northern Affairs Canada.

Simeone, T. (2008). The Arctic: Northern Aboriginal peoples. Library of Parliament INFOSERIES, Parliamentary Information and Research Service Publication PRB 08–10E, 24 October. Retrieved August 20, 2009, from: www.parl.gc.ca/information/library/PRBpubs/prb0810-e.pdf.

Sisco, A., Caron-Vuotari, M. Stonebridge, C., Sutherland, G. and Rhéaume G. (2012). *Lessons Learned: Achieving Positive Educational Outcomes in Northern Communities.* Ottawa: Conference Boards of Canada.

Sisco, A., and Stonebridge, C. (2010). *Toward Thriving Northern Communities.* Ottawa: The Conference Board of Canada.

Snipp, M. C. (2006). The Changing Political and Economic Status of the American Indians. *American Journal of Economics and Sociology* 45(2): 145–158.

Statistics Canada. (2006). *Harvesting and Community Well-being among Inuit in the Canadian Arctic: Preliminary Findings from the 2001 Aboriginal Peoples Survey – Survey of Living Conditions in the Arctic.* Ottawa: Government of Canada.

Stefánsdóttir, M. M. (2014). *Large Scale Projects in the Arctic: Socio-economic Impacts of Mining in Greenland.* Akureyi: University of Akureyri.

Steward, J. H. (1972). *Theory of Culture Change: The Methodology of Multilinear Evolution.* Chicago, IL: University of Illinois Press.

Storey, K. (2010). Fly-in/Fly-out: Implications for Community Sustainability. *Sustainability* 2(5): 1161–1181.

Tester, F. J. (2009). Iglutaasaavut (Our New Homes): Neither "New" nor "Ours": Housing Challenges of the Nunavut Territorial Government. *Journal of Canadian Studies* 43(2): 137–158.

Tester, F. J. and Kulchyski, P. T. (1994). *Mistakes: Inuit Relocation in the Eastern Arctic, 1939–1963.* Vancouver: UBC Press.

Todd, Z. (2010) *Food security in Paulatuk, NT – Opportunities and Challenges of a Changing Community Economy.* Department of Rural Economy, Faculty of Graduate Studies and Research. Edmonton: University of Alberta.

Truth and Reconciliation Commission. (2015) Ottawa: Government of Canada. Retrieved from: http://pm.gc.ca/eng/news/2015/12/15/final-report-truth-and-reconciliation-commission-canada.

Usher, P. J. (1993). Northern Development, Impact Assessment, and Social Change. In: Dyck, N. and Waldram, J. (eds), *Anthropology, Public Policy and Native Peoples in Canada.* Toronto: McGill-Queens Press, pp. 98–130.

Usher, P. J., Cobb, P., Loney, M., and Spafford, G. (1992). Hydro-Electric Development and the English River Anishanabe: Ontario Hydro's Past Record and Present Approaches to Treaty and Aboriginal Rights, Social Impact Assessment and Mitigation and Compensation. In: *Report for Nishanawbe Aksi Nation: Grand Council Treaty 03 and Tema-Augama Anishanabai.* Ottawa: P. J. Usher Consulting.

Usher, P. J. and Wenzel, G. (1987). Native Harvest Surveys and Statistics: A Critique of their Construction and Use. *Arctic*: 145–160.

Vick-Westgate, A. (2002). *Nunavik: Inuit-controlled Education in Arctic Quebec* (No. 1). Calgary: University of Calgary Press.

Waldram, J. (2004). *Revenge of the Windigo. The Construction of the Mind and Mental Health of North American Aboriginal People.* Toronto: University of Toronto Press.

Waldram, J. B., Herring, A. and Young, T. K. (2006). *Aboriginal Health in Canada: Historical, Cultural, and Epidemiological Perspectives*. Toronto: University of Toronto Press.

Walker, D. A., Forbes, B. C., Leibman, M. O., Epstein, H. E., Bhatt, U. S., Comiso, J. C. . . . and Kaplan, J. O. (2011). Cumulative Effects of Rapid Land-Cover and Land-Use Changes on the Yamal Peninsula, Russia. In: *Eurasian Arctic Land Cover and Land Use in a Changing Climate*. Dordrecht: Springer, pp. 207–236.

West, C. (2011). The Survey of Living Conditions in the Arctic (SLiCA): A Comparative Sustainable Livelihoods Assessment. *Environment, Development and Sustainability* 13(1): 217–235.

Wilson, K. (2003) Therapeutic Landscapes and First Nations Peoples: An Exploration of Culture, Health and place. *Health & Place* 9(2): 83–93.

Wilson, K. and Rosenberg, M. (2002). Exploring the Determinants of Health for First Nations Peoples in Canada: Can Existing Frameworks Accommodate Traditional Activities? *Social Science and Medicine* 55: 2017–2031.

Woolcock, M. (2001). The Place of Social Capital in Understanding Social and Economic Outcomes. *Canadian Journal of Policy Research* 2(1): 11–17.

Wootton, B. C. and Metcalfe, C. (2010). A Nice Cold Drink: Keeping Water Potable in Canada's North. *Water Canada 2019* (March/April): 22–24.

Young, T. K. and Mollins, C. J. (1996). The Impact of Housing on Health: An Ecologic Study from the Canadian Arctic. *Arctic Medical Research* 55(2): 52–61.

Zum Brunnen, C. (1997). A Survey of Pollution Problems in the Soviet and Post- Soviet Russian North. In: Alden Smith, E. and McCarter, J. (eds), *Contested Arctic: Indigenous Peoples, Industrial States, and the Circumpolar Environment*. Seattle, WA: University of Washington Press, pp. 88–121.

8 Resource revenue regimes around the circumpolar north

A gap analysis

Lee Huskey and Chris Southcott

The international press has heralded the new age of Arctic resource development. Arctic nations are optimistic about the prospects for resource development in the north. New resource projects are being discussed in areas of current production, like Alaska and Russia, as well as new regions, like Nunavut and Greenland (Rasmussen, 2011). While there are reasons to be cautious about this optimism, changing world demand for resources and changing conditions brought on by climate change may open up the long recognized Arctic resource storehouse. For some regions of the north resource development will introduce a new experience. For regions with a history of development, their experience presents a mixed history of success.

The governments of Arctic nations and major resource producers may view potential resource development positively in terms of additional jobs, revenues, and profits. Residents of northern regions are likely to have more mixed feelings. Past experience with booms and busts, limited local employment, or investment opportunities, and locally suffered costs may make residents wary of resource development opportunities. At the same time, limited economic opportunities mean northern residents may be searching for ways to insure that development of the region's resources results in prosperity and they are not simply left with "a hole in the ground" after the resource is removed. In particular they are looking for ways to ensure that a larger share of revenues remains in the region to build "linkages" that can be used to bring about a more sustainable economic future.

This chapter deals with the question of resource revenues. We present a survey of the literature on resource revenue regimes and the way they impact regions such as the north. The revenue that local communities receive from resource production is one way the prosperity associated with development can be shared with local residents. When these revenues are saved they provide a means of sustaining that prosperity beyond the end of resource production (Irlbacher-Fox and Mills, 2007). The chapter first discusses the economic relationship between regions and resource development. This is followed by a general discussion of the resource regimes across the circumpolar north. Based on the existing literature, the remainder of the paper asks five questions that can be used to guide research on finding ways to ensure a more sustainable future for these regions. Each question defines a key element for building an understanding of resource revenue regimes in the north.

Resources and the northern economy

Resource production is an important part of the economies of the circumpolar north. The economic base of the north's local and regional economies consists of three sectors; the international resource sector, the traditional or subsistence sector, and the public sector. Each of these provides real income for residents of the north. Local resources are the basis for the traditional economy which produces for local consumption throughout much of the region. Resources are also a major part of the circumpolar north's exports to the rest of the world. While the public sector in the north is primarily supported by transfers from higher levels of government, resource revenues may also provide funds to support the region's third key economic sector (Huskey, 2010).

The north's predicament is that while the social and environmental cost of resource development is concentrated in the north, the regional benefits of natural resource production are limited. The benefits of resource development for a northern region will depend on the portion of the resource rents and production costs that stay in the region. Unfortunately, the technical nature of most resource production for export means that most of the income earned by resource production flows to non-northern owners of the capital and technical know-how and skilled labor. These inputs are brought from outside the region. The large scale of many resource projects often means that even unskilled workers are brought into the region since the sparsely settled northern regions can't supply enough labor. The concentrated nature of natural resources and the enclave structure of production also limit the connection with communities. The structure of production means there are few linkages between resource production and the communities of the north. Most of the potential benefits flow from northern regions.

These observations are not new. There is an increasing realization among socio economic researchers that extractive industries production is often problematic for producing countries, regions, and communities. Building on earlier Canadian staples theory, researchers have shown that, despite an intuitive belief that natural resource development will increase the wealth, and therefore the well-being of producing regions, a resource curse exists. The paradox of the resource curse is that extractive resource development often leads to a decrease in development possibilities in these regions and produces other problems (Humphreys, Sachs, and Stiglitz, 2007; Collier, 2008).

Elsewhere, we have discussed the relevance of the resource curse discourse for the circumpolar north (Huskey and Southcott, 2016). We have noted that there is value in using staples theory as a conceptual framework to better understand how northern communities can avoid the resource curse. Staples theory was initially a conceptual framework to describe how Canada developed through its role as the producer of staples, or basic commodities, and how these commodities impacted the development of the country. It became a means of describing how countries could develop economically based on staples production when the Canadian economist Mel Watkins borrowed A. O. Hirschman's concepts of backward, forward, and final demand linkages to show how these linkages could encourage the spreading of benefits into other sectors of the economy (Watkins, 1963). Problems with the original conceptualization of linkages led Hirschman to introduce a forth type – fiscal linkages (Hirschman, 1977).

This staples framework allows researchers to understand how the development of a resource commodity can either assist a community or region to diversify into a sustainable economy or how it can provoke economic decline and produce a "staples trap" (Watkins, 2007). Backward, forward, and final demand linkages are important to understand the sustainability of northern communities. Potential new linkages are developed constantly that could result in new forms of economic sustainability, but these linkages are likely to be less important than the fourth and final type of linkage – fiscal linkages. There is an increasing recognition that resource rents should be used to mitigate any negative impacts of resource production and to assist marginalized populations in resource regions to develop sustainable communities (Pegg, 2006; Segal, 2011). In the past northern and Indigenous peoples had little or no access to resource rents as the ownership and control of the land rested with central governments and private industry who saw little need to share control of these rents (Watkins, 1977). Now, however, new comprehensive treaties and the devolution of political powers to northern regions means that northern communities have the ability to potentially access resource rents and use them to create more sustainable forms of development (Coates and Crowley, 2013; Einarsson et al., 2004; Irlbacher-Fox and Mills, 2007). Fiscal linkages are possible if resource rents can be directed towards local development and leakages are minimized.

While there is an increased belief that, properly managed, extractive industries can actually contribute to sustainability, there is still a lack of a coordinated effort of researchers on how exactly to do this. How can resource extraction be best used to create successful societies? What are the best ways to avoid the staples trap/resource curse? How can resource extraction best contribute to the well-being of producing regions and communities? What institutions best ensure that producing regions and communities benefit from resource extraction?

While these are complex questions that require complex answers, in situations where the proper institutions exist, an important part of resolving the staples trap/resource curse is ensuring a significant enough share of resource revenues stay in the region. When resource revenues are saved these savings provide a means of smoothing the benefits from development. The timing of the use of resource revenues may limit the boom-bust nature of development.

Resource revenue regime

Resource revenue regimes describe the fiscal rules of the game for resource production. Fiscal rules are the complement of the rules and regulations that govern how resources may be produced (ICMM, 2009). Taxes, royalties, and rents paid by resource companies are components of resource revenue regimes for government organizations. Local government and Indigenous organizations may also share in the income produced by resource production as owners through profits. Fiscal rules will vary not only across governments; they may also vary across types of resources.

The fiscal rules affect the amount of natural resource revenues a region receives in two ways. First, the rate and rules of a revenue regime will determine how much of the income earned from any given level of production will be claimed by the public sector. This is the direct effect. If the tax rate is 10 percent for every $100 resource production earns, $10 is kept by government. The indirect effect of taxes or other fiscal mechanisms operates through their impact on firms' production decisions. In the past resource development in some parts of the north might have been undertaken for strategic or political reasons, but the profit motive guides most production today. The fiscal rules of northern development will affect the profit potential of northern resource development and this will influence production decisions. An increase in tax rates will positively affect revenues from a given level of production but will have a negative effect on revenues if it discourages production.

In the north most subsurface resource rights are in public hands, and public ownership of natural resources remains the norm. Revenue regimes will reflect the attitude of northern government to resource ownership. If governments act like private owners they would control production decisions and hire a private resource company to manage the production and distribution of the natural resources. In this case, the public revenue flows to government in the form of profits from resource production.

A more common model in North America is the sale by the government of the right to use and profit from the use of a natural resource in a particular location. Under this ownership scheme public revenues from royalties, lease payments, and bonuses, rents may be set by law, through negotiation, or auction. Government may also tax resource producers. Taxes on production can be levied on units of output, income, or profits. Property taxes are based on the value of capital used in resource production or the value of resources in the ground. The manner of taxation will affect the production decisions of resource firms, the government's share in market and the timing of revenue flows.

The final determinant of the resource revenue regime in any part of the north is the center of control. The fiscal regime will probably depend more on where decisions are made than where the resources are located. Decisions about the resource use, revenues raised, and the distribution of natural resource revenues may be decided at the local, provincial, or national level. There may even be a mix in the locality of decision-making.

When higher levels of government control resource use and resource revenues in a region, the resource revenue regime for that region can be defined by that government's distribution policy. Local governments may receive impact funds as a type of compensation scheme for the cost borne in the region of production or they may receive funds distributed to all local governments in revenue sharing. It's likely that as the center of decision-making moves away from the resource location, more of the rents generated by production are lost to the region.

Throughout the north institutional changes have brought more control back to the regions affected by resource development. Land claims negotiations in Canada and the US placed the control of some of the north's natural resources in the hands of Indigenous groups by granting subsurface rights of ownership and regulatory authority for some resource uses. New local governments in Alaska, the North Slope Borough, and Greenland are examples of other types of institutional innovations that provide more fiscal control in the north.

Resource revenues provide the possibility that a resource producing region can prevent the reverse that historically has sometimes followed resource booms. Resource revenues provide a means for compensating the local residents for the cost of resource production. If resource revenues can be saved they also provide a way to spread out the benefits of resource development beyond the life of the project. Of course, the decision to save resource revenues is the decision not to spend the revenues. How the local area decides to spend its resource revenues can be considered as part of the fiscal regime.

How do resource revenue regimes vary across the circumpolar north?

The staples perspective seems a worthwhile lens to consider regional and national variations in resource revenue regimes. The notions of linkages and leakages allow us to more clearly see how communities can benefit from the revenue collected for natural resource extraction: which regimes allow for the greatest linkages (or stickiness) and which creates leakages.

Revenues that can be used for local linkages can come from several sources. One traditional source is the profits gained by the company that extracts the resource. At one time companies were locally based and as a result profits were mostly spent locally – allowing for important forward, backward linkages, and final demand linkages. Now however, most resource companies are international in nature and owned by shareholders in many areas of the world (United Nations, 2007). The large investment capital requirements to build the necessary infra-structure to support northern resource development means that returns on invest-ments would normally leave the region (Huskey and Morehouse, 1992).

Some new variations in this regard do exist however in that northern communities can become economic partners in resource ventures and as such share directly in the profits that are generated (as well as the losses). In Alaska, Native-owned development corporations, created under the terms of the Alaska Native Claims Settlement Act, are owners of resources and invest in resource developments. Perhaps the most well-known example of this is the Red Dog Mine in Northwest Alaska where the NANA Regional Corporation negotiated a joint venture lease arrangement with the Teck Mining Company by which they become economic partners in the development and share in the profits (Haley and Fisher, 2012). In Canada, Indigenous organizations, represented by the Aboriginal Pipeline Group, have partnered with international oil and gas companies to be a one-third owner of the proposed Mackenzie Gas Pipeline in the Northwest Territories (Nuttall, 2008). These organizations hope to share directly in the profits created by resource projects.

While other local benefits can come from things such as wages, increasingly fiscal linkages from resource rents are seen as the most immediate and likely source of benefits (Bridge, 1999). Under this situation, finding out how to best obtain, manage, and distribute these rents is a priority for northern communities and governments. Information on resource revenue regimes across the north generally exists in "grey literature" which describes who collects these resource rents, what mechanisms are used to collect these rents, and where these rents go (Taylor, Severson-Baker, and Winfield, 2004). These regimes often vary between those used for oil and gas development and those used for mining.

Generally, oil and gas regimes vary by the contractual form used by the state for relations with the developing company. There are many variations but the four main types are concession or license agreements, production-sharing agreements, joint venture agreements, and service agreements (Radon, 2007). In the north, oil and gas developments are primarily concession or license agreements where a government allows a company to explore, develop, and sell the resource in exchange for royalties which are generally based on production (Alaska Dept of Revenue, 2012). Concessions are usually obtained either through open bids or negotiations. Often agreements include up-front payments (a license fee or signing bonus) to the government, especially in the case of open bids. Governments retain these payments whether development occurs or not. As a result, concession agreements are usually low-risk for governments but could be high risk for companies (Radon, 2007: 100). The other types of contractual agreements are partnerships between governments and companies where governments allow companies to develop a resource in exchange for a share of production (production-sharing agreements) or as an actual business partner (joint venture agreements). Under service agreements governments hire companies to develop a resource in exchange for a set fee.

Resource regimes are often characterized by the fiscal tools used by governments to capture resource rents from resource production. These vary in a number of ways (Table 8.1). Profit-based tools capture a portion of any profit gained by the producing company while production-based tools take a certain amount based on the outputs of the project. Companies tend to prefer profit-based taxes since they are only taken when the project produced a profit while governments have a more secure guarantee of rents when production-based instruments are used. There is a fairly well developed discourse around the impacts of the various types of taxation on resource development (ICMM, 2009; Lund, 2009). Instruments include royalties, income based taxes, rent based taxes, state equity, export duties, import duties, and others (Goldsworthy and Zakharova, 2010: 7).

Mining regimes have been heavily influenced by globalization pressures over the past 25 years. A key study done by the World Bank noted that increasingly countries are harmonizing their revenue regimes as they seek access to

Table 8.1 Key fiscal instruments

Instrument	Advantages and disadvantages
Royalty	A fixed fee per unit produced or a percentage of production or gross revenue. Royalties provide a minimum payment for resources used, produce stable and early revenue, and are relatively easy to administer. However, beyond modest levels they can distort investment and production decisions because they are insensitive to costs. They are regressive.
Income based taxes	Corporate income taxes are less distortionary since they are based on revenue less cost. Foreign investors appreciate the fact that they give rise to foreign tax credits. However, they are relatively more complex to administer. Revenue is also delayed: by how much depends on capital depreciation allowances, which are often made generous to attract investment (i.e., provide faster payback).
Rent based taxes	Pure rent-based taxes are neutral since payment is only required after the investor has earned its required rate of return. However, in practice rent is approximated (Appendix II). Those based on a measure of achieved return are most effective but are also the most difficult to administer. The balance of risks is skewed towards the government.
State equity	Enables the government to share in the upside and is often viewed to increase the sense of national involvement. However, "paid" equity requires the government to contribute to initial capital outlays, and often gives rise to conflicts of interest arising from the government's role as regulator.
Export duties	Not very common. Export duties are relatively easy to administer but they distort the decision of whether to sell crude oil domestically or abroad and are insensitive to costs.
Import duties	Provides revenue even before royalties due to the import needs during project development. To mitigate the negative impact on investors, full or partial exemptions are often provided.
Other	Other instruments include: signature and production bonuses; land rental payments; withholding taxes on interest, dividends, and services; and value added tax, if applicable

Source: taken from Goldsworthy and Zahkarova, 2010: 7

increasingly internationalized mining investments (Otto, 2006). Countries with high royalty and taxation rates have tended to lower these while countries with low rates have increased their rates. Generally "the overall trend has been a reduction in many taxes applied to mining"(Otto, 2006: xiv). Despite this harmonization there is still a great deal of variation between countries. As the authors of the World Bank study indicate, there is no one best system or rate for governments to maximize resource rent returns as every region has unique social and political circumstances requiring differing policies.

Alaska

In Alaska, the resource revenue regimes for oil and gas are relatively well known as they have been established since the early 1970s. Most of the resource rents are captured by the state of Alaska which uses several vehicles to capture this rent. The federal government does capture some of these rents but mostly in the form of federal corporate income tax or with developments on federal-owned lands. Oil and gas revenues in Alaska generally fund just over half of the state budget and 90% of discretionary spending (Alaska Dept of Revenue, 2012). This rent is captured by the state through four main mechanisms: royalties, property taxes, state corporate income taxes, and production taxes. Royalties are paid by companies to the owners of lands that produce oil and gas. These royalties vary but generally are 12.5% of the value of production. Royalties generally represent around 30% of state oil and gas revenues. The federal government also collects royalties on land that it owns but shares a portion of this (often about 50%) with the state. The federal government controls offshore production but pays the state of Alaska 27% of royalties collected on properties that are between 3 and 6 miles offshore.

The state corporate income tax is 9.4% of income over $90,000 and currently represents around 10% of state oil and gas revenues. Companies also pay property taxes on the value of their properties. While this does not represent a large percentage of total oil and gas revenues it is significant that local governments can also collect property taxes on these properties (which are deducted from state property taxes). These have been used by northern communities such as the North Slope Borough to capture some of these rents and create fiscal linkages (Morehouse and Leask, 1980). The other vehicle which Alaska uses to capture resource rents is a production tax. This is currently the largest source of capture. It is often a complicated measure with credits for a variety of activities.

These four vehicles allow the Government of Alaska to capture enough rent that Alaskans do not have to pay any state income tax or state sales tax. These revenues are therefore used to pay for most of the services of the state government. In addition, Alaskans created a capital fund, or sovereign wealth fund, to which a certain portion of yearly oil and gas revenues flow. This fund, the Alaska Permanent Fund, discussed below, was created to "prudently take some of the non-renewable oil wealth and transform it into a renewable source of wealth for future generations of Alaskans" (Alaska Permanent Fund, 2009: 5). The fund is primarily a state trust fund for savings but also pays out an annual dividend to all residents of Alaska.

In terms of mining revenues, the regime in Alaska is quite different from that of oil and gas. While mining was the basis of the economic development of Alaska for the first half of the twentieth century, it currently contributes little to state finances. In 2012 the State of Alaska collected $41 million from the mining industry compared to the $6.1 billion collected from the oil and gas industry (Caldwell, 2013). A study of the mining industry done in 2006 noted that state mining revenue came from three main vehicles: annual claim rentals or mineral lease rentals on claims or leases on state lands, a royalty payment of 3% of the net income the company earns from the sale of minerals taken from state lands, and a mining license tax for mining operations on all lands in Alaska, whether private, state or federal (Rothe, 2006). Most revenues come from the mining license taxes which in 2004 represented $10,317,238 out of a total state mining revenue total of $13.7 million (2). Mineral rents and royalties represented $3 million of this total while corporate income tax was $323,000. The value of all mineral production in Alaska in 2004 was $1.06 billion.

While mining revenues represent a small contribution to state revenues, they are crucially important to the localities where mines are located. While mines contributed $13.7 million to the state in 2004, they also contributed $11 million to local governments through municipal or borough taxes. In 2012 the mining industry made $126 million in payments to Alaska Native corporations (Resource Development Council, 2013). The Red Dog Mine was the sole taxpayer to the Northwest Arctic Borough in 2012 and the payment that year was $13 million.

Canada

In Canada natural resources are primarily the responsibility of the provincial governments. In the Territories however it was the responsibility of the federal government until recently. Since the creation of Alberta and Saskatchewan in 1905, the territorial governments have slowly evolved towards having the same powers as provinces (Cameron and Campbell, 2009). This devolution of powers has increasingly been focused on the control of lands and natural resources since this is often seen as a primary focus of economic development and of financing for territorial initiatives. The devolution of political power from the federal government to the territories is furthest advanced in the Yukon. During the 1990s the federal government transferred responsibility for oil and gas develop-ment to the Yukon Territorial Government. Responsibility for other areas of resource development was transferred to the Yukon in 2003 (Alcantara, Cameron, and Kennedy, 2012).

The responsibility for land and resources in the Northwest Territories and Nunavut is still with the federal government although a devolution agreement was signed with the Government of the Northwest Territories in 2013 and is scheduled to be implemented later in 2014. A devolution agreement with Nunavut is currently being negotiated (Alcantara, 2013). While taxation and royalty regimes may change as territorial governments take over these new responsibilities, the experience with the Yukon is that territorial legislation will initial "mirror" federal legislation until increased capacity allows the territories to alter the situation (Alcantara, Cameron, and Kennedy, 2012: 333).

While oil and gas production in the Canadian Territories has much potential, current production is very limited and as such information about resource rents is also limited. Oil and gas development in the Northwest Territories and in Nunavut is the responsibility of the Northern Oil and Gas Branch of Aboriginal Affairs and Northern Development Canada. On their website the royalty regime is described as being "one of the most competitive in the world" (Aboriginal Affairs and Northern Development Canada, 2014). Royalties are listed as being "1%, rising by 1% every 18 months to a maximum of 5% until project payout." After development costs are paid royalties are "the greater of 30% of net or 5% of gross." In addition to royalty payments, federal corporate tax rates are 16.5% (15% after 2012) and territorial/provincial corporate tax rates are between 10% and 11.5% (Alaska Dept of Revenue, 2012). There are currently no provisions for specific property taxes on resource developments in any of the Canadian Territories or for production taxes. The existence of modern comprehensive land claims in the Canadian Territories means that revenues are often shared with Indigenous organizations based on terms of the various treaties. Indigenous groups also control development where they own subsurface rights (Irlbacher-Fox and Mills, 2007).

Mining royalties and taxation have been an issue of much discussion lately in Northern Canada. Recent studies have pointed out that these royalty rates are among the lowest in the world and offer few benefits to northerners (Cizek, 2005; Taylor et al., 2010). Another study notes the difference between Northern Canada and Alaska and Norway where they "consistently collect more economic rent from non-renewable resources than Canadian jurisdictions do" (Irlbacher-Fox and Mills, 2007: 17). In the Yukon the royalty regimes differ between placer gold and other minerals. Placer gold is subject to a basic royalty of 2.5% per unit of gold based on a fixed unit price of $15 per ounce. As a result the Yukon collects 37.5 cents for every ounce of gold produced. In 2005 $30 million worth of gold was mined in the Dawson Mining District for which the Yukon Territorial Government collected $26,370 (Irlbacher-Fox and Mills, 2007: 13), or less than .09% of the produced value. With the then existing revenue sharing agreement the region's First Nations would have received $600 as their share of gold mining in the Dawson area.

For hardrock, or quartz, mining, the Yukon royalty system is based on profits from the mine and is a sliding scale. For profits from $10,000 to $1 million, 3% of profits are paid. This increased gradually until profits exceed $35 million. At this point royalties are 12% of profit (KPMG, 2011). The Northwest Territories and Nunavut have royalties collected by the federal government with identical royalty schemes. For these jurisdictions the royalty is the lesser of 13% of the value of production or a sliding scale based on profits that start at 5% for amounts between $10,000 and $5 million and rise to 14% for amounts over $45 million.

Greenland

Greenland's experience with resource regimes has primarily concerned mining although oil and gas exploration in the country started in the 1970s and has increased over the past few years. As of today there are no commercial oil and gas operations there. Mining in Greenland started in the mid-nineteenth century and became important with the opening of the Ivittuut cryolite mine in the 1860s.

Indeed during the last part of the nineteenth century and the first part of the twentieth century this mine financed most state activities in Greenland (Sinding, 1992: 226). After World War II the expenses of the state surpassed the rent coming from mining but the Ivittuut mine continued to contribute until the 1980s. In the 1970s a lead-zinc mine at Maarmorilik was opened and operated until 1990.

As a colony of Denmark, all revenues from mining in Greenland flowed to Denmark until 1988. After Home Rule was implemented in 1978, a new fiscal arrangement was negotiated so that by 1989 resource management and revenue was to be shared equally between Greenland and Denmark (Sinding, 1992). Under the self-government agreement of 2009, Denmark handed over all control of resource development to the Greenlandic government (Harsem, Eide, and Heen, 2011).

Following Home Rule, Greenland and Denmark jointly produced a new resource development strategy and a new mining act was introduced in 1991 (Sinding, 1992). Under the new law the state could claim both a tax and permit fee on five-year prospecting permits. Ten-year exploration permits would require a fee, as would permits to allow a mine into production. Mines would pay a 35% income tax on profits after capital costs and other deductions are accounted for. Municipalities would not be allowed to tax mining companies.

Recently, the non-renewable resource sector has been the subject of much debate in Greenland (Areddy and Bomsdorf, 2013). In March 2013 a new government was elected largely due to a feeling that Greenland was not going to get enough benefit from several possible new mines that were hoping to open there. Under agreements with the previous government, capital costs and deductions meant that Greenland would not get any revenues from these mines for many years. A new government was elected and a new arrangement was negotiated with the proponent of the largest mining project in October 2013. According to this new agreement a royalty, based on sales, would be payable if withholding tax and corporate taxes were less than the calculated royalty in a particular year. This would mean that the government would be guaranteed to get some income from the project. The royalty was structured on a sliding scale with the first five years at 1% going up to 5% at 16 years and beyond (Kolver, 2013).

Norway

Norway's resource regime is probably the most well-known of Arctic states. Resource curse researchers often point to Norway's management of its oil and gas production as the best example of how to manage non-renewable resources (Listhaug, 2005; Thurber, Hults, and Heller, 2011). Unlike other countries, Norway has implemented a system where the state captures around 80% of rents from oil and gas development (Gylfason, 2001).

Norway's current resource regime is linked to a long-standing tradition of national state control of natural resources. The Norwegian state in highly central-ized and while recent changes concerning Finnmark have eased this centralized control, the resource regime is almost entirely controlled by the national government. The current legislation governing oil and gas development is derived from "Concession Laws" developed to control foreign involvement in hydro-electric development at the beginning of twentieth century (Ville and Wicken, 2013). While Norway lagged behind similar countries in terms of economic development through most of the twentieth century, the discovery of oil and gas in the North Sea offshore of southern Norway in the early 1970s transformed the economy turning it into one of the most wealthy in the world (Larsen, 2005).

The present system has evolved continuously since then but has been marked by the desire to both ensure that resource rents are captured and used to the benefit of all Norwegians and to avoid the Dutch Disease that first accompanied North Sea oil and gas development (Larsen, 2005). It is different from other jurisdictions in that it no longer captures rent through royalties. Instead it does so through taxing the profits of producing companies, through dividends from Statoil, the state oil and gas company, and by ensuring it is an equity partner in most projects (Persily, 2011).

Oil and gas companies in Norway are taxed at a normal business rate of 28% of profits. In addition, there is a special 50% tax on offshore oil and gas production meaning that this production is taxed at a marginal rate of 78% (Persily, 2011). All current oil and gas development in Norway is offshore. Tax concessions are sometimes awarded to encourage investment in difficult projects such as the Snohvit offshore development in the High North.

Companies may also pay property taxes on onshore infrastructure to municipalities. Whether there is property tax and what these rates may be vary by municipality.

In addition to taxes, the Norwegian state owns 67% of the Statoil oil and gas company, a company it created in 1972. Through Concession Law negotiations it has ensured that Statoil controls a large share of activities in Norway (Persily, 2011, Ville and Wicken, 2013). All of the dividends from Statoil flow to the government. In addition to Statoil, the Norwegian state has forced oil and gas companies to include them as an equity partner. Through another state-owned company, Petoro, Norway owns a percentage of most oil and gas developments above and beyond the share owned by Statoil. Indeed, this is currently the major source of oil and gas revenues for the state (Persily, 2011).

In order to save resource rents for future needs, and especially in order to avoid the economic problems associated with Dutch Disease, the Norwegian state transfers all of these rents into a fund called the Government Pension Fund – Global. Despite its name it is a sovereign wealth fund and not a pension fund. As this fund invests primarily outside Norway, the negative impacts that large resource rents can have on a domestic economy are lessened. The fund is often portrayed by both international financial institutions and the international media as the best example of dealing with oil and gas rents (Velculescu, 2008). The government can take out up to 4% of the asset value of the fund each year but unlike the Alaska Permanent Fund, no moneys are distributed directly to the public. The fund is currently the largest sovereign wealth fund in the world.

Unlike the oil and gas industry, the mining industry in Norway is not subject to any special taxes and pay normal corporate tax rates. Certain municipalities in northern Norway that are close to mining operation would like to see a new mining tax. It is hoped that proceeds from such a tax could go to local communities to mitigate increased demands on these communities (Karlsbakk, 2012).

Russia

Russia is one of the world's largest producers of natural resources. It is one of the largest exporters of crude oil with close to 13% of world production (Deloitte, 2012). It accounts for 18.4% of all natural gas production in the world of which 90% is exported. In terms of mining it is the world's largest producer of chromium, nickel, and palladium, and is the second largest producer of aluminum, platinum, and zirconium. In 2009 the mining and metallurgical industries accounted for almost 20% of the country's industrial output and around 15% of Russia's total exports (CountryMine, 2009).

The Russian resource regime has changed considerably over the past 20 years since the fall of the Soviet Union. Today it has a regime that is heavily controlled by the state (Deloitte, 2012). State controlled companies dominate the oil and gas sector and there are severe administrative restrictions (Thurber, Hults, and Heller, 2011). Apart from the diamond sector which is largely controlled by the state, mining in Russia is controlled by large private sector firms that are led by prominent wealthy Russian businessmen. Industry analysts contend that the Russian state can exercise "indirect control through some of the oligarch promoters who owe allegiance to the Kremlin" (Thomas White Global Investing, 2011: 8). Increasingly these firms are creating new relations with foreign investors and foreign firms in order to internationalize their operations.

The main regime in the oil and gas sector is a combination of taxes and royalties although a few developments are governed by production-sharing agreements (Goldsworthy and Zakharova, 2010: 9). Most state revenues come from a mineral extraction tax (MET) and an export tax. The MET is a volume-based royalty rate that in 2010 was 419 rubles per metric ton. This rate is adjusted periodically depending on prices in the world market. The MET is adjustable based on the source of the oil and gas. New and riskier fields have lower MET rates in order to encourage investment. The export duty is based on a sliding scale related to the market price of the oil and gas. In 2010 marginal rates were 35% for the market price in excess of $15 per barrel up to $20 per barrel; 45% for the excess over $20 per barrel to $25 per barrel; and 65% over $25 per barrel. A separate export duty exists for refined products. The corporate income tax on oil and gas in Russia is 20% with deductions for MET and export duties as well as for investments based on the maturity of the fields. This is also applicable to mining operations. Of this percentage 2% goes to the federal government and 18% to the regional government. It is possible for the regional government to reduce their amount by 4.5% to encourage development (PricewaterhouseCoopers, 2012).

A value added tax of 18% is paid on oil and gas sold domestically. Other taxes include a dividend withholding tax of 9% for residents and 15% for non-residents, an interest withholding tax of 20%, unified social tax of 2% to 26% of gross payroll, a property tax of up to 2.5% of assets, and import customs duties ranging from 5% to 30% (Goldsworthy and Zakharova, 2010: 11). All of this means that for every dollar worth of oil and gas exported from Russia, the government gets 90 cents when the price is over $25 per barrel.

In terms of the mining industry, in addition to the same corporate income taxes as oil and gas companies, production is subject to the Mineral Resources Extraction Tax. This is a royalty-like tax which is applied to the value of the extracted resource (PricewaterhouseCoopers, 2012). The tax varies by the type of material mined. Copper is taxed at 8%, gold at 6%, and iron ore at 4.8%. Most of these taxes go into the federal budget and are used to pay for government services across Russia. At the same time, Russia has created sovereign wealth funds that are used to pay off external debt, as future savings, and as a means of financial stabilization (Tsani, 2013).

In terms of local control over resource rents, historically Russia has been a very centralized system (Riabova, 2001). Following the fall of the Soviet Union there was a movement towards the decentralization of power towards regional governments but this has been reversed over the past 15 years (Libman, 2010: 7). In addition, the new private/public sector resource firms with strong connections to the federal government are increasingly influential in local and regional decision-making (Tynkkynen, 2007: 856). The currents system is highly centralized. Regional governors are no longer elected by regional populations are instead appointed by the federal president. At the same time, regions with large resource developments benefit from resource rents. They receive a share of the MET and other taxes and receive direct contributions from resource companies in their efforts to ensure good relations with the region (Libman, 2013: 86). These are used to provide public goods to those living in the region. As a result, despite the centralized nature of the Russian political system, those regions with resources often have higher level of services.

Sweden and Finland[1]

Neither Sweden nor Finland has important oil and gas developments but they both have relatively long histories of mining. Sweden and Finland have both been noted recently as being jurisdictions favored by the mining industry due to their low taxation policies (Wilson, McMahon, and Cervantes, 2013). One industry review of Sweden's taxation policy noted that companies there only have to pay 22% of gross profit and that this low rate is unsustainable (GMP Securities, 2013: 9). In Finland, mining companies are required to pay the common corporate tax rate of 20% of profits minus deductions. An additional mining tax is currently under discussion. Local communities in northern Norway, northern Sweden, and northern Finland are hoping that funds from a mining tax could be used to assist these communities deal with the impacts of mining in their regions (Karlsbakk, 2012).

Both Sweden and Finland have highly centralized national states. All resource taxes go directly to the national government and are then redistributed according to the desires of the central government.

Five key questions

New resource regimes offer possible new means of increasing fiscal linkages in northern regions. To increase the possibilities of this happening, key research questions have to be answered. This section outlines five key questions that research can help answer in order to increase the chances of resource development helping, rather than hurting, the sustainability of these communities.

Where does the money go in Arctic resource development?

The first area that we need to clarify is where exactly the money from Arctic resource development goes. If we want to find out about how to increase fiscal linkages we have to have a better idea of what money is produced by these operations and where this money goes. The brief discussion above concerning the resource regimes of the various Arctic regions gives us an idea but very little research exists in more detail. It would be wise to undertake a series of case studies that attempted to get a better idea of how to measure linkages and leakages in an Arctic context.

The ideal would be to develop a model of measuring linkages and leakages that could be used by Arctic communities and regional government so that they would have better idea of where the money from resource development goes.

Which resources regimes give the greatest fiscal linkages to northern communities?

Empowering communities and regions to be able to track resource development money would greatly assist them in trying to answer the next research question – how to increase linkages and decrease leakages. This is perhaps the central key question of this research area and one that would provide the most useful information. Which regimes best maximize fiscal benefits for the north and most limit leakages out of the region? The review above has revealed a great deal of variation based on a number of factors including the degree of political decentralization and the existence of new comprehensive land claims.

What are the best vehicles for saving resource revenues?

Linked to the question of maximizing fiscal linkages is a more specific question relating to the best vehicles for gathering and saving revenues. Several authors have looked at various types of funds and their relative value for resource dependent regions (Baena, Sévi, and Warrack, 2012; Humphreys and Sandbu, 2007; Tsani, 2013). This research often points out that two of the most successful are the Alaska Permanent Fund and the Norwegian Government Pension Fund – Global. Can these funds be replicated in other regions of the Arctic? What lessons can be learned from Norway and Alaska and other non-Arctic regions? Resource savings funds do not necessarily have to be large scale. Various Arctic communities have implemented different types of smaller funds that they attempt to use in order to stabilize development associated with natural resource exploitation. Often these are linked to Impact Benefit Agreements. Yet very little is known about what funds exist and what forms they take.

What are the best ways to distribute resource revenues to communities?

If research leads to a better understanding of how to increase linkages and decrease leakages, and if we find out new ways of saving the fiscal benefits of doing this, another key question is how best to spend these benefits. One of the most important objectives is to use this funding to try and transform the finite nature of natural resource exploitation into sustainable forms of development. In addition, funding may be necessary to mitigate negative impacts. What is the best way of doing this? Some research exists relating to the Alaska Permanent Fund and the Norwegian Government Pension Fund – Global on the best ways of doing this. Discussion is occurring around whether or not distributing benefits directly to individuals, as is done by the Alaska Permanent Fund, is a good thing or a bad thing (Baena, Sévi, and Warrack, 2012). Impact benefits agreements and similar arrangements between communities and industry often contain funding that is to be distributed to communities for a variety of reasons. Research on past experiences with varying forms of distribution can assist communities to find ways to maximize linkages created from this funding.

To what degree does resource regime structure affect production decisions?

Finally – if communities decide they want to see resource development occur, they have to know how decisions that they make concerning resource development will affect whether senior governments and industry proceed with production. More has to be known about how investment decisions are made – both by industry and senior governments. We are interested in the tradeoff between the direct and indirect effects of fiscal measures on resource revenues. When governments consider adjusting resource revenues to encourage more production or to increase revenues, will these adjustments work?

Note

1   Iceland has no large-scale mining and no current oil and gas developments. The national economy is still largely dependent on the fishing industry. Aluminum smelting is also important but most material used in this process is imported. The smelters exist in Iceland primarily due to the relatively cheap hydroelectric rates.

References

Aboriginal Affairs and Northern Development Canada (2014). Oil and Gas in Canada's North—Active Exploration and New Development. Retrieved from www.aadnc-aandc.gc.ca/eng/1100100037301/1100100037302.

Alcantara, C. (2013). Preferences, Perceptions, and Veto Players: Explaining devolution Negotiation Outcomes in the Canadian territorial North. *Polar Record*, *49*(249), 167–179.

Alcantara, C., Cameron, K., and Kennedy, S. (2012). Assessing Devolution in the Canadian North: A Case Study of the Yukon Territory. *Arctic*, *65*(3), 328–338.

Areddy, J. and Bomsdorf, C. (2013). Greenland Opens Door to Mining. *Wall Street Journal*. Retrieved from http://online.wsj.com/news/articles/SB10001424052702304682504579157323886693660.

Baena, C., Sévi, B., and Warrack, A. (2012). *Funds from Non-Renewable Energy Resources: Policy Lessons from Alaska and Alberta* (Vol. 51). Amsterdam: Elsevier.

Bridge, G. (1999). *Harnessing the Bonanza: Economic Liberalization and Capacity Building in the Mineral Sector*. Paper presented at the Natural Resources Forum.

Caldwell, S. (2013). Is Alaska Getting Its Fair Share from Mining? *Alaska Dispatch*. Retrieved from www.alaskadispatch.com/article/20130616/alaska-getting-its-fair-share-mining.

Cameron, K. and Campbell, A. (2009). The Devolution of Natural Resources and Nunavut's Constitutional Status. *Journal of Canadian Studies*, *43*(2), 198–219.

Cizek, P. (2005). *Plundering the North for Hyper-Profits: Non-Renewable Resource Extraction and Royalties in the Northwest Territories 1998–2004*, Yellowknife: Canadian Arctic Resources Committee.

Coates, K. and Crowley, B. (2013). *New Beginnings: How Canada's Natural Resource Wealth Could Re-shape Relations with Aboriginal People*. Ottawa: MacDonald-Laurier Institute.

Council, R. D. (2013). Alaska's Mining Industry. Retrieved from www.akrdc.org/issues/mining/overview.html.

CountryMine (2009). Mining in Russial. Retrieved from www.infomine.com/countries/russia.asp.

Deloitte. (2012). *Tax and Legal Guide to the Russian Oil & Gas Sector*. New York: Deloitte Touch Tomahmatsu Ltd.

Einarsson, N., Nymand Larsen, J., Nilsson, A., and Young, O. R. (2004). *The Arctic Human Development Report*. Akureyri: Stefansson Institute.

Freudenburg, W. R. (1992). Addictive Economies – Extractive Industries and Vulnerable Localities in a Changing World-Economy. *Rural Sociology*, *57*(3), 305–332.

Freudenburg, W. R. and Wilson, L. J. (2002). Mining the Data: Analysing the Economic Implications of Mining for Nonmetropolitan Regions. *Sociological Inquiry*, *72*(4), 549–575.

Fund, A. P. (2009). *An Alaskan's Guide to the Permanent Fund*. Anchorage: Alaska Permanent Fund Corporation.

Goldsworthy, B. and Zakharova, D. (2010). *Evaluation of the Oil Fiscal Regime in Russia and Proposals for Reform*. Washington, DC: International Monetary Fund.

Gylfason, T. (2001). Natural Resources, Education, and Economic Development. *European Economic Review*, *45*(4–6), 847–859.

Haley, S. and Fisher, D. (2012). *Shareholder Employment at Red Dog Mine*. Anchorage: Institute of Social and Economic Research.

Harsem, O., Eide, A., and Heen, K. (2011). Factors influencing Future Oil and Gas Prospects in the Arctic. *Energy Policy*, *39*(12), 8037–8045.

Hirschman, A. O. (1977). A Generalized Linkage Approach to Development, with Special Reference to Staples. *Economic Development and Cultural Change, 25*, 67–97.

Humphreys, M. and Sandbu, M. (2007). The Political Economy of Natural Resource Funds. In Humphreys, M., Sachs, J., and Stiglitz, J. E. (Eds.), *Escaping the Resource Curse* (pp. 194–233). New York: Columbia University Press.

Huskey, L. (2010). Globalization and the Economies of the North. In Heininen, L. and Southcott, C. (Eds.), *Globalization and the Circumpolar North.* Fairbanks: University of Alaska Press.

Huskey, L. and Morehouse, T. A. (1992). Development in Remote Regions: What Do We Know? *Arctic*, 128–137.

Huskey, L. and Southcott, C. (2016). "That's Where My Money Goes": Resource Production and Financial Flows in the Yukon Economy. *The Polar Journal, 6*(1), 11–29.

ICMM. (2009). *Minerals Taxation Regimes: A Review of Issues and Challenges in their Design and Application,.* London: International Council on Mining and Metals and the Commonwealth Secretariat.

Investing, T. W. G. (2011). *Sitting on a Gold Mine: Metals and Mining in Russia*, Chicago, IL: Thomas White International, Ltd.

Irlbacher-Fox, S. and Mills, S. J. (2007). *Devolution and Resource Revenue Sharing in the Canadian North: Achieving Fairness Across Generations.* Walter and Duncan Gordon Foundation.

Karlsbakk, J. (2012). Local Mayors Want Tax from Mining. *Barents Observer*. Retrieved from http://barentsobserver.com/en/politics/local-mayors-want-tax-mining-07-11.

Kolver, L. (2013). London Mining Awarded Exploitation Licence for Greenland Project *Mining Weekly.com.* Retrieved from www.miningweekly.com/article/london-mining-awarded-exploitation-licence-for-greenland-project-2013-10-24.

KPMG. (2011). *A Guide to Canadian Mining Taxation.* Toronto: KPMG Canada.

Larsen, E. R. (2005). Are Rich Countries Immune to the Resource Curse? Evidence from Norway's Management of Its Oil Riches. *Resources Policy, 30*(2), 75–86.

Libman, A. (2010). *Subnational Resource Curse: Do Economic or Political Institutions Matter?* Frankfurt: Frankfurt School of Finance & Management.

Libman, A. (2013). Natural Resources and Sub-national Economic Performance: Does Sub-national Democracy Matter? *Energy Economics, 37*, 82–99.

Listhaug, O. (2005). Oil Wealth Dissatisfaction and Political Trust in Norway: A Resource Curse? *West European Politics, 28*(4), 834–851.

Lund, D. (2009). Rent Taxation for Nonrenewable Resources. *Resource, 1*, 287–305.

Morehouse, T. A. and Leask, L. (1980). Alaska's North Slope Borough: Oil, Money and Eskimo Self-Government. *Polar Record, 20*(1), 19–29.

Nash, J. (1979). *The Mines Eat Us and We Eat the Mines: Dependency and Exploitation in Bolivian Tin Mines.* New York: Columbia University Press.

Nuttall, M. (2008). Aboriginal Participation, Consultation, and Canada's Mackenzie Gas Project. *Energy & Environment, 19*(5), 617–634.

Otto, J. E. A. (2006). *Mining Royalties: A Global Study of Their Impact on Investors, Government, and Civil Society.* Washington, DC: World Bank Publications.

Pegg, S. (2006). Mining and Poverty Reduction: Transforming Rhetoric into Reality. *Journal of Cleaner Production, 14*(3–4), 376–387.

Persily, L. (2011). *Norway's Different Approach to Oil and Gas Development*, Alaska Natural Gas Transportation Projects Office of the Federal Coordinator.

PricewaterhouseCoopers (2012). *Corporate Income Taxes, Mining Royalties and other Mining Taxes: A Summary of Rates and Rules in Selected Countries*, London: PricewaterhouseCoopers.

Radon, J. (2007). How to Negotiate an Oil Agreement. In Humphreys, M., Sachs, J., and Stiglitz, J. E. (Eds.), *Escaping the Resource Curse* (pp. 89–113). New York: Columbia University Press.

Rasmussen, R. O. (Ed.) (2011). *Megatrends*. Copenhagen: Nordic Council of Ministers.

Revenue, A. D. (2012). *Alaska's Oil and Gas Fiscal Regime*. Anchorage: Alaska Department of Revenue. Retrieved from www.dor.alaska.gov/Portals/5/Docs/Publications/acloserlook.pdf.

Riabova, L. (2001). Coping with Extinction: The Last Fishing Village on the Murman Coast. In Aarsæther, N. and Bærenholdt, J. O. (Eds.), *The Reflexive North*. Copenhagen: Nordic Council of Ministers.

Rothe, A. (2006). *A Review of Industrial Hard Rock Mining in Alaska*, Halycon Research. Report commissioned by Alaskans for Responsible Mining.

Securities, G. (2013). *Taxation Trends in the Mining Industry*. Toronto: GMP Securities Inc.

Segal, P. (2011). Resource Rents, Redistribution, and Halving Global Poverty: The Resource Dividend. *World Development*, *39*(4), 475–489.

Sinding, K. (1992). At the Crossroads: Mining Policy in Greenland. *Arctic*, *45*(3), 226–232.

Taylor, A., Grant, J., Holroyd, P., Kennedy, M., and Mackenzie, K. (2010). *At a Crossroads: Achieving a Win-Win from Oil and Gas Developments in the Northwest Territories*. Drayton Valley: Pembina Institute and Alternatives North.

Taylor, A., Severson-Baker, C., and Winfield, M. (2004). *When the Government is the Landlord: Economic Rent, Non-renewable Permanent Funds, and Environmental Impacts Related to Oil and Gas Developments in Canada*. Drayton Valley: Pembina Institute.

Thurber, M. C., Hults, D. R., and Heller, P. R. P. (2011). *Exporting the "Norwegian Model": The Effect of Administrative Design on oil Sector Performance* (Vol. 39). Amsterdam: Elsevier.

Tsani, S. (2013). Natural Resources, Governance and Institutional Quality: The Role of Resource Funds. *Resources Policy*, *38*(2), 181–195.

Tynkkynen, V.-P. (2007). Resource Curse Contested – Environmental Constructions in the Russian Periphery and Sustainable Development. *European Planning Studies*, *15*(6), 853–870.

United Nations (2007). *Transnational Corporations, Extractive Industries and Development*. New York: UN.

Velculescu, D. (2008). Norway's Oil Fund Shows the Way for Wealth Funds. *IMF Survey*.

Ville, S. and Wicken, O. (2013). The Dynamics of Resource-based Economic Development: Evidence from Australia and Norway. *Industrial and Corporate Change*, *22*(5), 1341–1371.

Watkins, M. (1963). A Staple Theory of Economic Growth. *Canadian Journal of Economics and Political Science*, *29*(2).

Watkins, M. (Ed.) (1977). *Dene Nation, the Colony Within*. Toronto: University of Toronto Press.

Watkins, M. (2007). Staples Redux. *Studies in Political Economy*, *79*(Spring), 213–226.

Wilson, A., McMahon, F., and Cervantes, M. (2013). *Survey of Mining Companies 2012/2013*. Vancouver: Fraser Institute.

9 Regional development in the circumpolar north

What else do we need to know?

Frances Abele

Mineral development is now and has long been a favoured aspect of northern development planning in most parts of the circumpolar north. Yet, while mineral development has brought some jobs and business opportunities, it has not brought development that can be termed sustainable. One step towards a better understanding of how resource development *could* contribute to the long-term sustainability of northern communities is consideration of development in its regional context. This modest reframing of the issue is the goal of this chapter.

Circumpolar regions differ in important ways, but they share a number of fundamental characteristics. Almost all are part of a larger polity in which non-northern populations –and externally sustained political systems – are dominant. Most northern regions are home to Indigenous populations who have worked to achieve a degree of collective self-determination and the means to protect their long-standing means of livelihood. All northern regions are now coming to terms with the implications of climate change, a global process of transformation that is affecting many parts of the north before it is felt elsewhere. All must adapt to the various effects of the warming climate, and ultimately, to the loss of markets for natural resources (assuming effective action towards meeting global carbon reduction targets). Circumpolar regions now face economic challenges and opportunities that differ in kind and to some extent in type from those occurring elsewhere on the planet.

Balanced development to enhance the resilience and vitality of northern Indigenous communities requires strategic attention to regional dynamics on several scales, starting with the local, along with bold efforts to create integrated and far-sighted development strategies for regions within national and sub-national jurisdictions. It is in this context that policies can be developed to better ensure benefits from resource development that can contribute to the long-term sustainability of northern communities.

Why focus on regions?

The disciplines of planning, economics and geography offer substantial research and discussion about the concept of region (Dawkins 2003). Some of this work is focused on the north (Petrov 2012). For present purposes, Dawkins' definition offers a good beginning. A region is:

> a spatially contiguous population (of human beings) that is bound either by historical necessity or by choice to a particular geographic location. The dependence on location may arise from a shared attraction to local culture, local employment centers, local natural resources, or other location-specific amenities.
>
> (Dawkins 2003: 134)

This way of conceiving of a region emphasizes the importance of a particular place –of land – to a human collectivity, and it envisions a variety of conditions that may bind a human collectivity to a place. It is a conception of region understood in terms of human purposes, experience and history. Such regions often also comprise an ecologically identifiable region, such as a coastal zone or a river system, as these are factors shaping human mobility and productive activity. For these reasons, Dawkins' definition is particularly appropriate to considerations of northern development, where Indigenous peoples' traditional territories define very long-standing zones of human activity,[1] and where the relationship between people and land is deep and intimate. In addition, now, Indigenous peoples' traditional territories are the sites upon which aspirations for the future are focused, even as all recognize that present and future are shared by descendants of the north's original inhabitants and more recent migrants.

But regions can be defined on different scales. Partly in response to growing interdependence and enhanced communication, and partly due to wider recognition of the extent to which the circumpolar basin has long been the 'Arctic Mediterranean' for Indigenous peoples, the basin itself is increasingly understood as an emerging geopolitical and economic region (Keskitalo 2007; Heininein and Southcott 2010; Glomsrød and Aslaksen 2009), with varying southern boundaries (Smith 2011). Further adding to the complexity, the circumpolar basin holds Indigenous regions on a different, and perhaps it is possible to say, more modern, scale. The Sami lands in Norway, Finland, Sweden and Northwest Russia are often identified as a distinct region of the Arctic (Henriksen 2010; Sillanpaa 1994), as are the Inuit circumpolar lands (e.g. Bjerregaard et al. 2004). The existence and activities of federated Indigenous peoples' organizations (the Sami Council, Inuit Circumpolar Council, Arctic Athabaskan Council) reinforce the conceptualization of these transnational regions (and they show the poor fit between Indigenous territories and contemporary country boundaries). Indigenous territories define regions within countries as well: Inuit Nunangat defines all of the Inuit lands in Canada while in other uses and quite commonly, Indigenous lands within provinces and territories are treated as regions. An example of this is found in the extensive literature on Nunavik, Inuit territories in the province of Quebec. A similar pattern prevails for Dene territory, which can be considered as one region, for some purposes, or as several, defined by modern institutions created by land claim agreements. An example of this latter is the region – defined by the Tlicho government, which exercises authority over Tlicho lands – developing land use plans and in other ways managing the lands as an economic region.[2]

All of this illustrates the way in which the meaning of the term 'region' depends upon institutions and purposes. It reminds us, as well, of the importance of attending to jurisdiction and sovereignty in regional economic development planning and assessment. The circumpolar north is overlain with the borders of modern nation-states, and within nation-states, of sub-national jurisdictional boundaries marking Indigenous and public governments. Increasingly, as well, national borders are traversed by circumpolar Indigenous organizations (such as the Inuit Circumpolar Council), of nation-states (Arctic Council) and of researchers (the most recently formed being EU-PolarNet).[3] None of the transnational organizations have state-like enforcement powers, but in addition to their individual missions, through discussion, knowledge development and dissemination they encourage perceptions of the circumpolar basin as a meaningful region.

In sum: thinking about healthy northern economic and social development in regional terms is intuitively appealing. It requires, though, that regions be identified purposefully, to make it possible to respect the reality of the communities in which people choose to live, while understanding change and formulating choices with appropriate territories and institutions in mind. A regional approach permits a focus on entire Indigenous territories, making it possible for peoples of those territories to make development plans that match the current and historical lands for which they are responsible. Regional approaches can also draw attention to entire areas affected by large natural resource developments. Contemporary jurisdictional boundaries, defined by constitution, legislation or international agreement, define the sites of authoritative decision-making and political debate at a range of scales.

In an increasingly interdependent and interactive world, it is important to think beyond nation-state boundaries, and to make decisions with relevant international authorities in mind. Yet most people do not live 'internationally'. They are born, grow up and make their lives in neighbourhoods, communities and sub-national regions. For that reason, the discussion that follows moves between a focus on communities and sub-national regions, the human-size regions about which democratic deliberation and decision-making is feasible, and the national and supranational scale where lie most of the financial resources and power.

What do we know so far?

The economies of regions around the pole can be considered in terms of four pillars: public investment, subsistence or land-based productive activity, mineral exploitation and a variety of 'diversifying', generally renewable, land-based economic activities such as tourism and artistic production. These vary in importance in different circumpolar regions, but a number of generalizations about each are possible.[4]

First, it is clear that outside the rare large industrial centres, the material base of community stability, economic vitality and cultural continuity depends upon there being available a blend of wage economy opportunities, harvesting and appropriately packaged transfer payments. As pointed out in Chapter 12 of this volume, despite long-standing predictions of the demise of the harvesting economy, or of the mixed economy of which it is now part,[5] northern Indigenous communities continue to seek sustenance from the land wherever possible. This seems to be true, remarkably, in almost all parts of the circumpolar north, despite important historical and institutional differences.

There are both cultural and practical reasons for this (Usher, Duhaime and Searles 2003; Natcher 2009; Abele 2009). Hunting, fishing and gathering, and traditional pastoral activities, are deeply respected and cherished activities, of intergenerational cultural value – and they are also the source of more nutritious food than is available by other means. Importantly, the capacity to hunt, fish and gather offers protection against the vicissitudes of the market economy; in the challenging circumstances of a globalizing north, having multiple sources of income and direct access to traditional food sources provides an important safety net. At the same time, alongside these important advantages, it is significant that the subsistence economy is not capable of generating taxable surplus – indeed, it is founded in an ancient set of economic practices whose purpose is precisely to avoid generating a surplus of any kind, least of all a cash surplus. Stable, enduring, highly valued and important to community resilience, the twenty-first century subsistence economy cannot stand on its own, nor can it fund public sector social investment.

For different reasons, this is also true of the second (also stable and enduring) pillar of the northern economy, public investment. For reasons of geography and population distribution, most parts of the circumpolar north are distant from manufacturing centres; they depend upon costly-to-transport imported goods and food. Energy costs are generally higher than is the case in more temperate climates, and all forms of construction, from housing to roads to communications systems, are also relatively costly. Because populations are small, the ratio of public service jobs to overall population size is large. Supporting governing institutions, and indeed, all human life in the northlands at a standard reasonably close to that of southern parts of the circumpolar nations has long required substantial public expenditure. Most of this has come from metropolitan governments located well outside the region, and per capita costs of governance and social provision are generally significantly higher than the same for southern regions.

Most public expenditure, too, has been invested on the expectation that eventually it would 'pay off' in the return of taxes or other revenue from resource development and income-taxed labour. It has rarely done so, but public investment in northern governance and infrastructure is unlikely to be withdrawn or radically reduced. Each circumpolar nation has reasons other than revenue for seeking to sustain resident northern populations. Although they are vulnerable to fluctuations in political direction, in overview public service jobs, social transfers and public investments in social and material infrastructure can be counted on as a stable feature of the northern economy for the foreseeable future.[6]

As the other chapters in this book illustrate, a great deal is known about the effects of the third pillar, mineral exploration and development, on northern communities and regional economies.[7] Certain negative local effects are common to all regions: pollution from the by-products of mineral extraction and refining, disruption of wildlife patterns and destruction of habitat, social disturbances related to sudden infusions of cash and the rigors of fly-in, fly-out commuting, and –eventually – the social disturbance following on the 'bust' part of the boom and bust resource cycle, whether this is due to depletion of the resource being exploited, or declining global prices. The severity of these effects varies substantially from region to region, depending upon the strength of public regulation and local control. The positive local and regional effects of mineral development are also well known: increased availability of well-paid (relatively) local employment, small business opportunities, stimulation of infrastructure improvements and resource revenues to public authorities and sometimes to local Indigenous governments, creating an opportunity for capital accumulation and its ultimate reinvestment. Again, the degree to which these potential benefits is realized depends upon the regulatory framework and taxation arrangements.

In contrast to the subsistence economy and public expenditure, resource-based development is inherently unstable, sensitive to international fluctuations in demand and subject to depletion. In the second case, it becomes necessary to find and open another mine or well; in the first, it is necessary to wait out the period of low prices before beginning again. In recognition of this instability, northern governments in many parts of the world have over the years provided some support for hunting and traditional pursuits, while aiming to promote economic diversification into such sectors as the fishery, forestry, tourism and arts and crafts.[8]

For the purposes of this analysis, this heterogeneous sector can be understood as the fourth pillar of northern economies. Although there are important differences among, for example, sale of art carvings, northern lights tourism and commercial fishery, they are similar in that they are based on exploitation of renewable resources, tend to be labour-intensive, require specialized skills and are unlikely to become the dominant source of economic growth in any given northern region.[9] All provide a reasonably stable and environmentally benign livelihood for some, have the potential to generate a taxable surplus, and are capable of contributing cash flow to the subsistence economy 'safety net'.[10]

Almost all northern regional economies are shaped by these four pillars, interacting in similar ways. The balance over time varies, something that is recognized by all northern governments, whether they are regional, local, Indigenous or the governments of nation-states. It remains, then, to consider what further research is needed to support citizens' and their governments in considering how best to build upon the pillars to face the climactic and demographic challenges that lie ahead.

What else do we need to know?

As the circumpolar north is opened to additional maritime traffic and mineral exploration, and as the northern impacts of climate change accumulate in a world where there will be many competing demands on national governments to respond to these impacts everywhere, it will be advisable for northern governments to use fiscal and human resources with wisdom, foresight and parsimony. The global demands will be great, and humanity's capacity to respond to them will be challenged, to say the least. It is a good moment to develop a plan.

It should be a plan based on a frank assessment of how earlier economic development schemes have worked out. Many northern countries, including Canada, have relied heavily upon the promotion of mineral discovery, development and export, on the expectation that jobs generated in this way would benefit the regional economy while revenue from resource exports would accrue to the national coffers and balance of payments. Contrary to expectations, nowhere except in Fenno-Scandia did promotion of mineral development stimulate the growth of a reasonably self-sustaining export-oriented resource-producing economy capable of generating substantial public revenue while providing adequate employment opportunities for the population. We have learned that the mineral exploitation 'pillar' cannot hold up the society on its own. In most of the circumpolar nations, the transfer of public funds has been important: governments were 'transplanted' to northern regions and have been financed by transfers from the centre. In Canada, Alaska and a few other countries, Indigenous peoples mobilized to interrupt this pattern, leading to redistributions of political power and some resource revenues to Indigenous organizations and governments. Indigenous people have maintained the subsistence economy all the while, though its viability has been severely challenged in certain locations. The expectation was for many decades that the subsistence sector would whither away, as wage income replaced harvesting. This has not happened; rather, the subsistence sector has demonstrated substantial flexibility in incorporating income from the other sectors. Many northern governments have recognized the importance of subsistence sector with small subsidy programs, while they have introduced an array of measures designed to support the 'diversifying' renewable sector.

If this general picture of how northern community economies have been working for the last few decades is roughly accurate, how stable is it? How can the people who live in such settings prepare for the challenges ahead? What knowledge do they require?

There is a need for knowledge about how governments and citizens may best create *mutually reinforcing* economic measures. In the north as in most parts of the world, there is a striking gap between public policies addressed to economic development and others directed to social development and environmental protection. 'Economic development' is usually understood to entail measures to create jobs and business opportunities. 'Social development' includes expenditures on education, health and poverty relief. 'Environmental protection' focuses on pollution and increasingly on climate change impacts. These are often treated as separate fields. For example, in Alaska, state-mandated regional development organizations (Alaska Regional Development Organizations – ARDORs) develop and implement regional economic development strategies. In the words of the

Bering Strait 2013–18 strategy, '[t]he ultimate goal of this process is to provide for sustainable and responsible development that benefits the people of the Bering Strait Region and improves the quality of life through economic opportunity' (Kawerak 2013: 4). There is no mention, however, of climate change mitigation or adaptation to impacts in the regional economic development plans. Rather, the state government operates the separate Alaska Climate Change Impact Mitigation Program, which supports communities to assess impacts and develop plans to respond. The Government of Greenland, similarly, has developed a thorough oil and mineral development strategy (Greenland 2014) that makes no mention of climate change, even though Greenland is the site of some of the most dramatic changes related to global warming, and the Greenland government has been active internationally on this issue. As Southcott concludes in Chapter 15, there is as yet scant evidence of the impacts of the changing climate on resource development, though more is known about its effect on subsistence production and infrastructure.

A similar pattern of isolated consideration of economic development, on one hand, and mitigation of and adaptation to climate change on the other, is evident throughout the circumpolar region. Given the eventual likely connection between economic activity –and particularly mineral extraction—and global warming, it is important that policies and state-mandated measures bring the two processes into the same conversation. In this regard, targeted multidisciplinary research can be helpful in identifying 'least worst' and 'most promising' measures.[11]

There is also a need for practical means to develop *public policies 'from the ground up'* – measures that respond to the dynamics of small regional and community economies. In the last decade, substantial knowledge has been generated by northern community members and associated scholars about the detailed dynamics of their community economies as they are affected by resource development projects (e.g. Gibson 2008; Parlee 2006; Parlee and Manseau 2005 and climate change (see http://climatechangenunavut.ca/en/community/arviat)). Increasingly, northern communities themselves are developing substantial local research capacities and sophisticated means of communicating among themselves (e.g. McGwinn 2015; Kennedy Dalseg and Abele 2015). It seems very likely, however, that some aggregation beyond the community level will be necessary for effective long-term development planning. Regional analyses will help to bring together the forces necessary for social and economic resilience and durability.

Given the weight of history – the legacy of colonization of the global northlands by southern-based institutions and governments – it is not self-evident that it is even possible to shift the policy onus from large exogenous forces to community and regional governments, or at least to put these authorities on an equal discursive and action basis with larger governments and corporations. It is obvious, though, that this would be highly desirable for protecting the integrity of the successful aspects of northern regional and community economies and for ensuring resource development provides greater benefits to these economies. The experience of federations (the European Union and federal countries) and of public regulatory hearings in various countries might be helpful bases for further study.

To date, there has been little scholarly attention devoted to the role that innovation might play in meeting the northern challenges ahead (Petrov 2005 is a rare exception). Formulating northern responses to growing accessibility, demographic change and the warming climate raises deep technological challenges. Responses must be synchronized with necessary and probably inevitable changes in the shape of national economies. It seems inevitable that nation-states will move away from emphasis on resource extraction towards remediation, conservation, alternative energy development and sophisticated recycling of precious material. Along the way, there is an opportunity to move towards technological leadership in what some have called 'the second economy' – a software-driven digital transformation of many aspects of the economy and social life.[12] Building upon such projects and opportunities requires a blend of social and technological innovation, and the development of durable linkages among economic sectors and across social purposes. Experience teaches that at least some of the necessary innovation will take place in northern communities (Petrov 2005; Abele 2009; Kennedy Dalseg and Abele 2015), but some of the impetus and certainly funding for research and development programs that build on this experience, adding scientific value, must come from national governments and international organizations.[13] An important question in this regard is how can the short-term benefits from extractive resource development be effectively used for these types of innovations, given the imperatives of electoral cycles and the power of corporate capital.

And then there is the question of *effective policy instruments.* In most regions, public sector expenditures are a relatively untapped tool for shaping economic growth. Little is known in detail about the impact of these important cash infusions. Purchasing policies have been developed in many regions to favour northern businesses, and often hiring policies favour northern or northern Indigenous residents. Also, in Nunavut, a decade-long experiment to use public sector decentralization as a means of economic decentralization and community development has been underway, without major success (Hicks and White 2015). These measures deserve further study, in light of what is known and what is not known about the impact of public investments of all kinds on the organization of social life, and economic activity, in northern communities and more generally, in northern regions. If sufficient coordinating capacity were available, necessary public expenditures could be rendered more effective if they were deployed in the context of an overall development strategy that coordinated use of all of the economic pillars discussed earlier.

Likely, also, more could be achieved by program redesign. To take a simple example from Canada's Inuit territories: infrastructure deficits play an important role in creating or sustaining social dysfunction in northern communities.[14] Crowded and inadequate housing, for example, leads to respiratory illness, poor school attendance, poor performance at work and other problems. These in turn add to stresses on families that cause mental and physical health difficulties. While comfortable, efficiently heated homes will not on their own solve the north's high school dropout rates, they will very likely help families come to terms with these. A coordinated research and development infrastructure strategy, involving experimental construction associated with coordinated education and training programs, is needed to address the north's housing problems and the medium-term civil engineering challenges that climate change adaptation will bring. Such a strategy would also provide education and employment opportunities in locations across the north. These would be stable positions requiring the skills of engineers, environmental scientists, social development specialists, educators, contractors and tradespeople. Currently a patchwork of ad hoc programs fund short-term worker training and housing construction in the north, while southern universities and federal government researchers conduct the necessary research. What is missing is continuity and coordination. Yet virtually no scholarly research effort has been devoted to understanding the dynamics of northern governments or northern governance, so that the reasons for their failure to develop such approaches can be better understood.[15]

Last thoughts

The older paradigms of development that characterized the period of colonization in all of the circumpolar countries are no longer adequate. In Russia, Canada, Greenland and Alaska, they amounted to massive social experiments that generated some (mostly exported) wealth, but also led to substantial disruptions and hardship. The legacy of these policies remain, in the locations and infrastructures of northern towns and cities, and in the lived experiences and memories of northern residents. And new challenges have appeared. It is time for all northern jurisdictions – Indigenous organizations and governments and territorial governments – to begin the important work of developing an innovative, far-sighted, community-based program for northern economic change. They must take into account the north's rapidly changing environmental circumstances, and impending structural changes in the world and the Canadian economies. In order to do this, a broad process of analysis and reflection is necessary, engaging northern citizens and the best research resources available – including the best insights of the global experience with regional economic development. It is in this type of a context that we can start to understand how to better utilize resource development to help northern communities become more sustainable.

Acknowledgements

I thank David Muddiman for superlative research assistance, and Sheena Kennedy Dalseg for her customarily perceptive comments on an early draft. Kennedy Dalseg, Josh Gladstone, Mary Ellen Thomas, Natan Obed and Deb Simmons have all nudged me along, though they may not agree with the resulting analysis, as have ReSDA colleagues David Natcher, Brenda Parlee and Chris Southcott. For his steadfast companionship and ever-intelligent questioning, I am as always deeply grateful to George Kinloch.

184   *Frances Abele*

Notes

1 For example, Betts 2005 offers a fascinating account of historical regions (which he refers to as focal economies), using archaeological research to arrive at an account of changing patterns of human use and residence in the centuries before European arrival.
2 www.tlicho.ca/tlichogovernment.
3 www.eu-polarnet.eu/about-eu-polarnet/. EU-PolarNet is a new and very large coalition of research institutions based in the European Union. For many years, Arctic researchers have formed working networks, some dating from the International Polar Year 2007–09, such as the Survey of Living Conditions in the Arctic (SLiCA – www.arcticliving conditions.org/). Others were formed to study particular problems, such as the Resources and Sustainable Development in the Arctic (ReSDA) network (http://yukon research.yukoncollege.yk.ca/resda/) and Arctic-FROST: Arctic FRontiers Of SusTainability: Resources, Societies, Environments and Development in the Changing North (https://arctic-frost.uni.edu/about/)
4 Except where otherwise noted, Glomsrød and Aslaksen 2009 provides a panoptic organization of the evidence for these statements. See also chapters in Heininen and Southcott 2010.
5 For present purposes, the terms 'harvesting' and 'subsistence' are used interchangeably. I do not emphasize the difference between subsistence and the mixed economy, since these are not found separately in contemporary life: subsistence requires a mixed economy, since it requires cash subsidy.
6 The post-soviet years in Russia are an exception to this generalization, where drastic reductions in formerly substantial levels of subsidy led to depopulation on a dramatic scale. Barring global economic collapse, it is unlikely that this experience will be repeated; indeed, as I shall argue late in this essay, with the changes wrought by climate change, it is likely that per capital northern expenditures will increase in most northern regions.
7 The literature on this question is massive. For useful syntheses of the Canadian experience, see Rodon and Levesque in this book; Keeling and Sandlos 2015; Gibson 2008.
8 For example, see the West Kitikmeot/Slave Study at www.enr.gov.nt.ca/sites/default/files/reports/wkssfinal.pdft, and the economic development strategies published by each territorial government.
9 An important exception to this is the northern fishery, which is some regions makes a large contribution to regional income and gross domestic product.
10 We have found no studies that analyse these sectors' overall contribution to community or regional economies.
11 In this regard, Norway is once again leading. As we go to press, the Commission on Green Competitiveness released its report, envisioning use of financial and other instruments for the transition to a thriving low carbon economy. www.regjeringen.no/en/aktuelt/mottok-rapport-fra-ekspertutvalget-for-gronn-konkurransekraft/id2518126/
12 See Wolfe 2016 for a discussion of the second economy and Canadian development.
13 There is also substantial potential for trans-northern policy transfer and mutual learning. In 2014, Norway published a new northern development strategy. Two of the priorities for this strategy are building a knowledge economy, and strengthening infrastructure, measures that entailed a significant increase in government spending in the Norwegian High North.
14 See for example Battle and Torjman 2013; I am not arguing, of course, that these are the sole cause of distressing health and other indicators in some northern communities; infrastructure improvements, however, are likely to be an important force for improvement, because of their direct impact on individual, family and community well-being.
15 Major exceptions to this are the publications of Graham White and Jack Hicks. See for example Hicks and White 2015; Hicks 2001; Cameron and White 1995.

References

Abele, F. (2009). 'The Past, Present and Future of Northern Development.' In F. Abele, T. J. Courchene, F. L. Seidle and F. St-Hilaire (Eds.), *Northern Exposure: Peoples, Powers and Prospects in Canada's North.* Institute for Research on Public Policy and McGill-Queen's University Press.

Abele, F. (2016). 'The North in New Times: Revising Federal Priorities.' In J. Higginbotham and J. Spence (Eds.), *North of 60: Toward A Renewed Canadian Arctic Agenda* (pp. 5–11). Waterloo: Centre for International Governance Innovation.

Betts, M. W. (2005) 'Seven Focal Economies for Six Focal Places: The Development of Economic Diversity in the Western Canadian Arctic.' *Arctic Anthropology*, 42 (1): 47–87.

Bjerregaard, P., T. K. Young, E. Dewailly and S. O. Ebbesson (2004). 'Indigenous Health in the Arctic: An Overview of the Circumpolar Inuit Population.' *Scandinavian Journal of Public Health*, 32 (5): 390–395.

Cameron, K. and G. White (1995). *Northern Governments in Transition: Political and Constitutional Development in the Yukon, Nunavut and the Western Northwest Territories.* Montreal: Institute for Research on Public Policy.

Dalseg, S. K. and F. Abele (2015) 'Language, Distance, Democracy: Development Decision Making and Northern Communications.' *Northern Review*, 41.

Dawkins, C. J. (2003) 'Regional Development Theory: Conceptual Foundations, Classic Works, and Recent Developments.' *Journal of Planning Literature*, 18(2): 131–172.

Gibson, V. V. (2008). 'Negotiated Spaces: Work, Home and Relationships in the Dene Diamond Economy.' Doctoral dissertation. University of British Columbia.

Glomsrød, S. and I. Aslaksen (Eds.) (2009). *The Economy of the North 2008.* Oslo: Statistics Norway. www.ssb.no/a/publikasjoner/pdf/sa112_en/sa112_en.pdf.

Greenland, Government of. (2014). *Greenland's Oil and Mineral Strategy 2014–2018.*

Haalboom, B. and D. C. Natcher (2012). 'The Power and Peril of "vulnerability": Approaching Community Labels with Caution in Climate Change Research.' *Arctic*, 65(4): 319–327.

Heininein, L. and C. Southcott (Eds.) (2010). *Globalization and the Circumpolar North.* Fairbanks: University of Alaska Press.

Henriksen, J. B. (Ed.) (2010). 'Sami Self-determination: Autonomy and Economy – the Authority and Autonomy of the Samediggi in the Health and Social Services Sector.' *Galdu Cala – Journal of Indigenous Peoples Rights*, 2. Kautokeino: Resource Centre for the Rights of Indigenous Peoples.

Hicks, J. and G. White (2015). *Made in Nunavut: An Experiment in Decentralized Government.* Vancouver: UBC Press.

Keeling, A. and J. Sandlos (2015). *Mining and Communities in Northern Canada: History, Politics and Memory.* Calgary: University of Calgary Press.

Kawerak Community Planning and Development Program and The Bering Strait Development Council (2013). *Bering Strait Comprehensive Economic Development Strategy 2013–2018.*

Keskitalo, C. (2007). 'International Region-building: Development of the Arctic as an International Region.' *Cooperation and Conflict*, 42(2): 187–205.

McBeath, J. and C. E. Shepro (2007). 'The Effects of Environmental Change on an Arctic Native Community: Evaluation Using Local Cultural Perceptions.' *The American Indian Quarterly*, 31(1): 44–65.

McGwinn, K. (2015). 'Resilience in the Face of Threat.' *The Arctic Journal*, November 22, 2016.

Natcher, D. C. (2009). 'Subsistence and the Social Economy of Canada's Aboriginal North.' *Northern Review*, 30: 83–98.

Neffke, F., M. Henning and R. Boschma (2011). 'How Do Regions Diversify Over Time? Industry Relatedness and the Development of New Growth Paths in Regions.' *Economic Geography*, 87(3): 237–265.

Parlee, B. L. (2006). Dealing with Ecological Variability and Change: Perspectives from the Denesoline and Gwich'in of Northern Canada. Doctoral dissertation, University of Manitoba.

Parlee, B. L. (2015). 'Avoiding the Resource Curse: Indigenous Communities and Canada's Oil Sands.' *World Development*, 74: 425–436.

Parlee, B. L. and M. Manseau. (2005). 'Using Traditional Knowledge to Adapt to Ecological Change: Denésǫné Monitoring of Caribou Movements.' *Arctic*, 58(1): 26–37.

Petrov, A. N. (2012). 'Redrawing the Margin: Re-examining Regional Multichotomies and Conditions of Marginality in Canada, Russia and Their Northern Frontiers.' *Regional Studies*, 46(1): 59–81.

Sandlos, J. and A. Keeling (2012). 'Claiming the New North: Development and Colonialism at the Pine Point Mine, Northwest Territories, Canada.' *Environment and History*, 18(1): 5–34.

Sillanpaa, L. (1994). *Political and Administrative Responses to Sami self-Determination: A Comparative Study of Public Administrations in Fennoscandia on the Issue of Sami Land Title as an Aboriginal Right.* Helsinki: Finnish Society of Sciences and Letters.

Smith. L. C. (2011). 'Agents of Change in the New North.' *Eurasian Geography and Economics*, 52(1): 30–55.

Stöhr, W. B. (1996). 'Local Synergy as an Explanation for Innovation in Peripheral and Frontier Areas.' In Y. Gradus and H. Lithwick (Eds.), *Frontiers in Regional Development.* Lanham, MD: Rowman and Littlefield.

United Nations (1987). *Our Common Future – Brundtland Report.* Oxford: Oxford University Press.

Usher, P. J., G. Duhaime and E. Searles. (2003). 'The Household as an Economic Unit in Arctic Aboriginal Communities, and Its Measurement by Means of a Comprehensive Survey.' *Social Indicators Research*, 61: 175.

White, G. (2001). 'And Now for Something Completely Northern: Institutions of Governance in the Territorial North.' *Journal of Canadian Studies*, 35(4): 80.

Wolfe, D. A. (2016). *A Policy Agenda for the Digital Economy.* Munk School of Global Affairs. Policy White Paper Series Paper No. 2016–02. http://munkschool.utoronto.ca/ipl/files/2016/02/IPL-White-Paper-No-2016–2.pdf.

Wolfe, R. J. and R. J. Walker (1987). 'Subsistence Economies in Alaska: Productivity, Geography, and Development Impacts.' *Arctic Anthropology*, 24(2): 56–81.

Zimmerling, S. (2016). A Review and Analysis of Canadian Indigenous Mining Policy. Master's Thesis, University of British Columbia.

10 Knowledge, sustainability, and the environmental legacies of resource development in Northern Canada

Arn Keeling, John Sandlos, Jean-Sébastien Boutet, Hereward Longley, and Anne Dance

Since the early twentieth century, grand visions of industrial development in the circumpolar Arctic have at times captured the attention and imagination of policy makers and the general public. The periods of high salience for northern resource development issues are many: explorer Vilhjalmur Stefansson's attempt in the 1920s to convince the Canadian government and the public that a "Polar Mediterranean" could be built in the Arctic based on agriculture and the expansion of cities; post-WWII endeavors by successive national governments to build their own version of a modern north; the resulting hydrocarbon and mining developments in Arctic locations ranging from Alaska to Russia and Scandinavia; and more recent comprehensive planning exercises such as Quebec's Plan Nord promoting a bright future for the region based on the exploitation of non-renewable resources. By the 1970s, questions surrounding the economic and environmental impact of these projects on northern communities, particularly Indigenous communities, came to the forefront in North America in part due to the public airing of the issue that emerged from Justice Thomas Berger's Mackenzie Valley Pipeline Inquiry, as well as from the growing environmental movement in the south (Berger, 1977; Nassichuk, 1987; Page, 1986; Sabin, 1995).

In Canada, the idea that development might not be wholly positive for northerners evolved (in the academic realm, at least) into an Innisian critique suggesting that the economic benefits of northern development were largely exported out of the North, leaving behind economic disruption (due to impacts on Indigenous hunting and trapping activities) and environmental devastation as its legacy (Rea, 1968; Zaslow, 1988). If this early discourse suffered from its tendency to render Indigenous people as passive victims of development, unable to adapt creatively to changing environmental and economic circumstances, it did permanently disrupt the idea that large-scale resource development embodies progress, with no short- or long-term environmental and social consequences for Indigenous and northern communities. Drawing a rough mixture of the staples thesis, dependency theory, and political ecology frameworks, authors have conceptualized resource peripheries in Canada and throughout the circumpolar north as "deeply contested spaces," with intersecting industrial, environmental, Indigenous, and geopolitical dimensions at play (Hayter, Barnes, and Bradshaw,

2003: 15; Keeling and Sandlos, 2009). Environmental historians joined the critique of the "industrial assimilation" (Piper, 2009) of northern environments around the globe in the twentieth century, highlighting the links between resource exploitation, environmental degradation, and the social and economic dislocation of local communities (e.g., Coates, 1991; Josephson, 2002; Morse, 2003; Sandlos and Keeling, 2012a).

Despite the wealth of research on development impacts in Northern Canada, one major gap we highlight is the paucity of studies examining the long-term environmental legacies of development, their ongoing impacts on communities, and the involvement of northerners—especially Indigenous communities—in their mitigation, management, and remediation. In order to invite comparisons with historical and community impacts of resource development across the circumpolar world, here we analyze the wider literature on the long-term environmental legacies of industrial development in the Arctic, particularly hydrocarbon extraction, abandoned mines, and other developmental impacts. Rather than conceptualize the legacies of modern development as engineering problems requiring technical solutions (as predominate policy and scientific approaches do), we propose that more work is needed on how to incorporate local community and Indigenous perspectives and participation into the remediation challenges of these environmental hot spots.

Technical knowledge tends to dominate and orient practices of environmental remediation—activities that are, as in the exploration and exploitation stages, deemed to be largely the domain of Western expertise. Whether a particular development is ongoing (like the Norilsk metallurgical complex in Siberia or oil production at Norman Wells in the Northwest Territories), completed (like the closed mines on Svalbard or in the Canadian High Arctic), or ephemeral (like the Canol pipeline from Norman Wells to Alaska), the "post-industrial landscape scars" (Storm, 2014) of historic resource activities may remain evident in the landscape and environment for long after. This is because of the slow recovery rates of many Arctic ecosystems, but also because of the material persistence of the environmental changes themselves (for instance, toxic contaminants, which may move and accumulate in the environment and in the Arctic biota [AMAP, 1997; 2002]).

Considerable research activity in recent years, whether independent scientific studies or research undertaken for assessment and regulatory processes, has focused on the planning, construction, and operational phases of large-scale developments (see chapters in this volume). We suggest that scholars have been less inclined to address the *long-term environmental legacies* of such developments, whether concluded or continuing, particularly the extent to which local and Indigenous knowledge and community perspectives are included in environmental remediation processes. Answering these questions will provide insight into the ongoing environmental challenges associated with historic resource development, particularly extractive industries, and help guide decision-making in the future. In this chapter, we explore first the origin, nature, and scope of environmental legacies and remediation challenges from extractive development. We also discuss the shortcomings of the policies and practices of remediation proposed for these sites, particularly highlighting their inattention to local and Indigenous knowledges associated. A final section examines potential interpretative and conceptual avenues for integrating community experiences and concerns, highlighting research initiatives from around the circumpolar region including frameworks focused on cumulative effects assessment, resilience and vulnerability, and the politics of knowledge in remediation.

Environmental legacies of northern resource development

The long-term environmental legacies of resource use and development in the Arctic pose important management and research challenges, particularly through the interaction of different resource uses and impacts at various scales. At the landscape scale, the environmental effects of development include surface disturbance from extractive activities and infrastructure, including roads, shipping lanes, pipelines, mines, drilling sites, impoundment facilities, and dams. In addition to disrupting fragile Arctic vegetation, these surface works may impact permafrost (thermokarst), alter hydrological patterns, and disrupt freshwater or even sea ice formation, resulting in increased erosion and/or sedimentation from inundation, flooding, slumping, or other processes (Walker et al., 1987). As Forbes, Ebersole and Strandberg (2001) note, although the overall spatial extent of these anthropogenic disturbances may be small, they nevertheless generate a patchy landscape that may significantly affect local wildlife and vegetation patterns, and persist for long periods in this low-energy, slow-recovering ecosystem. Linear developments such as roads and pipelines have significantly larger direct effects along their corridors, but also beyond as wildlife, hunters, and herders seek to either avoid these installations, or to use roads for increased access to fish and wildlife resources (Klein and Magomedova, 2003; Herrmann et al., 2014; LeClerc and Keeling, 2015). Physical environmental legacies at closed developments abound in the Arctic, in the form of settlements (depopulated or nearly so), defunct production and transport facilities, and abandoned resource sites (and their wastes), from the Kennecott Mine in Alaska to the Polaris and Nanisivik mine sites in the Canadian High Arctic to the Russian mining ghost town of Pyramiden on Svalbard.

The waste products generated by resource exploration and development activity have also left a significant environmental legacy in parts of the Arctic. Mining activities generate vast amounts of waste in the form of overburden, waste rock, and tailings, which may be stored at the surface or deposited in nearby water bodies (Lottermoser, 2010). Mine tailings have the potential to pollute local watercourses through physical erosion, acidification, and the release of heavy metals or trace process chemicals. The controversial practice of using lakes or fjords for tailings disposal, such as at Maarmorilik in Western Greenland, may result in long-term contamination of these environments (Sondergaard et al., 2011). Hydrocarbon exploration and production produces significant polluting wastes, including both hydrocarbons themselves as well as chemicals in muds and brines used in drilling, or waters (often saline and/or sulfurous) brought to the surface during drilling (Walker et al., 1997). Small spills from hydrocarbon developments have been commonplace in the Arctic, and pipeline leaks also release hydrocarbons (AMAP, 1997, Poland, Riddle and Zeeb, 2003). Contamination has also affected the local environment near mineral processing and oil and gas installations, including air pollution, fuel spills, and community wastes, such as at the notoriously polluting Norilsk and Kola Peninsula smelters in Russia (AMAP, 2007; 1997). Mineral resource exploration and development processes also generate wastes, via detritus such as drill cores, seepage from drill holes, fuel dumps, and abandoned equipment (Duhaime and Comtois, 2003; Duhaime, Bernard and Comtois, 2005). Although contaminated sites in the Arctic may be relatively fewer in number, as Poland et al. (2003: 377) concluded, "It is clear that the Arctic has very seriously polluted sites that are as bad as sites anywhere else in the world."

Three key issues related to these environmental legacies stand out as warranting further investigation. First, there are opportunities to explore the intersection of past environmental changes associated with large-scale development, and those associated with other natural resource and environmental issues in the Arctic. There is a well-established scientific literature examining the impacts of resource exploitation on wildlife, particularly on species important to northern communities, and making conservation recommendations (e.g., Johnson et al., 2005; Klein and Magomedova, 2003; Herrmann et al., 2014). But often, these studies are restricted to a single development and/or target species (referred to in environmental assessment processes as Valued Ecosystem Components); less frequently are broader questions around the legacies of historic resource utilization for sustainable regional development considered (Chance and Andreeva, 1995; Caulfield, 2000; Klein, 2000). Yet these legacies not only influence environmental conditions in the present, but also offer "examples and experiences of value in assessing proposed new projects and for designing guidelines for their development to avoid unnecessary environmental, social and economic impacts" (Klein, 2000: 92). Research examining the interaction of reindeer herders and the oil industry in the Yamal Peninsula region of Russia (discussed further below) provides a useful example of research into the legacies of resource conflict and coexistence (Stammler, 2002). Another recent collection, *Mining and Communities in Northern Canada* (Keeling and Sandlos, 2015), explored Indigenous community responses to mineral development at a series of sites across the Canadian Arctic and subarctic.

Second, attention to the cross-scale effects of these environmental legacies would provide greater insight into the challenges of their management. Although the long-term environmental legacies of any given resource use may be localized, as noted above these local impacts interact with other resources at different temporal, ecological, or management scales. For instance, pipeline development impacts include their interaction with complex, long-range caribou migration cycles and spatial patterns (Nuttall, 2010), while mines may be voracious consumers of energy and renewable resources far beyond the mine site itself (Keeling, 2010; Piper, 2009). In terms of temporal scale, long-term legacies interact in complex ways with changing resource management and use practices and other drivers of environmental change (including climate change), presenting difficult management challenges in themselves. For instance, the extremely long-term burden of care and monitoring of tailings management facilities, particularly those posing toxic or radiological hazards, presents unique environmental management challenges. In an Arctic context, climate change threatens to disrupt permafrost regimes and runoff from in tailings facilities, resulting in potential physical instability and the release of contaminants (Pearce et al., 2011; Prowse et al., 2009). Similarly, at the Giant Mine in Yellowknife, Northwest Territories, the stabilization and long-term storage of 237,000 tons of toxic arsenic trioxide waste from decades of gold mining poses a perpetual care scenario with complex technical and financial dimensions; nevertheless, the environmental hazard posed by arsenic exposures is borne by the local community (Sandlos and Keeling, 2012b). The critical question remains: how do local communities and/or national authorities address the legacy effects of resource exploitation that appear distant in time and space from the original development?

The problem of managing the cross-scale effects of development legacies such as the Giant Mine is further reflected in a third key challenge, that of environmental remediation and resource (re)development at legacy sites. The cyclical and volatile nature of Arctic resource economies means that particular resource sites may be subject to sudden closure and abandonment, often leaving behind considerable environmental problems (Keeling, 2010). The environmental legacies of former resource sites include contemporary responses to their impacts, including cleanup of abandoned/contaminated sites. Both new and legacy sites pose difficult technical and governance challenges; efforts to remediate them are shaped by jurisdictional confusion and inertia, site-specific considerations, and reclamation goals and strategies that are vague, contested, and difficult to finance and execute (Dance, 2015). Although there is a thriving technical literature on environmental remediation in the Arctic (Olsen, 2001; Udd and Bekkers, 2003; Poland et al. 2003; Udd and Keen, 1999), the related political and socio-economic questions remain poorly documented (Pepper, Roche and Mudd, 2014; Dance, 2015). Indeed, the scope of the remediation challenges in the Arctic is not well understood: inventories of abandoned mines, exploration sites, and toxic hotspots have appeared (Mackasey, 2000, Worrall et al., 2009, Alaska Department of Natural Resources, 2012), but a better sense of the national and circumpolar scope of the problem is needed (AMAP, 1997: 2002).

The incorporation of local knowledge and participation in remediation policy and practice, particularly involving Indigenous people, represents an important but understudied aspect of this issue (Assembly of First Nations, 2001; McBeath and Shepro, 2007; NOAMI, 2003; Sistili et al., 2006).[1] Some scholars argue that environmental assessment and monitoring of all stages of large-scale resource development projects can in many cases confine Indigenous knowledge primarily to wildlife impacts, placing emphasis on the generation of usable statistics from TEK data rather than considering Indigenous governance, cultural values, and cosmologies, thus reinforcing an inequitable distribution of political power between developers and Indigenous communities (Ellis, 2005; White 2006; Houde, 2007; O'Faircheallaigh and Corbett, 2005; Stevenson, 1996). Government and industry too often regard large-scale, for-profit, capital-intensive activities as falling outside the realm of competency typically attributed to Indigenous people. Such a constricted view of Indigenous knowledge thus "makes it easy for scientists and resource managers to disregard the possibility that Indigenous peoples might possess distinct cultural perspectives on modern industrial activities such as logging and mining" (Nadasdy, 2003: 120–121). How to incorporate Indigenous communities more effectively as co-managers of resource projects at the remediation stage—whether through the application of TEK and/or local political control over projects—has become an even more significant research question as increasing numbers of environmental legacy sites are produced and identified throughout the Canadian North (Tsuji et al., 2001; Parlee, 2012; Haalboom, 2014; 2016; Sandlos and Keeling, 2016a). This disconnect between the knowledge held by scientists and community members, resulting from divergent worldviews or public understandings of technical issues, can lead to conflict and uncertainty, undermining the public acceptability or "social license" for remediation solutions.

Alongside remediation, redevelopment may bring new life to formerly derelict or degraded resource landscapes in the Arctic. Public and private expenditures on remediation represent a significant form of economic activity at these sites, while periods of rising commodity prices have in recent years attracted capital investment to former resource sites that had been considered exhausted or uneconomic (Bouw and Ebner, 2011; Kivinen, 2017). Whether subject to remediation, redevelopment, or both, these reanimated "zombie" resource sites (Keeling and Sandlos, 2017) may cause conflict with local communities and land users (Boutet, 2015). They also raise conceptual questions around the categories of waste and value at abandoned sites, as well as a series of important research questions concerning the legacies of past development, including: What are the socioecological implications of renewed disturbance/development activities at resource sites? What are the environmental conflicts and challenges engendered by restoration and/or remediation activities? What are the best practices and policy mechanisms of cleaning up Arctic developments? How might these activities positively/negatively interact with other resources and local communities?

Extractive industries and environmental legacies: interpretive frameworks

The existing literature does suggest several potential frameworks for tackling the critical issues related to resource development, environmental legacies, and communities in the Arctic. First, taking a concept from practices in environmental assessment (see Noble, this volume), the analysis of cumulative effects of resource development may provide a framework for exploring the intersections of current and proposed projects with the existing environmental and social legacies of development. The practice of cumulative effects assessment and management, now just "growing out of its infancy" (Canter and Ross, 2010: 267), typically aims at the assessment and monitoring of multiple development projects at a resource site and/or at the regional scale, and their projected impacts on valued ecosystem and socio-economic components such as wildlife (eg., Walker et al., 1997, Johnson et al., 2005, Spaling, Zwier and Ross, 2000). Although typically focused on immanent and/or future development patterns and impacts, the notion of cumulative effects might be extended backwards in time, to account for the socioecological legacies of past development at resource sites (such as the "zombie mines" or abandoned sites noted above). This is not to suggest researchers simply replicate existing or proposed cumulative effects monitoring practices for the Arctic; rather, the *concept* of cumulative effects itself may be critically evaluated and applied to understanding the social and ecological legacies of *past* developments and their implications for present and future development. In doing so, it will be critical to incorporate not only those metrics and indicators of physical environmental changes typically used in environmental assessment, but also analysis of the cumulative socio-cultural impacts of resource development and landscape change on local communities, including and especially Indigenous communities (Angell and Parkins, 2011; Ehrlich, 2010; Paci and Vilebrun, 2005; Klein, 2000). Such an approach should also incorporate Indigenous knowledge as a substantive and meaningful contribution to understanding, monitoring and managing both environmental remediation and the cumulative effects of past and present developments (Sandlos and Keeling, 2016a).

One compelling model of the interdisciplinary investigation of cumulative past, current, and future effects of large-scale development is the research undertaken in the Yamal-Nenets region under the Environmental and Social Impacts of Industrial Development in Northern Russia (ENSINOR) project. Combining quantitative ecological studies of environmental impact and change related to oil and gas developments in the region with qualitative research involving Indigenous herders and oil industry workers, this project (and its successor studies) has provided important insights into how "historical experience and current Nenets agency could serve as a stable basis to continue the decades-old co-existence of industrial development and nomadic pastoralism" (Forbes et al., 2009: 22047; see also Kumpala et al., 2011; Stammler, 2005). Notably, this collaborative research also resulted in the crafting and adoption by representatives of industry, local authorities, herders, and researchers of a "declaration of coexistence" outlining in general the terms of social and environmental co-operation in the region ("Declaration . . .," 2007). Such interdisciplinary and collaborative explorations of the cumulative environmental legacies of development would doubtless contribute significantly to the understanding of their complex effects in other regions, such as Northern Alaska or the mining regions of the Northwest Territories and the Quebec-Labrador peninsula.

A similar interdisciplinary exploration of cumulative, long-term environmental legacies was undertaken by the LASHIPA (Large-Scale Historical Exploitation of Polar Areas) Project, which deployed geographers, historians, and archaeologists in an investigation of whaling, mining, and other historical resource developments in the far North and South Atlantic Ocean. For instance, the LASHIPA team con-nected material landscape change (including human settlements and environmental impacts) at resource sites on Svalbard (Spitsbergen) with an analysis of the political economy of development, colonial and geopolitical strategies, and far-flung networks of actors and technologies (Avango et al., 2011; Hacquebord, 2012; Avango, 2017). The LASHIPA project's approach (and that of its successor project, Resource Extraction and Sustainable Arctic Communities [REXSAC]) points to the fruitful possibilities of comparative studies of industrial legacies in the circumpolar Arctic (cf. Espiritu, 1997; Midgley, 2012; REXSAC, 2017).

As the work of the ENSINOR project in particular suggests, given the deep interconnections between the Arctic environment and the society, culture and well-being of northern people, perspectives on the vulnerability, and resilience of socioecological systems could provide important insights into how local communities have coped or are coping with the legacies of development (Forbes et al., 2009; Chapin III et al., 2004; Stammler, 2002; 2005). This framework links questions of anthropogenic environmental change with questions of local/Indigenous knowledge, socio-cultural adjustment, and adaptation to change (for instance, by highlighting changing uses of post-industrial landscapes). While attempts to model adaptation and vulnerability conditions may be of dubious utility in the widely varying and dynamic contexts of Arctic communities and environments (cf. Robards and Alessa, 2004), there are nevertheless potential insights to gain from the elaboration of these concepts in the literature on environmental change. As Nuttall observes, historical research on resource development can provide "greater understanding of the vulnerability of small-scale societies and whether peoples have developed successful adaptative [sic] strategies to meet social, economic, political and environmental challenges as the global economy ebbs and flows" (Nuttall, 2000: 402). Boutet et al. (2015) use historical research to explore this dynamic of local Indigenous adaptations to mineral development and closure in Northern Canada, including the socio-economic and environmental legacies of such development (for instance, impacts on subsistence harvesting practices—see LeClerc and Keeling, 2015; Rixen and Blangy, 2016).

In engaging with ideas of resilience and adaptation, however, it is vitally important to inject the critical perspectives of Emilie Cameron and others who argue these concepts may overemphasize local agency, to the potential "exclusion of the social, cultural, political, economic, and historical geographies of colonialism, and particularly the ways in which colonial formations in the region interweave with both past and present interests in securing natural resources" (Cameron, 2012: 109). Similarly, an emerging critique examines the incorporation of these concepts into managerialist policy discourses aimed at "accommodating change, rather than contesting it" (O'Brien, 2012). Critical perspectives on adaptation and resilience also highlight their tendency (as discussed above) to problematically contain Indigenous knowledge and agency to the scale of the local (Perramond, 2007; Thompson et al., 2006).

As Cameron suggests, by examining the legacies of environmental change within broader critical frameworks, we can begin to better understand how these legacies are related to processes of socioecological change at different scales. Recent popular and scholarly work on the historical political economy of the Arctic highlights the intersection of "local" environmental changes in the region with global-scale processes such as colonial expansion and capitalist resource exploitation, from whaling to the fur trade to energy resources (Avango et al., 2011; Cameron, 2011; Emmerson, 2010; Nuttall, 2010). The legacies of these earlier rounds of resource development, both social and environmental, not only persist in the region today, but in many ways helped create the conditions for contemporary and future development, both materially (for instance, through colonial territorial expansion, (re)settlement, and infrastructure development [Sandlos and Keeling, 2012a]) and ideologically (through the reinforcement of the notion of the Arctic as a resource frontier [Nuttall, 2010; Stuhl, 2013]). Leigh Johnson has provocatively dubbed the role of historic anthropogenic environmental changes (in this case, climate change) in fostering the conditions for renewed or further Arctic resource development as "accumulation by degradation" (2010). While questionable as a strategy *per se* of resource appropriation, this framing highlights something of the complex intersection of environmental degradation, remediation, and redevelopment characteristic of some legacy sites such as the zombie mines noted above, and the political economy of Arctic resource development (Caulfield, 2000; Midgley, 2012).

These critical insights point to a further promising set of connections and avenues of investigation surrounding the "materiality" and "ongoingness" of resource development impacts, both at resource sites and beyond (cf. Lepawsky and Mather, 2011). Recent scholarship on the geographies and political ecology of resource development seeks to question not only the taken-for-granted status of nature as a "resource," but also the suite of material flows and connections surrounding resource and energy developments, including the complex metabolism of waste, energy, and ecological processes at resources sites and beyond (Bakker and Bridge, 2006; Bridge, 2009). By tracing the environmental-historical networks that constitute resources as commodities—what William Cronon (1992) evocatively describes as "the paths out of town"—social scientists can better connect the legacies of local environmental degradation and resource conflicts with socioeconomic and ecological processes at national and global scales, including: changing patterns of resource demand and consumption, global energy and resource flows, environmental change, and the uneven geographical development that constructs the Arctic as a resource periphery. Nuttall, for instance, suggests a focus on the social anthropology of pipelines, "including pipelines as complex and interdependent technological, social, economic and political systems and networks" (2010: 21).

At the most intimate scale, Richardson and Weszkalnys (2014: 20) argue, "resource exploitation is a process where bodies, technologies, infrastructures, and substances become entangled throwing the porosity between human bodies and their resource environments into sharp relief." Certainly, the "slow violence" of toxic threats posed to northern peoples over the very long term through toxic exposures provides disturbing examples of such bodily entanglements (Sandlos and Keeling, 2016b). For instance, decades of pollution from uranium extraction near Elliot Lake in Canada undermined the downstream Serpent River

Anishnaabek people's use of and confidence in freshwater fish and beaver—impacts which persisted long after the mine's closure (Leddy, 2013). Seeing these relations as ongoing and extending beyond resource extraction activities themselves emphasizes the dynamic temporality and spatiality associated with resource developments and their environmental legacies for northern communities, even long after a particular development activity has ceased. This perspective also provides important openings to consider a wider range of experiences and knowledges associated with both the historical impacts and subsequent repair of post-industrial landscapes.

Conclusion

This chapter has examined literature on the history and environmental legacies of extractive development in the Arctic (with an emphasis on the Canadian experience), and the integration of community concerns into industrial remediation projects. In the context of the current and forecasted rapid development of extractive industries in the Arctic, we suggest it is critical to understand the full scope and extent of these environmental legacies, including ecosystem disturbance, waste disposal, and toxicity, as well as contemporary efforts to mitigate or remediate their ongoing effects. This is an especially important concern for Arctic and northern communities themselves, as they seek to maximize the benefits of resource development while mitigating or eliminating the negative impacts. Extractive industries, typically large scale, transformative, and disruptive in their environmental and social impacts, pose a particular challenge in this regard.

Arturo Escobar states, in a South American context, that ultimately "the environment is cultural and symbolic construction, and the way it is constructed has implications for how it is used and managed." Consequently, he asks, "How would sustainability and conservation look if approached from the perspective of the world construction of the black groups of the Pacific?" (Escobar, 2008: 120). Analogous questions formulated with regards to the use of non-renewable resources and economic development in the circumpolar north would certainly merit more attention. In particular, such questions can serve to bring to the forefront local understandings of environmental and personal health, the impacts of contaminants or pollution, and the environmental legacies of development on local landscapes and resources which may reflect northern experiences of colonialism and marginalization that are not easily effaced through the provision of more scientific "expertise."

For Indigenous communities, remediation and redevelopment activities may reawaken or reproduce the negative effects of past rounds of industrial development. Further, while local communities and Indigenous groups are increasingly consulted or even participate in remediation work, environmental reclamation processes remain dominated by technical experts and government agencies, often to the exclusion of local or traditional knowledge. Currently, the ability of Indigenous people to influence large-scale development projects is limited by the confinement of TEK to renewable resource issues and the imbalance of political power among state, corporate, and Indigenous actors within the northern development triumvirate. Nonetheless, because many northern Indigenous communities have been exposed historically to decades of development ranging from hydroelectric damming, to mining, hydrocarbon extraction, and militarization, they maintain significant knowledge of the environmental impacts of industrial development and its long-term legacies. Indigenous knowledge is crucial to identifying remediation goals and standards acceptable to the community, as well as to assessing cross-scalar and cumulative impacts and improving the governance and long-term care of contaminated or remediated landscapes. There are also opportunities to enhance the benefits for local communities, beyond an ameliorated local environment, but these opportunities must be generated through dialogue with communities throughout the life of a remediation project (and indeed, as part of new development proposals). Although it may seem a cliché to speak of approaching development (and remediation) "as if community mattered" (Ross and Usher, 1986), more research is needed on the incorporation of these community perspectives as a foundation for the sustainable management of non-renewable resources and their environmental legacies in the Arctic.

Our review also explored potential conceptual frameworks for further research into the wider environmental legacies of extractive development. The importance of recognizing these legacies when considering contemporary development proposals in northern regions could be addressed through an expanded notion of *cumulative effects*, which would account not only for the effects of multiple concurrent resource developments, but also address the ongoing environmental and social effects of past extractive activities. In terms of accounting for community-level impacts and engagements, perspectives on the *vulnerability and resilience of socioecological systems* could provide important insights into how local communities have coped or are coping with the legacies of development. "Adaptationist" approaches, however, must remain attuned to the wider political economy and scaled politics of development and remediation that frame local manifestations of "resilience" in the face of extractive-industrial impacts. Finally, critical attention to *resource materialities* points to the entanglements of human bodies, biophysical systems, and the networks of governance, infrastructure, and economy that produce ongoing environmental injustice (for instance, through toxic exposures) at sites of extraction and remediation, even long after the putative end of a particular development.

Note

1 The question of the intersection of Indigenous or Traditional Ecological Knowledge in resource development is dealt with at greater length in an earlier version of this paper: Keeling, A., Sandlos, J., Boutet, J.-S., and Longley, H., "Managing Development? Knowledge, Sustainability and the Environmental Legacies of Resource Development in Northern Canada," ReSDA Gap Analysis Report #12, Yukon College, Whitehorse, YT.

References

Alaska. Department of Natural Resources (2012). Abandoned Mine Lands Program. Retrieved from http://dnr.alaska.gov/mlw/mining/aml/. 26 Feburary 2013.

Angell, A. C. and Parkins, J. R. (2011). Resource Development and Aboriginal Culture in the Canadian North. *Polar Record, 47*(240), 67–79.

Arctic Monitoring and Assessment Program (AMAP) (1997). Arctic Pollution Issues: A State of the Arctic Environment Report. Oslo: AMAP.

Arctic Monitoring and Assessment Program (AMAP) (2002). Arctic Pollution 2002: Persistent Organic Pollutants, Heavy Metals, Radioactivity, Human Health, Changing Pathways. Oslo: AMAP.

Arctic Monitoring and Assessment Program (AMAP). (2007). Arctic Oil and Gas 2007. Oslo: AMAP.

Assembly of First Nations and MiningWatch Canada. (2001). After the Mine: Healing Our Lands and Nations—A Workshop on Abandoned Mines. MiningWatch Canada Report, Ottawa, Canada.

Avango, D. (2017). Remains of Industry in the Polar Regions: Histories, Processes, Heritage. *Entreprises et Histoire, 87*, 133–149.

Avango, D., Hacquebord, L., Aalders, Y., De Haas, H., Gustafsson, U., and Kruse, F. (2011). Between Markets and Geo-politics: Natural Resource Exploitation on Spitsbergen from 1600 to the Present Day. *Polar Record, 47*(240), 29–39.

Bakker, K. and Bridge, G. (2006). Material Worlds? Resource Geographies and the "Matter of Nature". *Progress in Human Geography, 30*(1), 5–27.

Berger, T. (1977). Northern Frontier, Northern Homeland: The Report of the Mackenzie Valley Pipeline Inquiry. Ottawa: Supply and Services Canada.

Boutet, J. S. (2015) The Revival of Québec's Iron Ore Industry: Perspectives on Mining, Development, and History. In Keeling, A. and Sandlos, J. (eds.), *Mining and Communities in Northern Canada: History, Politics, and Memory* (pp. 169–206). Calgary: University of Calgary Press.

Boutet, J. S., Keeling, A., and Sandlos, J. (2015). Historical Perspectives on Mining and the Aboriginal Social Economy. In Southcott, C. (ed.), *Northern Communities Working Together: The Social Economy of the Canadian North* (pp. 198–227). Toronto: University of Toronto Press.

Bouw, B. and Ebner, D. (2011). The Commodity Cycle Speeds Up. *Globe and Mail*, 14 January. Retrieved from www.theglobeandmail.com/report-on-business/industry-news/energy-and-resources/the-commodity-cycle-speeds-up/article1871363/. 14 May 2012.

Bowes-Lyon, L.-M., Richards, J. P., and McGee, T. M. (2009). Socio-economic Impacts of the Nanisivik and Polaris Mines, Nunavut. In Richards, J. P. (ed.), *Mining, Society, and a Sustainable World* (pp. 371–396). Heidelberg: Springer-Verlag.

Bridge, G. (2009). Material Worlds: Natural Resources, Resource Geography and the Material Economy. *Geography Compass, 3*(3), 1217–1244.

Cameron, E. (2011). Copper Stories: Imaginative Geographies and Material Orderings of the Central Canadian Arctic. In Baldwin, A., Cameron L., and Kobayashi A. (eds.), *Rethinking the Great White North: Race, Nature, and Whiteness in Canada* (pp. 169–190). Vancouver: University of British Columbia Press.

Cameron, E. (2012). Securing Indigenous Politics: A Critique of the Vulnerability and Adaptation Approach to the Human Dimensions of Climate Change in the Canadian Arctic. *Global Environmental Change, 22*, 103–114.

Canter, L. and Ross, B. (2010). State of Practice of Cumulative Effects Assessment and Management: The Good, the Bad and the Ugly. *Impact Assessment and Project Appraisal*, *28*(4), 261–268.

Caulfield, R. (2000). The Political Economy of Renewable Resource Management in the Arctic. In Nuttall, M. and Callaghan T. V. (eds.), *The Arctic: Environment, People, Policy* (pp. 485–514). Amsterdam: Harwood Academic Publishers.

Chance, N. A. and Andreeva, E. N. (1995). Sustainability, Equity, and Natural Resource Development in Northwest Siberia and Arctic Alaska. *Human Ecology*, 23(2), 217–240.

Chapin III, F. S., Peterson, G., Berkes, F., Callaghan, T. V., Angelstam, P., Apps, M., . . . and Whiteman, G. (2004). Resilience and Vulnerability of Northern Regions to Social and Environmental Change. *AMBIO: A Journal of the Human Environment*, *33*(6), 344–349.

Coates, P. A. (1991). *The Trans-Alaska Pipeline Controversy: Technology, Conservation, and the Frontier*. London, UK: Associated University Press.

Cronon, W. (1992). Kennecott Journey: The Paths Out of Town. In Cronon, W., Miles, G., and Gitlin, J. (eds), *Under an Open Sky: Rethinking America's Western Past*. New York: W. W. Norton.

Dance, A. (2015). Northern Reclamation in Canada: Contemporary Policy and Practice for New and Legacy Mines. *The Northern Review*, *41*, 41–80.

"Declaration on coexistence of oil & gas activities and indigenous communities on Nenets and other territories in the Russian North" (2007). Retrieved from www.arcticcentre.org/? DeptID=10165. 27 February 2013.

Duhaime, G., Bernard, N., and Comtois, R. (2005). An Inventory of Abandoned Mining Exploration Sites in Nunavik, Canada. *Canadian Geographer*, *49*(3), 260–271.

Duhaime, G. and Comtois, R. (2003). Abandoned Mining Equipment in Nunavik. In Rasmussen, R. and Koroleva, N. (eds.), *Social and Environmental Impacts in the North: Methods in Evaluation of Socio-economic and Environmental Consequences of Mining and Energy Production in the Arctic and Sub-Arctic*. Dordrecht: Kluwer Academic.

Ehrlich, A. (2010). Cumulative Cultural Effects and Reasonably Foreseeable Future Developments in the Upper Thelon Basin, Canada. *Impact Assessment and Project Appraisal*, *28*(4), 279–286.

Ellis, S. C. (2005). Meaningful Consideration? A Review of Traditional Knowledge in Environmental Decision Making. *Arctic*, *58*(1), 66–77.

Emmerson, C. (2010). *The Future History of the Arctic*. New York: Public Affairs.

Escobar, A. (2008). *Territories of Difference: Place, Movements, Life*. Durham, NC: Duke University Press.

Espiritu, A. (1997). Aboriginal Nations: Natives in Northwest Siberia and Northern Alberta. In Smith, E. A. and McCarter, J. (eds.), *Contested Arctic: Indigenous people, Industrial States, and the Circumpolar Environment* (pp. 41–67). Seattle, WA: University of Washington Press.

Forbes, B. C., Ebersole, J. J., and Strandberg, B. (2001). Anthropogenic Disturbance and Patch Dynamics in Circumpolar Arctic Ecosystems. *Conservation Biology*, *15*(4), 954–969.

Forbes, B. C., Fresco, N., Shvidenko, A., Danell, K., and Chapin III, F. S. (2004). Geographic Variations in Anthropogenic Drivers that Influence the Vulnerability and Resilience of Social-ecological Systems. *AMBIO: A Journal of the Human Environment*, *33*(6), 377–382.

Forbes, B. C., Stammler, F., Kumpula, T., Metchtyb, N., Pajunen, A., and Kaarlejärvi, E. (2009). High Resilience in the Yamal-Nenets Social-ecological System, West Siberian Arctic, Russia. *Proceedings of the National Academy of Sciences*, *106*(52), 22041–22048.

Ford, J. D. and Smit, B. A. (2004). A Framework for Assessing the Vulnerability of Communities in the Canadian Arctic to Risks Associated with Climate Change. *Arctic, 57*(4), 389–400.

Hacquebord, L. (ed.) (2012). *LASHIPA: History of Large Scale Resource Exploitation in Polar Areas.* Groningen: Barkhuis Publishing.

Hacquebord, L. and Avango, D. (2009). Settlements in an Arctic Resource Frontier. *Arctic Anthropology, 46*(1–2), 25–39.

Haalboom, B. (2014). Confronting Risk: A Case Study of Aboriginal Peoples' Participation in Environmental Governance of Uranium Mining, Saskatchewan. *The Canadian Geographer, 58*(3), 276–290.

Haalboom, B. (2016). Pursuing Openings and Navigating Closures for Aboriginal Knowledges in Environmental Governance of Uranium Mining, Saskatchewan, Canada. *The Extractive Industries and Society, 3*(4), 1010–1017.

Hayter, R., Barnes, T. J., and Bradshaw, M. J. (2003). Relocating Resource Peripheries to the Core of Economic Geography's Theorizing: Rationale and Agenda. *Area, 35*(1), 15–23.

Herrmann, T. M., Sandström, P., Granqvist, K., D'Astous, N., Vannar, J., Asselin, H., . . . and Cuciurean, R. (2014). Effects of Mining on Reindeer/Caribou Populations and Indigenous Livelihoods: Community-based Monitoring by Sami Reindeer Herders in Sweden and First Nations in Canada. *The Polar Journal, 4*(1), 28–51.

Houde, N. (2007). The Six Faces of Traditional Ecological Knowledge: Challenges and Opportunities for Canadian Co-management Arrangements. *Ecology and Society, 12*(2), [online] www.ecologyandsociety.org/vol12/iss2/art34.

Johnson, C., Boyce, M. S., Case, R. L., Cluff, H. D., Gau, R. J., Gunn, A., and Mulders, R. (2002). Cumulative Effects of Human Development on Arctic Wildlife. *Wildlife Monographs, 160*(1), 1–36.

Johnson, L. (2010). The Fearful Symmetry of Arctic Climate Change: Accumulation by Degradation. *Environment and Planning D: Society and Space, 28*, 828–847.

Jorgenson, M. T., Kidd, J. G., Carter, T. C., Bishop, S., and Racine, C. H. (2003). Long-term Evaluation of Methods for Rehabilitation of Lands Disturbed by Industrial Development in the Arctic. In Rasmussen, R. and Koroleva, N. (eds.), *Social and Environmental Impacts in the North: Methods in Evaluation of Socio-economic and Environmental Consequences of Mining and Energy Production in the Arctic and Sub-Arctic* (pp. 173–190). Dordrecht: Kluwer Academic.

Josephson, P. R. (2002). *Industrialized Nature: Brute Force Technology and the Transformation of the Natural World.* Washington, DC: Island Press.

Keeling, A. (2010). "Born in an Atomic Test Tube": Landscapes of Cyclonic Development at Uranium City, Saskatchewan. *The Canadian Geographer, 54*(2), 228–252.

Keeling, A., Sandlos, J., Boutet, J-S., and Longley, H., "Managing Development? Knowledge, Sustainability and the Environmental Legacies of Resource Development in Northern Canada," ReSDA Gap Analysis Report #12, Yukon College, Whitehorse, YT.

Keeling, A. and Sandlos, J. (2009). Environmental Justice Goes Underground? Historical Notes from Canada's Northern Mining Frontier. *Environmental Justice, 2*(3), 117–125.

Keeling, A. and Sandlos, J. (eds.) (2015) *Mining and Communities in Northern Canada: History, Politics, and Memory.* Calgary: University of Calgary Press.

Keeling, A. and Sandlos, J. (2017). Ghost Towns and Zombie Mines: The Historical Dimensions of Mine Abandonment, Reclamation and Redevelopment in the Canadian North. In Bocking, S. and Martin, B. (eds.), *Ice Blink: Navigating Northern Environmental History.* (pp. 377–420). Calgary: University of Calgary Press.

Kivinen, S. (2017). Sustainable Post-Mining Land Use: Are Closed Metal Mines Abandoned or Re-Used Space? *Sustainability, 9*(10), 1705.

Klein, D. R. (2000). Arctic Grazing Systems and Industrial Development: Can We Minimize Conflicts? *Polar Research, 19*(1), 91–98.

Klein, D. R. and Magomedova, M. (2003). Industrial Development and Wildlife in Arctic Ecosystems. In Rasmussen, R.and Koroleva, N. (eds.), *Social and Environmental Impacts in the North: Methods in Evaluation of Socio-Economic and Environmental Consequences of Mining and Energy Production in the Arctic and Sub-Arctic* (pp. 35–56). Dordrecht: Kluwer Academic.

Kumpula, T., Pajunen, A., Kaarlejärvi, E., Forbes, B. C., and Stammler, F. (2011). Land Use and Land Cover Change in Arctic Russia: Ecological and Social Implications of Industrial Development. *Global Environmental Change, 21*(2), 550–562.

LeClerc, E. and Keeling, A. (2015). From Cutlines to Traplines: Post-industrial Land Use at the Pine Point Mine. *The Extractive Industries and Society, 2*(1), 7–18.

Leddy, L.C. 2013. Poisoning the Serpent: The Effects of the Uranium Industry on the Serpent River First Nation, 1953–1988. In Hele, K. S. (ed.), *The Nature of Empires and the Empires of Nature: Indigenous Peoples and the Great Lakes Environment* (pp. 125–148). Waterloo: Wilfrid Laurier University Press.

Lepawksy, J. and Mather, C. (2011). From Beginnings and Endings to Boundaries and Edges: Rethinking Circulation and Exchange Through Electronic Waste. *Area, 43*(3), 242–249.

Lottermoser, B. G. (2010). *Mine Wastes: Characterization, Treatment and Environmental Impacts*, third edition. Berlin: Springer.

Mackasey, W. O. (2000). Abandoned Mines in Canada. Unpublished report prepared for MiningWatch Canada by WOM Geological Associates, Inc.

McBeath, J. and Shepro, C. E. (2007). The Effects of Environmental Change on an Arctic Native Community: Evaluation Using Local Cultural Perceptions. *The American Indian Quarterly, 31*(1), 44–65.

Midgley, S. (2012). Co-producing Ores, Science and States: High Arctic Mining at Svalbard (Norway) and Nanisivik (Canada). Master's thesis. Memorial University, St John's, Canada.

Morse, K. (2003). *The Nature of Gold: An Environmental History of the Gold Rush*. Seattle, WA: University of Washington Press.

Nadasdy, P. (2003). *Hunters and Bureaucrats: Power, Knowledge, and Aboriginal-State Relations in Southwest Yukon*. Vancouver: University of British Columbia Press.

Nassichuk, W. W. (1987). Forty Years of Northern Non-renewable Natural Resource Development. *Arctic, 40*(4), 274–284.

National Orphaned and Abandoned Mines Initiative (NOAMI) (2003). Lessons Learned on Community Involvement in the Remediation of Orphaned and Abandoned Mines. NOAMI report.

Nuttall, M. (2000). Indigenous Peoples, Self-determination and the Arctic Environment. In Nuttall, M. and Callaghan, T. V. (eds.), *The Arctic: Environment, People, Policy* (pp. 377–409). Amsterdam: Harwood Academic Publishers.

Nuttall, M. (2010). *Pipeline Dreams: People, Environment, and the Arctic Energy Frontier*. Copenhagen: International Working Group for Indigenous Affairs.

O'Brien, K. (2012). Global Environmental Change II: From Adaptation to Deliberate Transformation. *Progress in Human Geography, 36*(5), 667–676.

Olsen, H. K. (ed.) (2001). *Mining in the Arctic: Proceedings of the Sixth International Symposium on Mining in the Arctic*. Lisse: A. A. Balkema.

O'Reilly, K. and Eacott, E. (2000). Aboriginal People and Impact and Benefit Agreement: Summary of the Report of a National Workshop. *Northern Perspectives, 25*(4). Retrieved from www.carc.org/pubs/v25no4/2.htm. 20 February 2013.

Paci, C. and Villebrun, N. (2005). Mining Denendeh: A Dene Nation Perspective on Community Health Impacts of Mining. *Pimatisiwin: A Journal of Aboriginal and Indigenous Community Health, 3*(1), 71–86.

Page, R. (1986). *Northern Development: The Canadian Dilemma.* Toronto: McClelland & Stewart.

Parlee, B. (2012). Finding Voice in a Changing Ecological and Political Landscape: Traditional Knowledge and Resource Management in Settled and Unsettled Claim Areas of the Northwest Territories, Canada. *Aboriginal Policy Studies, 2*(1), 56–87.

Pearce, T., Ford, J. D., Prno, J., Duerden, F., Pittman, J., Beaumier, M., . . . and Smit, B. (2011). Climate Change and Mining in Canada. *Mitigation and Adaptation Strategies for Global Change, 16*, 347–368.

Pepper, M., Roche, C. P., and Mudd, G. M. (2014). Mining Legacies – Understanding Life-of-Mine Across Time and Space. In *Proceedings, Life-of-Mine 2014.* Brisbane: Australasian Institute of Mining and Metallurgy (AusIMM).

Perramond, E. (2007). Tactics and Strategies in Political Ecology Research. *Area, 39*(4), 499–507.

Piper, L. (2009). *The Industrial Transformation of Subarctic Canada.* Vancouver: University of British Columbia Press.

Poland, J. S., Riddle, M. J., and Zeeb, B. A. (2003). Contaminants in the Arctic and the Antarctic: A Comparison of Sources, Impacts, and Remediation Options. *Polar Record, 39*(4), 369–383.

Prowse, T. D., Furgal, C., Chouinard, R., Melling, H., Milburn, D., and Smith, S.L. (2009). Implications of Climate Change for Economic Development in Northern Canada: Energy, Resource, Transportation Sectors. *Ambio, 38*(5), 272–281.

Rea, K. J. (1968). *The Political Economy of the Canadian North.* Toronto: University of Toronto Press.

Resource Extraction and Sustainable Arctic Communities [REXSAC]. (2017) REXSAC Annual Report. Stockholm: REXSAC.

Richardson, T. and Weszkalnys, G. (2014). Resource Materialities. *Anthropological Quarterly, 87*(1), 5–30.

Rixen, A. and Blangy, S. (2016). Life After Meadowbank: Exploring Gold Mine Closure Scenarios with the Residents of Qamini'tuaq (Baker Lake), Nunavut. *The Extractive Industries and Society, 3*(2), 297–312.

Robards, M. and Alessa, L. (2004). Timescapes of Community Resilience and Vulnerability in the Circumpolar North. *Arctic, 57*(4), 415–427.

Ross, D. and Usher, P. (1986). *From the Roots Up: Development as if Community Mattered.* Croton on Hudson, NY: Bootstrap Press.

Sabin, P. (1995). Voices from the Hydrocarbon Frontier: Canada's Mackenzie Valley Pipeline Inquiry, 1974–1977. *Environmental History Review, 18*(1), 17–48.

Sandlos, J. and Keeling, A. (2012a). Claiming the New North: Mining and Colonialism at the Pine Point Mine, Northwest Territories, Canada. *Environment and History, 18*(1), 5–34.

Sandlos, J. and Keeling, A. (2012b). Giant Mine: Historical Summary. Report submitted to Mackenzie Valley Environmental Review Board public registry.

Sandlos, J. and Keeling, A. (2016a). Aboriginal Communities, Traditional Knowledge, and the Environmental Legacies of Extractive Development in Canada. *Extractive Industries and Society, 3*(2), 278–287.

Sandlos, J. and Keeling, A. (2016b). Toxic Legacies, Slow Violence, and Environmental Injustice at Giant Mine, Northwest Territories. *Northern Review*, *42*, 7–21.

Sistili, B., Metatawabin, M., Iannucci, G., and Tsuji, L. J. S. (2006). An Aboriginal Perspective on the Remediation of Mid-Canada Radar Line Sites in the Subarctic: A Partnership Evaluation. *Arctic*, *59*(2), 142–154.

Sondergaard, J., Asmund, G., Johansen, P., and Rigét, F. (2011). Long-Term Response of an Arctic Fiord System to Lead-Zinc Mining and Submarine Disposal of Mine Waste (Maarmorilik, West Greenland). *Marine Environmental Research*, *71*, 331–341.

Spaling, H., Zwier, J., and Ross, W. (2000). Managing Regional Cumulative Effects of Oils Sands Development in Alberta, Canada. *Journal of Environmental Assessment Policy and Management*, *2*(4), 501–528.

Stammler, F. (2002). Success at the Edge of the Land: Past and Present Challenges for Reindeer Herders in the West Siberian Yamalo-Nenetskii Autonomous Okrug. *Nomadic Peoples*, *6*(2), 51–71.

Stammler, F. (2005). *Reindeer Nomads Meet the Market: Culture, Property and Globalisation at the End of the Land*. Muenster: Litverlag.

Stevenson, M. G. (1996). Indigenous Knowledge in Environmental Assessment. *Arctic*, *49*(3), 278–291.

Storm, A. (2014). *Post-industrial Landscape Scars*. London: Palgrave Macmillan.

Stuhl, A. (2013). The Politics of the "New North": Putting History and Geography at Stake in Arctic Futures. *The Polar Journal*, *3*(1), 94–119.

Thompson, A., Robbins, P., Sohngen, B., Arvai, J., and Koontz, T. (2006). Economy, Politics and Institutions: From Adaptation to Adaptive Management in Climate Change. *Climatic Change*, *78*, 1–5.

Tsuji, L., Kataquapit, J., Katapatuk, B., and Iannucci, G. (2001). Remediation of Site 050 of the Mid-Canada Radar Line: Identifying Potential Sites of Concern Utilizing Traditional Environmental Knowledge. *Canadian Journal of Native Studies*, *21*(1), 149–160.

Udd, J. E. and Bekkers, G. (eds.) (2003). Mining in the Arctic: Proceedings of the 7th International Symposium on Mining in the Arctic. Montreal, Quebec: Canadian Institute of Mining, Metallurgy and Petroleum.

Udd, J. E. and Keen, A. J. (eds.) (1999). Mining in the Arctic: Proceedings of the 5th International Symposium on Mining in the Arctic. Lisse: A. A. Balkema.

Walker, D. A., Webber, P. J., Binnian, E. F., Everett, K. R., Lederer, N. D., Nordstrand, E. A., and Walker, M. D. (1987). Cumulative Impacts of Oil Fields on Northern Alaska Landscapes. *Science (New Series)*, *238*(4828), 757–761.

White, G. (2006). Cultures in Collision: Traditional and Euro-Canadian Governance Knowledge Processes in Northern Land-Claim Boards. *Arctic*, *59*(4), 401–414.

Worrall, R., Neil, D., Brereton, D., and Mulligan, D. (2009). Towards a Sustainability Criteria and Indicators Framework for Legacy Mine Land. *Journal of Cleaner Production*, *17*, 1426–1434.

Zaslow, M. (1988). *The Northward Expansion of Canada, 1914–1967*. Toronto: McClelland and Stewart.

11 Impact and benefit agreements and northern resource governance

What we know and what we still need to figure out

Ben Bradshaw, Courtney Fidler, and Adam Wright

Mineral exploration and extraction in areas like Northern Canada has long been recognized as a source of economic activity and opportunity. In 2011, a recent peak year, exploration expenditures were $4.2 billion across the country, while the total production of minerals, metals, and coal was valued at $50.9 billion (Intergovernmental Working Group on the Mineral Industry, 2016). During such times, the sector has generated substantial wealth for mining firms, governments, servicing companies, employees, and, increasingly, communities proximate to a mine site. For example, in 2007, the owners of the Raglan Mine in northern Quebec issued a payment of $16.7 million to the Inuit of Nunavik through Makivik Corporation. This payment was made as per the requirements of the Raglan Agreement, a contract signed in 1995 by the (then) owners of the mine, Falconbridge, and regional Inuit organizations on whose traditional lands the mine was built.

Since this time, equivalent agreements, commonly termed Impact and Benefit Agreements (IBAs), have increasingly been established across Canada, Australia, and, most recently, Greenland. IBAs are negotiated directly between mine developers and Indigenous communities with limited state interference. Indeed, outside of a few jurisdictions where IBAs are mandatory as a result of a completed land claim (e.g. Nunavut), their use is voluntary from a regulatory perspective. Nevertheless, IBAs have become institutionalized as firms have come to recognize that it is in their commercial interest to address evident gaps in the regulatory processes used to permit mine developments (Bradshaw and McElroy, 2014). Consistent with this observation, their de facto purpose is to deliver enhanced impact mitigation and tangible benefits to communities in exchange for community acceptance of a project (Kennett, 1999). Routinely, IBAs provide benefits such as payments, training, and employment, and preferential treatment of community businesses; less routinely, IBAs offer a means for would-be impacted communities to influence project design and operation. For example, as a condition of their support for Inco's Voisey's Bay nickel mine, the Labrador Inuit Association secured limits on winter shipping to and from the mine in order to protect winter ice travel by Inuit hunters.

The growing use of IBAs by Canadian Indigenous communities over the past two decades should not be read as evidence of their unequivocal embrace or trouble-free use. In some instances, an IBA is simply not viewed as sufficient to mitigate or offset the anticipated impacts of mineral extraction. Such was the case for the Lutsel K'e Dene First Nation who, in the mid-2000s, expressed opposition to uranium exploration in the Thelon region of the Northwest Territories (NWT) even if supported by negotiated agreements. In instances where IBAs have indeed been used, community signatories have sometimes been frustrated by perceived withholding of benefits by the developer or some other failure of IBA implementation (O'Faircheallaigh, 2002); when this occurs, community protests and project slow-downs are possible. Such was the case for DeBeers Canada's Victor mine, whose supply lines were disrupted by an 18-day roadblock in February 2009 staged by members of the Attawapiskat First Nation. The protest developed because certain portions of the community felt DeBeers Canada was not living up to the terms of their IBA. Community discontent was also evident for reasons that would seem to exceed the scope of the IBA. Though not explicitly a subject of negotiations, it was expected that local incidence of substance abuse, domestic violence, and other social ills would be reduced as average community incomes grew via mine-related employment; four years after IBA ratification, these social issues were still evident.

Hence, notwithstanding their growing use, IBA signatories and analysts alike regard IBAs as an imperfect governance solution. Of particular concern is: the uncertain position of IBAs in mine permitting, especially relative to regulatory processes like Environmental Assessment (EA) and the execution of the Crown's consultation obligations; the perception that Indigenous community well-being is not sufficiently improving through IBA-enabled mine developments, especially given limited use of adaptive management to address social impacts as they emerge within IBA-signatory communities; and growing concerns that IBAs represent a partial measure that allows the Crown to continue to avoid its obligations and thereby perpetuate ongoing injustice. Though these concerns represent a practice deficit, some also represent a deficit in our collective knowledge of IBAs, which ideally would be addressed through future research. In support of this end, this chapter reviews select scholarship focussed on IBAs to highlight what is known about IBAs and identify knowledge gaps, especially in Northern Canada. What follows is a descriptive overview of three dominant areas of IBA scholarship: the relationship of IBAs to existing regulatory processes; IBA negotiation, implementation and effectiveness; and IBAs and the pursuit of social justice.

IBAs and regulatory processes

In one of the first papers to systematically assess the emergence of IBAs in the context of existing regulatory processes, Galbraith, Bradshaw and Rutherford (2007) suggested that IBAs be seen as "supraregulatory" governance instruments in that they necessarily function alongside, though are seldom officially a part of, public regulatory systems. The relationship of IBAs to regulatory processes such as EA, the consultation obligations of the Crown, and, most recently, legislation that aims to increase the transparency of payments made by extractive sector firms to governments, including Indigenous ones, is a complicated one, but one that has attracted scholarship.

For many scholars, it is noteworthy that IBAs have emerged in two jurisdictions, Canada and Australia, where regulatory processes governing mining, such as EA, are relatively progressive. Evidently such systems, even in places like the Mackenzie Valley, NWT where EA is conducted through a co-managed board, offer insufficient protection and opportunity for Indigenous communities. This assertion is supported by the content of – and even more obviously the label – an IBA. Before an Indigenous community will support and enable a mine develop-ment, it must reach an agreement with the developer (an *Impact Agreement*) on how the impacts of mine construction and operation will be managed beyond that required through regulatory processes. This can be done through, for example, mutually agreed-upon changes in project design, enhanced impact mitigation provisions, and enabling community involvement in environmental monitoring. Though some of these provisions may seem routine and even disingenuous (see Noble and Birk, 2011), the establishment of an agreement, and hence ongoing relationship, between a mine developer and a local Indigenous community to limit, measure, and manage impacts is significant in that it serves to overcome a disconnected permitting process: one process and administrative body deter-mines the suitability of a proposed mine development; and a second process and administrative body monitors outcomes following project approval. This is problematic in itself, but made more so by the *ex ante* nature of EA, which requires that a binary decision (yes/no) be made based on probabilistic forecasts (O'Faircheallaigh, 1999). The continuity created by an agreement ensures that unforeseen changes are properly identified and managed, and thereby provides more assurance to a community than does a disjointed regulatory system; this seems to hold true even with the advent of more stringent follow-up provisions in EA.

Of equal importance, an Indigenous community has no reason to support and enable a mine development without coming to an agreement with the developer (a *Benefit Agreement*) on the delivery of benefits to locals, such as employment and training, mine business contracts, and direct financial payments. Though other benefits may ultimately deliver more wealth to a community, payments tend to attract the greatest attention in negotiations. Payments can take a variety of forms, including equity interest, royalties based on profits, royalties based on the value of production, royalties based on the volume of outputs, and fixed payments; often, communities will secure fixed payments to cover their costs of IBA administration, coupled with some form of profit-sharing (Gibson and O'Faircheallaigh, 2010). Rather than complement public regulatory systems, this second aim of an IBA makes up for these systems' most glaring limitation: though locals inevitably experience a disproportionate share of project impacts, regulatory systems govern-ing resource extraction do not privilege locals in terms of capturing project benefits, even where those locals hold special rights. For example, the creation or enhancement of the positive impacts of development or not mentioned in the *Canadian Environmental Assessment Act*, 2012; its concern is merely the avoidance, mitigation, or compensation of adverse ones.

Many scholars have pointed to the need to better define the relative roles of IBAs and regulatory processes such as EA (e.g. Kennett, 1999; Fidler and Hitch, 2007; Galbraith et al., 2007; Diges, 2008), and some have even offered guidance for integrating the two processes (e.g. Lukas-Amulung, 2009; Gibson and O'Faircheallaigh 2010, Noble and Birk, 2011). These calls and prescriptions stem from a recognition that the public conduct of EA and private negotiation of IBAs regularly overlap and can sometimes interact in problematic ways. Lukas-Amulung (2009) notes from her research of development processes in the Northwest Territories that EA and IBA negotiations tend to overlap in three key areas: the scoping stage; the deliberation stage; and the resolution stage (see Figure 11.1).

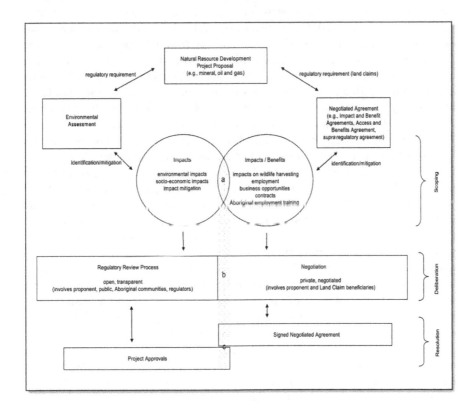

Figure 11.1 A conceptualization of the overlapping relationship between Negotiated Agreements (NAs) and Environmental Assessment (EA) in the Northwest Territories

Note: Potential overlaps are indicated by the shaded area. The overlaps in the EA and NA processes exist in: (a) the scoping stage, when project impacts and benefits are identified; (b) the deliberation stage, when regulatory review of the EA and the negotiation of the NAs occur; and (c) the resolution stage, when project approvals are granted and NAs are signed (Lukas-Amulung, 2009: 32).

In some instances, interaction is overt. For example, Shanks (2006) identifies cases where EA panels have recommended that IBAs be in place as a condition of regulatory approval. Similarly, Fidler (2009) notes that an explicit element of the IBA signed in support of the proposed Galore Creek mine in northern BC outlined how the Tahltan First Nation and mine proponent would collaborate to achieve EA approval.

To address problematic interactions between IBAs and EA, scholars have conceptualized normative models of integration, and considered the place of IBAs in regional planning. For example, Lukas-Amulung (2009) proposes the establishment of a bilateral IBA-EA coordination plan at the early stages of a proposed project to enable information sharing between the processes, and the coordination of monitoring provisions post-approval. Echoing this suggestion, Noble and Birk (2011: 19) discuss the potential benefits of linking "negotiated agreements and associated monitoring programs with EIA-based follow-up practices in support of improved community engagement and project impact management". For Gibson and O'Faircheallaigh (2010), a critical challenge for communities lies in ensuring information arising from IBA negotiations and the EA process is iterative, with one informing the other in a manner that leverages optimal benefits in a time-sensitive environment. Certainly, as argued by Caine and Krogman (2010), IBA negotiations and especially ratification should be informed by EA, especially with respect to the identification of the likely impacts of a proposed project impacts; however, contemporary EA processes should also be informed by the deliberations that ideally precede IBA negotiations through which an Indigenous community crafts a vision for the future that they hope to realize through partnering on a development. This argument is consistent with those of O'Faircheallaigh (2007; 2010) and Siebenmorgen and Bradshaw (2011) who see benefit in linking or at least placing IBA negotiations within larger scale, collaborative land-use planning processes. In a related vein, O'Faircheallaigh (2010) draws attention to the wider, problematic implications of agreement-making for Indigenous communities, including: limiting access to judicial systems to exercise Indigenous or even basic citizen rights because of binding commitments in a signed IBA; dissuading participation in a political campaign or strategy that runs counter to the interests of the project partner; and altering a community's relationship with the Crown with respect to funding of programs and fulfillment of the Crown's consultation and accommodation duties.

This latter issue is especially complex, and remains an area in need of further analysis. The Crown's duty to consult and accommodate stems from the Supreme Court of Canada's 2004 ruling in Haida Nation v British Columbia (Minister of Forests), which established that the Crown is obliged to consult and seek to accommodate the needs of Aboriginal peoples when it contemplates actions or decisions (e.g. the permitting of a mine) that may affect an Aboriginal person's Aboriginal or treaty rights. Consistent with the Court's ruling, Gogal, Riegert and Jamieson (2005) note that a mine proponent does not owe an independent duty to consult Aboriginal peoples; however, the Crown may delegate the procedural aspects of consultation to industry, which they have routinely done (Garton, 2009;

Cameron and Levitan 2014); an IBA is an obvious form of accommodation as was made explicit in the 2012 signing of an *Accommodation Agreement* by Avalon Minerals and the Deninu Kue First Nation in support of the Nechalacho Rare Earth Elements Project; and it is ultimately in the commercial interest of firms to ensure that these duties are fulfilled given that, should the Crown fail to do so, a proposed project can be stalled.

Mining firms' commercial interest in IBAs is made clear in Lapierre and Bradshaw (2008), who, in addition to identifying the business case for IBAs, argue that the industry has moved from reluctant acceptance of the need to negotiate an IBA to embracing the opportunity to partner with Indigenous communities proximate to a proposed mine site because of their need for infrastructure and labour, their desire to avoid protests, and their larger, often global, reputation. Most recently, relations between firms operating mines and their partner Indigenous communities in Canada have been tested by the passing of the *Extractive Sector Transparency Measures Act* (2014), which requires Canadian companies to disclose payments to Aboriginal governments, formed through confidential IBAs. This legislation has not been well received by Indigenous governments in Canada (see NRCAN 2014 Public Consultation Comments) and creates new tension and questions pertaining to the role of government in IBAs.

Notwithstanding considerable attention to understanding better the relationships between IBAs and state-led regulatory processes, further research is needed to address a number of pertinent questions, such as:

- How do communities understand their legal rights? How does this impact their approach to IBA negotiations?
- How does industry approach IBA negotiations? Is it to have a rights-based discussion or is it based on "how much money will it take to get the project the green light?"? Does this matter?
- Does legal context from region to region impact the negotiation and content of an IBA?
- Are IBA negotiations recognized as de facto replacements for the Crown's Duty to Consult and Accommodate? If so, what are the implications for Indigenous communities and industry?
- How are legal and cultural norms around the concept of FPIC changing in Canada, and what might this mean for IBA negotiations?
- Is there the potential to harmonize EA and IBA processes? Is there interest?

IBA negotiation, implementation, and effectiveness

When members of the Attawapiskat First Nation blocked the supply route to the Victor Mine in February, 2009, they expressed outwardly what members of IBA-signatory communities have routinely felt inwardly. Though IBAs have undoubtedly filled some notable gaps in existing regulatory systems, there is widespread recognition in the world of IBA practice and the IBA research community that

Table 11.1 Factors that contribute to successful IBA implementation as identified in scholarship

Factors	Description	Key sources
1. Clear goals	Precision and clarity in the way that goals and intended outcomes are stated in the agreement is critical in ensuring smooth implementation. Is there ambiguity? Is language precise?	O'Reilly 2000, Sosa and Keenan 2001, Weitzner 2006, O'Faircheallaigh 2007, Prno 2007, Woodward and Company 2009, Gibson and O'Faircheallaigh 2010, Prno et al. 2010, Siebenmorgen and Bradshaw 2011, Wright 2013, Dylan et al. 2014
2. Institutional structures for implementation	Implementation cannot be expected to occur as an add-on role for the parties to the agreement. New structures such as committees specifically tasked with implementation will be necessary to drive the agreement forward.	Keeping 1998, Sosa and Keenan 2001, Galbraith et al. 2007, Lewis 2009, Gibson and O'Faircheallaigh 2010, Wright 2013
3. Clear allocation of responsibilities	Clearly state who is responsible for doing what, and make sure that the responsible person or organization has the authority required. Essential to have both company and community champions of the agreement.	Isaac and Knox 2004, Klein et al. 2004, Weitzner 2006, Fidler and Hitch 2007, O'Faircheallaigh 2007, Diges 2008, Lewis 2009, Gibson and O'Faircheallaigh 2010, Noble and Fidler 2011, Siebenmorgen and Bradshaw 2011
4. Adequate resources	Dedicated resources, including: a) Funds to hire, e.g., technical experts to review environmental reports; b) Resources may include staff and access to experts or information; c) Training & Capacity to engage in policy work.	O'Reilly 2000, Sosa and Keenan 2001, O'Faircheallaigh 2007, Diges 2008, Lewis 2009, O'Faircheallaigh 2010, Woodward and Company 2009, Gibson and O'Faircheallaigh 2010, Wright 2013, Caine and Krogman 2010, Prno et al. 2010, Noble and Fidler 2011, Siebenmorgen and Bradshaw 2011

5. Penalties and incentives for compliance	This type of adaptive management is about ensuring appropriate responses if the intent of the agreement is not met.	O'Reilly 2000, Gogal et al. 2005, Hitch 2005, Weitzner 2006, Caine and Krogman 2010, Gibson and O'Faircheallaigh 2010
6. Monitoring	Provisions for monitoring might include a requirement for regular reporting of data and a list of performance indicators that will be used to track the progress under the agreement.	Public Policy Forum 2005, Woodward and Company 2009, Prno et al. 2010, Gibson and O'Faircheallaigh 2010, Caine and Krogman 2010, Noble and Birk 2011, Siebenmorgen and Bradshaw 2011, Wright 2013
7. Review mechanisms	Periodic review as well as commitments to fund the review and a process for the findings to be considered and acted upon.	O'Reilly 2000, Prno 2007, Lewis 2009, Woodward and Company 2009, Gibson and O'Faircheallaigh 2010, Caine and Krogman 2010, Siebenmorgen and Bradshaw 2011
8. Capacity for amendment	Process for amendment that is not too onerous or problems will not be addressed. Amendments can be important for a few reasons: a) the relationship between the company and community is dynamic, which can result in unanticipated situations that need to be addressed; b) the body of knowledge, understanding and approaches to IBAs is frequently changing; c) the alternative to amendment, dispute resolution, is expensive and disruptive.	Keeping 1998, Weitzner 2006, Diges 2008, Knotsch and Warda 2009, Lewis 2009, Prno et al. 2010, Gibson and O'Faircheallaigh 2010, Caine and Krogman 2010, Wright 2013

IBAs could be doing more to advance community interests. For example, Prno, Bradshaw and Lapierre (2010) reported on the effectiveness of a number of IBAs negotiated in support of three diamond mines in the Northwest Territories. While the IBAs were generally found to be meeting their broad objectives, especially with respect to the delivery of benefits, the authors identified some deficiencies, especially around building trust with mining firms, relieving capacity strains, and enabling follow-up and adaptive management when problems arise. As might be expected, the authors discovered highly varied opinions on the effectiveness of the IBAs from agreement to agreement, community to community, and sometimes focus group to focus group within a single community. For example, the region's first agreements, signed in haste in support of the Ekati mine, were viewed with pride by some focus group participants and barely veiled contempt by others (e.g. "[They exist] to shut us up ... They threw [the IBA] at us; take it or leave it!" (Prno, Bradshaw and Lapierre, 2010: 5)).

Efforts to make sense of IBA effectiveness, and especially problematic performance, have rightly directed attention to IBA negotiation and implementation. Reflecting the former, O'Faircheallaigh and Corbett (2005) showed that negotiations can severely impact the terms and, ultimately, progressiveness of resultant IBAs; indeed, the authors revealed too many instances where negotiated agreements contained provisions of less significance than existing regulatory requirements! These kinds of extreme outcomes are less likely given efforts to improve the knowledge base and bargaining power of communities seeking to negotiate an agreement (see Gibson and O'Faircheallaigh, 2010). Nevertheless, many scholars argue that power imbalances at the negotiating table will always be problematic (e.g. Caine and Krogman, 2010; Szablowski, 2010; St-Laurent and Le Billon, 2015), and factors beyond the negotiating table will continue to affect an IBA's content. This latter point is well shown in O'Faircheallaigh's (2015) comprehensive review of the practice of IBA negotiation, which, among other things, reveals how economic, social, and political contextual variables like institutional frameworks and the state of regional economies can impact IBA content.

IBA implementation has similarly been a subject of research and prescription. One of the earliest contributions, from O'Faircheallaigh (2002), drew attention to the many causes of weak IBA implementation, such as inadequate resources, weak language and institutional arrangements, an absence of specific penalties for non-compliance, staff turnover, etc. Many of these concerns were used by Gibson and O'Faircheallaigh (2010) to develop prescriptions for improved IBA implementation. More specifically, the authors identified eight factors internal to agreements that are known to contribute to their successful implementation, and consequently the effectiveness of an IBA: 1) clear goals; 2) institutional structures for implementation; 3) clear allocation of responsibilities; 4) adequate resources; 5) penalties and incentives for compliance; 6) monitoring; 7) review mechanisms; and 8) capacity for amendment. Though the scholarly basis of these eight factors is healthy (see Table 11.1), few of the articles are empirical, with many authors repeating one another's claims.

O'Faircheallaigh (2015) addresses this empirical deficit somewhat and draws attention to some other considerations for successful implementation of company-community agreements including measures to: address mine ownership changes or simply managers' changing priorities; maintain good community-company relations; and ensure senior managers focus on implementation issues on a regular basis.

While weak negotiation and implementation clearly explain some communities' frustrations with IBAs, some of these frustrations can also be attributed to communities' growing expectations of benefit delivery, coupled with their failure to make explicit a number of implicit expectations (Siebenmorgen and Bradshaw, 2011). For communities, the decision to sign an IBA, and thereby tacitly embrace the mineral economy, is a significant one; such decisions are only made because community members anticipate improving outcomes over time, not just with respect to employment and income, but also community well-being (Knotsch and Warda, 2009). Research on the experiences of IBA-signatory communities has revealed a number of community well-being concerns, especially as a result of new monies and the rotational work schedule, such as increased family stress, substance abuse and crime, and mental health impacts such as depression (Gibson, 2008; Peterson, 2012; Shandro et al., 2011). IBA negotiators are evidently aware of these issues and IBAs increasingly make reference to community well-being as an overarching goal of the agreement; however, few if any IBAs contain concrete provisions that ensure that such issues can be effectively managed (Jones and Bradshaw, 2015). Impact monitoring is improving (Klinck et al. 2016), but adaptive management of problematic outcomes is often insufficient (Jones and Bradshaw, 2015).

Though, in response to calls (e.g. Prno and Bradshaw, 2008; Caine and Krogman 2010), some important research has been completed on IBA negotiation, implementation, and effectiveness, it remains an area in need of attention if IBAs continue to be promoted as a way to resolve competing interests, obtain consent, and deliver beneficial outcomes (Gibson and O'Faircheallaigh 2010). In particular research is needed to address questions such as:

- How variable are IBA negotiations and implementation?
- How much information sharing occurs among communities and among companies, and how does this impact IBA negotiations?
- What are contemporary and historical constraints to effective IBA negotiation and implementation, and how can they be addressed?
- What are the governance and capacity issues that have to be addressed to improve IBA negotiation and implementation?
- Are IBAs benefiting communities? Are they meeting their explicit and implicit aims, especially around community well-being?
- Are IBAs benefiting industry? Are they meeting their explicit and implicit aims?

IBAs and social justice

For IBA scholars, and even more so for Indigenous community leaders making use of IBAs as a means of asserting interests and securing benefits, a fundamental question is never far from mind: are IBAs redressing historical injustices or perpetuating further injustice? Given the significance of past injustices experienced by Indigenous peoples as a result of mineral exploration and mine construction, operation, and abandonment (see, for example, Gibson and Klinck, 2005; Sandlos and Keeling, 2015), the use of IBAs evidently constitutes progress over a dismal past with respect to impact mitigation, benefit delivery, and self-determination. Indeed, for some (e.g. Sosa, 2011; Boreal Leadership Council 2012; 2015; Bradshaw and McElroy, 2014; Nunatsiavut Government, 2016), an IBA even has the potential to serve as an expression of an Indigenous community's free, prior, and informed consent (FPIC) to major projects proposed within its territories.

This *relative* progress argument was adopted by O'Faircheallaigh (1999: 76) in an early review of IBAs in which the author noted that:

> Social Impact Assessment and negotiation processes [e.g. IBAs] . . . can help shift the balance of power toward those who are relatively weak . . . [These] processes involve providing communities with information and a capacity to apply that information, and this can significantly enhance their relative power.

The author recently repeated the *relative* progress claim in O'Faircheallaigh (2013). Nevertheless, critics have raised important challenges about their use. One set of criticisms reflects problems of IBA practice that give rise to injustice, and yet holds out the possibility for – indeed seeks to enable – refined usage; another set of criticisms places IBAs at the centre of an evolving and unjust neoliberal governance regime and, in even worse light, within a longer history of colonialism.

Exemplifying the former, research has drawn attention to a number of problematic aspects of IBAs from a social justice perspective, such as: their confidentiality, which limits knowledge sharing between communities impacted by mining and those contemplating local mine development, and sometimes even limits communications within a community negotiating an IBA; power imbalances that manifest in unjust tactics used in their negotiation; the inevitable emergence of conflict and division within communities, and between communities and their larger governance institutions; the marginalization of certain voices in communities; the challenge of disbursing monies within collective societies; their lack of dynamism relative to the desire of communities to offer ongoing consent; the failure to manage what are now well-known impacts of mine development for Indigenous community well-being; and their failure too often to meet community expectations, which can lead communities to question their decision to negotiate an IBA (Sosa and Keenan, 2001; O'Faircheallaigh, 2006; 2010; 2012; Caine and Krugman, 2007; Gibson and O'Faircheallaigh, 2010; Prno, Bradshaw and Lapierre, 2010; Siebenmorgen and Bradshaw, 2011; Jones and Bradshaw, 2015; St-Laurent and Le Billon, 2015).

In addition to drawing attention to problematic elements within IBAs, Caine and Krogman (2010) and O'Faircheallaigh (2010) effectively move discussions of IBAs into the larger context of community sovereignty, Crown-community relations, and even meanings of democracy. This larger context is given more attention by those critics of IBAs who have identified their use as part of neoliberal governance and a colonial history marked by cultural erosion, environmental dispossession, and loss of autonomy (e.g. Cameron and Levitan, 2014; St-Laurent and Le Billon, 2015). The question of autonomy is a particularly compelling one. Whereas O'Faircheallaigh (2010: 71) extols IBAs for enabling "access to mining income [which] can provide Aboriginal groups a degree of autonomy from the state . . . adding to their negotiating power in dealing with the state in relation, for instance, to service delivery, land title and management, and governance," Cameron and Levitan (2014) suggest that IBAs have simply shifted Indigenous communities' dependence to corporations and allowed the state to retreat even further from fulfilling its fiduciary obligations; for the authors, IBAs have allowed the honour of the Crown to remain unfulfilled.

These latter arguments echo those of, for example, Slowey (2001) and Coulthard (2014), who draw attention to state-enabled processes of Indigenous self-determination that create "neocolonization" rather than decolonization. Such writings offer important new terrain for IBA scholarship and create new opportunities to problematize IBA practice. To this end, Jones and Bradshaw (2015) acknowledge the complicity of IBAs in enabling mine developments and perpetuating historical injustices, and fault IBA negotiators for their lack of attention to Indigenous conceptions of health and especially its historical determinants. In response, however, they challenge these negotiators, as well as other relevant governance actors, to understand better, and then account for within IBAs and other contemporary governance processes, the complexities that inform Indigenous well-being, and especially legacies of colonialism. Problematically, the means by which this might be done are still in their infancy.

Intuitively, then, a number of knowledge gaps exist with respect to IBAs and social (in)justice, none of which are easily addressed, such as:

- Are IBAs a counterbalance to historical injustices or do they perpetuate them? Is autonomy from the state welcome?
- Do communities feel pressured to negotiate an IBA? Are community members aware of IBA negotiations when they happen? Do they have sufficient opportunities to contribute?
- What are alternatives to IBAs that might create greater opportunities for communities in terms of both economic development and social justice?

Conclusions

To repeat the introduction of this chapter, the growing use of IBAs by Canadian Indigenous communities over the past two decades should not be read as evidence of their unequivocal embrace or trouble-free use. And though considerable scholarship has developed over the past two decades around IBAs, many questions surrounding their use remain. This chapter has aimed to reveal these in the hope of driving further research and refined practice. This dual aim fits well with ReSDA's overall goal of examining ways to build the potential for sustainable resource development in the Canadian north for years to come.

References

Birk, J. and Noble, B. (2011). Comfort Monitoring? Environmental Assessment Follow-up Under Community–Industry Negotiated Environmental Agreements. *Environmental Impact Assessment Review*, 31(1): 17–24.

Boreal Leadership Council (2012). *Free, Prior, and Informed Consent in Canada*. Ottawa: Boreal Leadership Council.

Boreal Leadership Council (2015). *Understanding Successful Approaches to Free, Prior, and Informed Consent in Canada. Part I*. Ottawa: Boreal Leadership Council.

Bradshaw, B. and McElroy, C. (2014). Company–Community Agreements in the Mining Sector, in C. Louche and T. Hebb (eds.), *Socially Responsible Investment in the twenty-first Century: Does it Make a Difference for Society?* (Critical Studies on Corporate Responsibility, Governance and Sustainability, Volume 7). Bingley, UK: Emerald Group Publishing, pp. 173–193.

Caine, K. J. and Krogman, N. (2010). Powerful or Just Plain Power-Full? A Power Analysis of Impact and Benefit Agreements in Canada's North. *Organization & Environment*, 23(1): 76–98.

Cameron, E. and Levitan, T. (2014). Impact and Benefit Agreements and the Neo-liberalization of Indigenous-State Relations and Resource Governance in Northern Canada. *Studies in Political Economy*, 93: 29–56.

Coulthard, G. (2014). *Red Skin, White Masks: Rejecting the Colonial Politics of Recognition*. Minneapolis, MN: University of Minnesota Press.

Diges, C. (2008). Sticks and Bones: Is Your IBA Working? Amending and Enforcing Impact Benefit Agreements, McMillan Binch Mendelsoh LLP. Retrieved at: at www.cbern.ca/content/uploads/sites/4/2016/03/Sticks-and-Bones_-Is-your-IBA-Working-Amending-and-Enforcing-Impact-Benefit-Agreements.pdf.

Fidler, C. (2009). Increasing the Sustainability of a Resource Development: Aboriginal Engagement and Negotiated Agreements. *Environment, Development and Sustainability*, 12(2): 233–244.

Fidler, C. and Hitch, M. (2007). Impact and Benefit Agreements: A Contentious Issue for Environmental and Aboriginal Justice. *Environments*, 35(2): 50–69.

Galbraith, L., Bradshaw, B., and Rutherford, M. B. (2007). Towards a New Supraregulatory Approach to Environmental Assessment in Northern Canada. *Impact Assessment and Project Appraisal*, 25(1): 27–41.

Garton, B. S. (2009). "Top Ten" Accommodation Agreement. Vancouver: Aboriginal Law: Solicitors' Issues – 2009 Update Paper 5.2.

Gibson, G. (2008). Negotiated Spaces: Work, Home and Relationships in the Dene Diamond Economy (Ph.D). The University of British Columbia, Vancouver.

Gibson, G. and Klinck, J. (2005). Canada's Resilient North: The Impact of Mining on Aboriginal Communities. *Pimatisiwin*, 3(1): 116–139.

Gibson, G. and O'Faircheallaigh, C. (2010). IBA Community Toolkit: Negotiation and Implementation of Impact and Benefit Agreements. Walter & Duncan Foundation. Retrieved at: http://gordonfoundation.ca/resource/iba-community-toolkit/.

Gogal, S., Riegert, R., and Jamieson, J. (2005). Aboriginal Impact and Benefit Agreements: Practical Consideration. *Alberta Law Review*, 43(1): 129–157.

Intergovernmental Working Group on the Mineral Industry (2016). Mining Sector Performance Report 2006–2015. Prepared for submission to the Energy and Mines Ministers' Conference in August 2016 in Winnipeg, Manitoba.

Jones, J. and Bradshaw, B. (2015). Addressing Historical Impacts Through Impact and Benefit Agreements and Health Impact Assessment: Why it Matters for Indigenous Well-Being. *Northern Review*, 41: 81–109.

Kennett, S. A. (1999). *A Guide to Impact and Benefit Agreements*. Calgary: Canadian Institute of Resources Law.

Klinck, R., Bradshaw, B., Sandy, R., Nabinacaboo, S., Mameanskum, M., Guanish, M., . . . and Pien, S. (2016). Enabling Community Well-Being Self-Monitoring in the Context of Mining: The Naskapi Nation of Kawawachikamach. *Engaged Scholar Journal*, 1(2): 114–130.

Knotsch, C. and Warda, J. (2009). *Impact Benefit Agreements: A Tool for Healthy Inuit Communities?* Ottawa: National Aboriginal Health Organization.

Lapierre, D. and Bradshaw, B. (2008). Why Mining Firms Care: Determining Corporate Rationales for Negotiating Impact and Benefit Agreements. Canadian Institute of Mining, Metallurgy and Petroleum Conference.

Lukas-Amulung, S. (2009). The Rules of Engagement? Negotiated Agreements and Environmental Assessment in the Northwest Territories, Canada (Thesis). Royal Roads University.

Nunatsiavut Government, 2016 Characterization of Voisey's Bay Impact and Benefit Agreement. Retrieved at: www.nunatsiavut.com/department/co-management/

Noble, B. and Birk, J. (2011). Comfort Monitoring? Environmental Assessment Follow-up under Community–Industry Negotiated Environmental Agreements. *Environmental Impact Assessment Review*, 31(1): 17–24.

NRCAN 2014 Public Consultation Comments. Retrieved at: http://open.canada.ca/en/consultations/mandatory-reporting-standards-for-extractive-sector-what-we-heard

O'Faircheallaigh, C. (1999). Making Social Impact Assessment Count: A Negotiation-based Approach for Indigenous Peoples. *Society and Natural Resources*, 12(1): 63–80.

O'Faircheallaigh, C. (2002). Implementation: The Forgotten Dimension of Agreement Making in Australia and Canada. *Indigenous Law Bulletin*, 5(20): 14–17.

O'Faircheallaigh, C. (2006). Mining Agreements and Aboriginal Economic Development in Australia and Canada. *Journal of Aboriginal Economic Development*, 5(1): 74–91.

O'Faircheallaigh, C. (2007). Environmental Agreements, EIA Follow-Up and Aboriginal Participation in Environmental Management: The Canadian Experience. *Environmental Impact Assessment Review*, 27: 319–342.

O'Faircheallaigh, C. (2010). Aboriginal-Mining Company Contractual Agreements in Australia and Canada: Implications for Political Autonomy and Community Development. *Canadian Journal of Development Studies*, 30(1–2): 69–86.

O'Faircheallaigh, C. (2013). Women's Absence, Women's Power: Indigenous Women and Negotiations with Mining Companies in Australia and Canada. *Ethnic and Racial Studies*, 36(11): 1789–1807.

O'Faircheallaigh, C. (2013). Extractive Industries and Indigenous Peoples: A Changing Dynamic? *Journal of Rural Studies*, 30: 20–30.

O'Faircheallaigh, C. (2015). *Negotiations in the Indigenous World: Aboriginal Peoples and Extractive Industry in Australia and Canada.* New York: Routledge.

O'Faircheallaigh, C. and Corbett, T. (2005). Indigenous Participation in Environmental Management of Mining Projects: The Role of Negotiated Agreements. *Environmental Politics*, 14(5): 629–647.

Peterson, K. C. (2012). Community Experiences of Mining in Baker Lake, Nunavut (Masters dissertation). University of Guelph, Guelph.

Prno, J. and Bradshaw, B. (2008). Program Evaluation in a Northern Aboriginal Setting: Assessing Impact and Benefit Agreements. *The Journal of Aboriginal Economic Development*, 6(1): 59–75.

Prno, J., Bradshaw, B., and Lapierre, D. (2010). Impact and Benefit Agreements: Are They Working? CIM Conference, (pp. 1–9). Vancouver.

Sandlos, J. and Keeling, A. (2015). *Mining and Communities in Northern Canada: History, Politics, and Memory.* Calgary: University of Calgary Press.

Siebenmorgen, P. and Bradshaw, B. (2011). Re-conceiving Impact and Benefit Agreements as Instruments of Aboriginal Community Development in Northern Ontario, Canada. *Oil, Gas & Energy Law*, 4. Retrieved at: www.ogel.org.

Shandro, J. A., Veiga, M. M., Shoveller, J., Scoble, M., and Koehoorn, M. (2011). Perspectives on Community Health Issues and the Mining Boom–Bust Cycle. *Resources Policy*, 36(2): 178–186.

Slowey, G. A. (2001). Federalism and First Nations: Finding Space for Aboriginal Governments, in Ian Peach (ed.), *Constructing Tomorrow's Federalism: New Routes to Effective Governance.* Winnipeg: University of Manitoba Press, pp. 157–170.

Sosa, I. (2011). License to Operate: Indigenous Relations and Free Prior Informed Consent in the Mining Industry. Sustainalytics. Toronto. Retrieved at: www.sustainalytics.com/sites/default/files/indigenouspeople_fpic_final.pdf.

Sosa, I. and Keenan, K. (2001). Impact Benefit Agreements Between Aboriginal Communities and Mining Companies: Their Use in Canada. 1–29. Retrieved at: www.cela.ca/publications/impact-benefit-agreements-between-aboriginal-communities-and-mining-companies-their-use.

St-Laurent, G. P. and Le Billon, P. (2015). Staking Claims and Shaking Hands: Impact and Benefit Agreements as a Technology of Government in the Mining Sector. *The Extractive Industries and Society*, 2(3): 590–602.

Szablowski, D. (2010). Operationalizing Free, Prior and Informed Consent in the Extractive Industry Sector? Examining the Challenges of a Negotiated Model of Justice. *Canadian Journal of Development Studies*, 30(1–2): 111–130.

12 Normalizing Aboriginal subsistence economies in the Canadian North

David Natcher

The traditional economies of Aboriginal peoples in Northern Canada have changed dramatically over the past century. Once reliant solely on the procurement of wildfoods, Aboriginal peoples adjusted their residency, land use, and social organization according to the seasonal and spatial availability of foods harvested from the land. For Aboriginal peoples, their use of wildlife resources was nested within a cultural system, where food procurement encompassed a complex array of social, spiritual, ecological, and economic dimensions. Soon following contact with Settler populations, the subsistence-based economies that long sustained the cultures and economies of Aboriginal peoples underwent irreversible change. While the intensity of these changes was experienced differently, by region and over time, the impacts on Aboriginal peoples have been very much the same, a slow yet consistent transition from subsistence-based to capitalistic-based forms of livelihood.

Accompanying this transition has been a notion of what constitutes the "real" economy. In contrast to the so called informal, or subsistence based economy of Aboriginal peoples, the real economy involved capitalistic enterprises, market transactions, and a reliance on wage labour opportunities. Northern policy documents are replete with evolutionary characterizations of the transitioning northern economy, along with depictions of Aboriginal peoples as non-progressive, and requiring government assistance. Government assistance was justified on grounds that Aboriginal peoples required support to participate more fully in the real economy so they too could benefit from the economic and social transformation of the North.

While the Canadian Government, and the various departments that assumed responsibility for the administration of Aboriginal affairs, can be most directly implicated in this transformative process, academics have also contributed by giving form and credibility to the "real" economy. This includes suggestions of the inevitability and evolutionary supremacy of market-based activities. Perhaps well intentioned, the scholarship of some academics has unwittingly advanced the separation (Polanyi, 1957) of Aboriginal culture from economy, and in so doing helped set a trajectory of economic 'modernization' of northern Aboriginal economies. Guided by theoretical and empirical analyses, a plethora of public policies, program, and services were introduced to hasten the transformation of

Aboriginal economies and to help Aboriginal peoples prepare for unprecedented economic change. Theories of acculturation and modernization in particular were advanced by leading scholars, and were then used by the Canadian Government to justify northern expansion of the real economy through extractive resource development. The notion of the real economy has become so ingrained in northern policy that natural resource development is now ubiquitous as the economic policy for the North. For northern communities to benefit directly from resource development, a better understanding of how subsistence harvesting can be integrated into a more inclusive notion of the northern economy is needed.

In this chapter I examine how subsistence economies have been treated by academics, and in particular how theories of acculturation and modernization have shaped government policies in ways that have proven detrimental to Aboriginal peoples to this day. I then critique the methods that have dominated the analysis of subsistence production, with their overemphasis on methodological individualism and their treatment of the household as an autonomous economic unity devoid of historical particularity. By way of conclusion I make a call for research into the normalization of subsistence-based economies and encouragement of equitable forms of public support for the subsistence-based livelihoods of Aboriginal peoples.

Defining subsistence

The term subsistence has received varied and uneven treatment by the social sciences. For some, subsistence has been characterized as the minimum resources necessary to support life (Lonner, 1986), and conceptualized as static and minimalist in terms of material needs. This conception has in effect entrenched a belief that subsistence represents a meagre economic existence, and a relic of the past (Wheeler and Thornton, 2005: 70). Yet Williamson (1997: iv) notes that subsistence is not simply an economic activity, but rather only a facet, albeit a central one, of a way of life laden with values that connect Aboriginal peoples to the lands they occupy. To the Nunatsiavutmiut, subsistence does not imply poverty, but rather its practice indicates wealth, freedom, and wholeness (Williamson, 1997: iv). Encapsulating both economic and cultural attributes, Kishigami (2008) argues that subsistence involves a series of food-obtaining activities (harvesting, processing, sharing, consuming, disposing) that are informed by norms, social relationships, technology, worldview, identity, and environmental knowledge that are all embedded in food procurement systems. I too have emphasized the relational integration and complimentary unity of subsistence, where economic and social interactions elude dualistic representations (Natcher, Castro and Felt, 2015). In this way subsistence represents a seamless whole, where culture, economy, and environment overlap, and boundaries become blurred.

Notwithstanding this broader and more holistic view, subsistence is, more often than not, characterized in the literature and public policy as purely an economic activity, and a means of household provisioning. This rather myopic view of subsistence is found, for example, in comprehensive land claims agreements, where subsistence is characterized as "the non-commercial means of providing food and other household necessities from the land" (LILCA, 2004: 161) or simply as "the taking of wildlife into possession, and includes hunting, trapping, fishing, netting, egging, picking, collecting, gathering, spearing, killing, capturing or taking [wildlife] by any means" (Nunavut Land Claims Agreement, 1993). While these definitions were agreed to by Aboriginal negotiators, and ultimately ratified by Aboriginal peoples themselves, these definition nonetheless fail to capture the cultural dimensions of wildlife harvesting. It is this definition, and variations thereof, that more often than not reflects how Aboriginal subsistence economies have come to be treated in research and northern public policy.

Theorizing subsistence

Prior to the Second World War, the consensus among academics, if not the general Canadian public, was that the traditional cultures of Aboriginal peoples had been irrevocably changed, with any remnants being disadvantageous to their adaption to the changing North. Like other indigenous populations of the world, Aboriginal peoples in the Arctic and sub-Arctic were seen as victims of progress (Bodley, 1975), and if unable to assimilate into Canadian society, would simply disappear (Buchanan, 2006: 93). The general treatises on economic development in the early half of the twentieth century condemned hunters and gatherers as bad examples for their adherence to traditional ways and their devotion to subsistence-based economies (Sahlins, 1971). The need to assimilate, or risk cultural extermination, was advanced in large part through social theories of acculturation and modernization. Acculturation—or changes that occur from sustained contact between cultures—and modernization—processes by which cultures are forced to accept the traits of others—dominated the social sciences during the first half of the twentieth century. Stemming from these theoretical frameworks, an era of scholarship emerged that concentrated on how social and cultural patterns inhibit economic change and the conditions that could best foster economic assimilation. In this regard, anthropologists and other social scientists in Canada began to play an influential role.

As early as the nineteenth to the early twentieth century, the Government of Canada played a significant role in shaping academic traditions. Serving as virtually the sole source of research funding at the time, the Canadian Government was in the position to shape the intellectual content of anthropological research and implant its own political interests. Being intellectually and professionally beholden to political administrations, some have even accused the discipline of anthropology of being complicit in the government's desire to eradicate Aboriginal culture in Canada (see Avrith-Wakeam, 1993). Whether or not this criticism is justified, it is true that during this period the professional goals of anthropology, and other social sciences, were defined to a large extent by the interests of government rather than the intellectual quest to conduct "pure" research. It was during this time that social scientists were employed by the Federal Government, for example by the National Museum of Canada, to carry out ethnographic analyses of Canada's Aboriginal peoples, the results of which were to inform the policies of Indian Affairs in Ottawa. As noted by Dyck (2006: 81) it was in the first half of the twentieth century that anthropologists first began step outside the relative safety of their disciplinary confines and began offering actual advice to policy makers concerning the future prospects of Aboriginal populations, who at the time were considered to be on the verge of demographic extinction. Policy makers were particularly interested in the economic conditions of Aboriginal peoples and relied on the informed research of social scientists to gain this perspective (Hedican, 1995: 116–117). It was during this period that practitioners of "salvage anthropology" set out to document the fleeting subsistence lifestyles of Aboriginal peoples, before those systems were lost completely to the inescapable wave of modernization. While a wealth of knowledge was generated from their efforts, their conclusions gave credence to the acculturation theories that were openly embraced by the Canadian Government. While examples abound, the reports of Tanner (1966) are indicative of the general findings of the time, which concluded that the traditional lifestyles, namely hunting, fishing, and trapping activities of Aboriginal peoples, had all but been supplanted by the lure of the city and the quick money to be made through other vocations. Preference for wage labour was further motivated by "the harsh life on the land, and the loss skills that have made traditional pursuits unattractive" (Tanner, 1966). Lotz (1970: 92–93) similarly suggested that as a result of economic change, Yukon Indian peoples have a greater propensity for wage labour and even more so for the welfare payments that could be found in settlement areas.[1]

Parallel predictions were made for Inuit where a perceived breakdown of traditional institutions was cast as inevitable change in the wake of modernization. Perhaps most noteworthy is the work of Diamond Jenness who, in *The Economic Situation of the Eskimo*, stated that the "economic prosperity . . . of an Eskimo community today is roughly proportional to the amount of wage employment it obtains, and not, as formerly, to the wildlife resources that exist in its neighborhood" (Jenness, 1978: 144). Jenness further concluded that those "Eskimos" that failed to reorient their forms of economy would ultimately lose their dignity and any measure of independence. Having been demoralized through their contact with Europeans, Jenness argued that the only remaining option was to assume appropriate wage labour occupations (in Buchanan, 2006: 97). However, Jenness also cautioned that Aboriginal peoples would in the early stages of their acculturation be inefficient and unreliable wage labourers, and would likely demonstrate little aptitude for new economic activities. Notwithstanding the challenges, Jenness was adamant that the Canadian Government should take all necessary steps to assimilate Aboriginal peoples into the national society through their education in the modern economic world, in so far as features of their traditional culture would not interfere with their economic progress. Therefore the ultimate goal of all Indian policy should be assimilation, with the ultimate objective being the preparation of Aboriginal peoples to contribute to the social and economic institutions of the modern Canadian state. Due to his scholarly expertise on northern and indigenous affairs, the recommendations of Jenness were treated as informed testimony on the economic conditions of Aboriginal peoples and proved influential in setting the future direction of public policy in Northern Canada.

While the Government of Canada sponsored numerous studies on economic change of northern Aboriginal peoples, Canadian universities were also promoting research programs of their own to aid in the social and economic development of Aboriginal peoples. Illustrative of this agenda is the 1968 Statement of Purpose of the Arctic Institute of North American at the University of Calgary, which reads:

> The Northern Indians and Eskimos are faced already with adaption to a strange way of living which eventually will absorb them and extinguish their own cultures. Research is needed on how best to ease their problems in becoming adapted to conditions that require them to work in time controlled, wage earning economy, and to accept life in a developed community.

This statement is indicative of much of the research that was being done in Northern Canada at this time. A noteworthy example can be found in the Cree Development Change Project, directed by Professor Norman Chance, and funded by the Canadian Department of Forestry and Rural Development (1966–1969). The goals of Cree Development Change Project were to increase the understanding of the process of economic, social, and political change among the Cree of James Bay and then arrive at a formula that could accelerate the social and economic development of Cree communities (Chance, 1968: 3). The lessons learned could then be applied to other Aboriginal communities that were also in need of social and economic advancement. Theories that guided the project were firmly grounded in acculturation and modernization and were premised on the belief that the effective development of the Cree "requires the adoption of modern attitudes and values such as self-reliance and need achievement, a willingness to accept new ideas, a readiness to express opinions, and future time-orientation that involved greater concern for planning, organization and efficiency" (Chance, 1968: 6). A key finding of the Cree Development Project was that "it is necessary to increase income and standard of living to a point whereby family members can maintain a sense of self-respect between generations, among peers, and in their contacts with outsiders" (Chance, 1968: 29). Furthermore, this income should be of sufficient duration to sustain generational change. Only when enough steady income gained from regular wage earning employment is achieved will there be incentive enough for Aboriginal people to choose training in skilled professions over other traditional vocations (Chance, 1968). However, it was also recognized that the Cree, as well as other Aboriginal peoples for that matter, should not be expected to make these changes willingly or on their own. But rather "will depend on the ability of others to assist [them] in maintaining a sense of self-respect" (Chance, 1968: 29). To achieve these outcomes the Cree would need to be appropriately motivated and enticed "to behave in more modern ways" (Weiner, 1966: 13).

To motivate such change, the federal and territorial governments prioritized a number of intensive job training programs across the North in order create a skilled

Aboriginal workforce. In their review of these training programs, Young and McDermott (1988: 201) found that most were implemented first and foremost to induce rapid cultural change among Aboriginal trainees, and ultimately among all northern Aboriginal peoples. These programs, informed in part by the conclusions of leading scholars of the day, were also based on government's own objectives to modernize the North, which initiated an era of profound social change for Aboriginal peoples (Kulchyski and Tester, 2008). The most overt expression of government's modernizing efforts was the actual relocation of some Aboriginal communities. The relocations were justified on the basis of education and the need to provide better training opportunities for Aboriginal peoples in ways that would facilitate their entrance into the modern industrial economy (Wynn, 2007: xix). By being relocated to more accessible regional centres, "Inuit could receive so-called rehabilitation and employment training that would in theory allow them to adopt modern livelihoods as miners" which in turn would transform Aboriginal peoples into passive workers in a modern capitalist economy (Sandlos, 2007: 239). Most government-sponsored training programs were delivered in such a way as to stimulate rapid acculturation, through extended periods of relocation, the exclusive use of English, and adherence to fixed schedules (Young and McDermott, 1988: 195). Hobart (1982: 54) also found that "[v]irtually every feature of northern training programs was designed to enhance the socialization of trainees: they were cut off from the role models, the reference group, and the significant others which sustain their distinctive patterns of motivation, interest, and activities." The explicit goal of the training programs was to induce acculturation (Young and McDermott, 1988) where no allowances for the inclusion of Aboriginal culture would be made (Hobart, 1982).

Peter Usher (1993: 104) notes that sole paradigm for understanding government's propagation of the social and economic change of Aboriginal peoples was based on flawed theories of acculturation and modernization. Wheeler and Thornton (2005) echo Usher's conclusions and suggest that the evolutionary perspectives of social and economic development were readily embraced by government in order to advance large-scale resource development projects under the guise of Aboriginal social development. Used in this way, the social welfare of Aboriginal peoples was conjoined to modernization and the industrialization of the North. From this perspective the development of an industrial northern economy could take precedence over other forms of livelihood as part of the natural evolution of society and Canada's northern economy (Hedican, 1995: 117). In this way the conclusions of scholars served as powerful impetus for government interventions that invited development schemes aimed at improving the economic conditions of Aboriginal peoples—often with disastrous effects and at the cost of traditional subsistence-based economies.

Researching subsistence

In many ways the subsistence research conducted in the North can be classified into two general groupings, theoretical and applied studies. The more theoretical grouping includes those studies that utilize subsistence data to advance social theory, for instance the theories of acculturation and modernization that were discussed above. Other noteworthy contributions include a number of small-scale theoretical studies carried out by Chabot (2003) and Gombay (2010) in Nunavik, and Dombrowski (2007) and Thornton (2001) in Alaska. These and other similar studies provide important insights on the cultural and political changes occurring in Aboriginal economies as seen through the lens of subsistence. Wenzel's (1991; 2005) research in Clyde River, Nunavut is particularly illustrative of this type of research, yet is unique in that it offers a detailed and longitudinal account of the social organization of Inuit subsistence harvesting over time (Harder and Wenzel, 2012).

General nutrition studies that address the food habits and nutrition of Aboriginal peoples in the North have long been conducted (Wein and Freeman, 1995; Duhaime et al., 2002; Kuhnlein, 2009) and more recently traditional ecological knowledge studies that include subsistence and environmental monitoring data have become increasingly common (Gilchrist, Mallory and Merkel, 2005; Ferguson and Messier, 1997). A more recent emergence in the area of subsistence studies can be found in food security literature. Here the works of Duhaime and Bernard (2008), Furgal et al. (2012), and Ford et al. (2009) are representative of this important research. A comprehensive review of Aboriginal food security in Northern Canada can be found in the Canadian Council of Academies Report on the *State of Aboriginal Food Security in the Canadian North* (2014).

The second category, or "applied subsistence studies," includes close-range studies of food procurement in regions, communities, or for specific species. This category includes harvest studies that are designed to estimate the harvest of fish, wildlife, and plants by Aboriginal peoples. Perhaps the most significant contribution to this area is the research conducted by the Alaska Department of Fish and Game, Division of Subsistence. Since the 1980s the Division of Subsistence has carried out research on the economic aspects of subsistence hunting and fishing by Alaska Natives. As noted by Fall (1990), the Division of Subsistence has focused its efforts on understanding the who, what, when, where, how, and how much of wildlife harvesting. For example, research conducted by the Division of Subsistence found that 75% to 98% of all Alaska Native households harvest wildfoods (Fall, 2014). Collectively this harvest results in an annual consumption of approximately 52,114,490 pounds of wildfoods by Alaska Native households (Fall, 2014).

In Canada, one of the earliest harvest studies was conducted in Nunavik. In September 1975, the Northern Quebec Inuit Association initiated a seven-year study entitled *Research to Establish Present Levels of Native Harvesting*. The study set out to determine the extent of Inuit harvesting, the results of which would be used to establish a best estimate of harvest levels by species and community (JBNQRMC, 1988: v). The objective of the harvest study was to provide data needed to establish guaranteed harvesting levels for Inuit households.

Since the completion of the Nunavik study, other land claims regions have carried out their own harvest assessments. The Inuvialuit Harvest Study was conducted from 1988 to 1997. The object of the IHS was to obtain a continuous, long-term record of Inuvialuit harvest levels for each of the six regional communities. Harvest data are to be used by co-management boards and other wildlife and fisheries agencies to determine and recommend subsistence quotas. Environmental screening and impact review boards also use harvest information to fulfill their role in dealing with resource development and for determining compensation in cases of loss or damage.

The Gwich'in Harvest Study (GHS) was a requirement of the Gwich'in Comprehensive Land Claim Agreement (1992). The objective of the GHS was to record the number of animals, fish, and birds harvested by Gwich'in within the Settlement Area. These harvest levels would then be used to calculate Gwich'in Minimum Need Levels for Gwich'in households and would inform the management efforts of the Gwich'in Renewable Resources Board (GRRB) and other government partners.

The Sahtu Settlement Harvest Study was required under the Sahtu Land Claim Agreement (1993). Administered by the Sahtu Renewable Resource Board, the Study recorded the total number of fish and wildlife harvested by Sahtu Dene and Métis between 1998 and 2003. Those harvest estimates were then used to also establish the "minimum need levels" of Sahtu Dene and Métis and were used for wildlife management purposes in the Sahtu region.

The Nunavut Wildlife Harvest Study (NWHS) was mandated by the Nunavut Lands Claim Agreement (NLCA) and carried out under the direction of the Nunavut Wildlife Management Board (NWMB). Harvest data were collected monthly from Inuit hunters between June 1996 and May 2001. The purposes of the Harvest Study were to determine current harvesting levels and patterns of Inuit use of wildlife resources, aid in the management of wildlife resources of Nunavut, and once again to establish "basic needs levels" (BNLs).

With the settlement of the Labrador Inuit Land Claims Agreement (LILCA –2005), Inuit of Nunatsiavut secured clearly defined rights to a 72,500km^2 land-base and a 48,690km^2 of coastal zone. Within the settlement region, Inuit residents have the right to harvest wildlife resources in order to meet their domestic needs or, as defined by the LILCA, Inuit Domestic Harvest Limits (IDHLS). Domestic need is defined as the amount of resources necessary to satisfy individual non-commercial use. The use of domestic harvest levels as a basis for wildlife harvesting policy was promoted by the federal and provincial governments for its ability to set clearly defined harvest limits and facilitate effective monitoring and enforcement capabilities. Since its settlement, the Nunatsiavut Government has implemented a community harvest study program that is establishing IDHLs for 138 different species and resources used by Inuit residing within the Nunatsiavut Settlement Region.

My reason for describing each of these harvest studies in relative detail is to demonstrate how subsistence has come to be characterized in land claims agreements—agreements that were designed in part to protect the harvesting rights and livelihood interests of Aboriginal peoples. In each of these cases, wildlife harvesting studies have been designed to establish minimal need levels of subsistence resources for Aboriginal households. While satisfying the interests of federal, territorial, and provincial governments, this approach reduces subsistence to a regulatory issue, where conservation of wildlife receives *a priori* concern. In each of these studies, household harvesting data are collected, probability statistics are run, replete with their confidence intervals and other factors of probability, and are then used to chart population dynamics for any given species in order to allocate harvesting rights to Aboriginal peoples. The seeming legitimacy of this process has been so compelling that Aboriginal peoples, who continue their struggle to regain control over their lands and resources, agree to participate, if not fully embrace this approach, even though they often times struggle with how such approaches can co-exist with their own understanding of subsistence as an overarching cultural system.

Admittedly our knowledge of subsistence economies has advanced considerably through the conduct of harvest studies. For example, through these studies we know that between 1976–1981 the Nunavik communities of Kangiqsualujjuaq, Inukjuaq, and Quaqtaq collectively harvested and consumed an estimated 2,019,064 kg of wildfood (JBNQRMC, 1988). We also know that between 1996–2001, communities in Nunavut harvested and consumed an estimated 6,622,522 kg of wildfood (NWMB, 2004). Based on these numbers it seems clear that wildlife harvesting remains critical to the health and well-being of these Inuit communities. However, these studies were designed to identify the number of species harvested during a defined period of time, which would then be used to determine minimum need levels. While useful for those purposes, these studies make no attempt to uncover how people's livelihood choices are embedded in culture and history, nor the economic and political settings at which subsistence now occurs. This has been one of the major shortcomings of previous subsistence studies in that they most often fail to acknowledge the structural barriers that influence the options available to Aboriginal harvesters. Lost in these studies is the fact that subsistence represents a cultural system, which cannot be reduced simply to an economic activity or regulatory process.

Equally problematic has been that with few exceptions (e.g., Wenzel, 1991; Harder and Wenzel, 2012; Collings, 2014) these studies have employed methodologies that are more or less consistent with what Halperin (1994: 144) defines as "householding." While making some allowances for inter-household cooperation, this approach generally treats the household as an autonomous socio-economic unit that engages in a variety of capitalist and non-capitalist opportunities in different combinations. As valuable as this approach might be, it remains firmly grounded in methodological individualism, which focuses on the household or individuals rather than the structures or institutions that continue to influence the transformation of Aboriginal subsistence economies. Too often obscured are the historical, cultural, and institutional contexts that shape Aboriginal subsistence economies. Narrowly focused, researchers tend to concentrate on the methodological challenges—response/strategic bias, over- and under-reporting, memory recall, sampling strategies—rather than the structural conditions that shape and provide cultural meaning to subsistence production. By ignoring the context, and attaching the same assumptions to the behaviour of all individuals or households, the variability that exists between individuals, communities, and regions is impossible to discern. Failing to account for the complexity and the context in which subsistence harvesting now occurs, we are left to simply tabulate the number of species harvested over a period of time, and calculate the economic and nutritional contribution in ways that can inform future allocations (i.e., minimum need levels). The belief that individual productivity must be measured and quantified through harvest studies is the most obvious example of how subsistence is now treated in the North.

Normalizing subsistence

Nearly 40 years ago Justice Thomas Berger (1977: 123) stated that:

> [i]t is self-deception to believe that large-scale industrial development would end unemployment and under-employment of Native people in the North. In the first place, we have always overstated the extent to which Native people are unemployed and underemployed by understating their continued reliance on the land. Secondly, we have never fully recognized that industrial development has, in itself, contributed to social, economic, and geographical dislocation among Native peoples.

Despite the impact of the Berger Report (1977), these words seem to have been lost on the Government of Canada who remain steadfast in their belief that the most expedient route for improving the social and economic conditions of Aboriginal communities in the North is through resource extraction. Such devotion is reflected in the Canada's 2009 Northern Strategy (Our North, Our Heritage, Our Future) (AAND, 2009) where mining and other major resource development projects were identified as "the cornerstones of sustained economic activity in the North and the key to building prosperous Aboriginal and Northern communities." As of December 2013, there were over 30 major industrial projects underway or moving through the permitting processes across the three northern territories. With a federal commitment of $25 billion in capital investment, the number of industrial projects in the North is expected to double by 2020. To meet the human resource needs of these industries, the Federal Government has committed to developing the skills, knowledge, and credentials of an Aboriginal workforce in order to keep pace with the North's fast-changing economy (AAND, 2009).

One example of the North's fast-changing economy can be found in northern Quebec, where the Quebec government's *Plan Nord,* promises "the orderly, respectful, and socially-responsible development for Northern Québec's natural resources over the next 25 years" (Government of Quebec, 2012). According to the Quebec government, an anticipated $80 billion in public and private investments will be directed to mining projects that in return will bring about substantial economic and social developments for northern Quebec's Aboriginal communities. Notwithstanding government's optimism, *Plan Nord* has been met with considerable opposition by Inuit and Cree communities who have heard such promises before only to "toil for a couple of dollars a day like slaves to help mining companies get rich off the development of Inuit lands" (George, 2011). As in the past, Canada's policies regarding economic development, and integration of Aboriginal communities into the "real" northern economy, remain premised on modernization schemes that often fall short of the promises made, and fail to consider other viable and culturally relevant forms of economy that still exist.

This is not to suggest that resources development is necessarily a bad thing. Nor is it possible or even preferable to return to a purely subsistence-based livelihood. Rather, if the impacts and benefits are managed effectively and equitably in terms of Aboriginal interests, then resource extraction can prove positive and even supporting of subsistence harvesting. For northern communities to benefit from resource development, however, we must not abandon the land-based economies of Aboriginal peoples by believing the only viable economic alternative for the North are vocations in the industrial economy. What I am calling for then is the normalization of subsistence, where Aboriginal peoples, if they choose, are provided the same support and opportunities as are being made available through resource extraction. The normalization of subsistence economies would necessarily involve a range of institutional support systems, alongside other forms of economy, that provide Aboriginal peoples the opportunity to exercise culturally acceptable norms and practices that can facilitate land-based livelihoods. The Hunter Support Programs (HSPs) that are used in some regions of the North are but one example. With funding provided by Aboriginal governments, and in some cases industry, HSPs aim to sustain local livelihoods, support the nutritional needs of community members, mitigate the costs associated with accessing traditional lands and resources, and support culturally important land-based traditions that might otherwise be hard to maintain. In these cases Aboriginal governments have implemented HSPs to ensure the subsistence economy continues for economic reasons but equally so for maintaining the cultural values embodied in subsistence harvesting. In this way, the normalization of Aboriginal subsistence-based economies involves adapting to new economic conditions without having to abandon cultural and economic traditions. For too long Canada's northern policies have been directed to "poor economic outcomes in lagging communities and regions" and have typically been undertaken in the absence of any consideration for the subsistence economies of Aboriginal communities. Yet it is the subsistence economy of Aboriginal peoples that could benefit most readily from government support. By normalizing subsistence economies, though HSPs or other institutional mechanisms, an economic basis could be provided that would invigorate local institutions and perpetuate the traditional values that have long been embedded in wildlife harvesting.

Conclusion

As early as the 1940s, social scientists in Canada began to step outside their traditional ethnographic confines by offering government counsel on how best to transition Aboriginal peoples to more productive forms of economy in the face of demographic extinction (Dyck, 2006: 81). With few exceptions, early anthropological research assumed evolutionary undertones and employed theories of acculturation and modernization to explain the rapid changes occurring in Aboriginal society. The influential role of anthropology, as well as other social sciences, was further spurred by large-scale resource development projects, where concepts of modernization and industrialization were presented as synonymous. Yet by conceptualizing subsistence as merely food-getting activities rather than an integral part of complex cultural systems, academics did much to advance the notion that the industrialization of the North was inevitable, and the expert advice of some social scientists was used by governments as endorsement that northern resource development was a "natural" and inevitable alternative to subsistence production.

Fortunately the theories of acculturation and modernization that were once so influential have been rejected and more or less relegated to the past. However, other theories have stepped in, and are in many ways reminiscent of the social theory that once justified the intensive involvement of government in the lives of Aboriginal peoples. While I have made a similar argument elsewhere (see Haalboom and Natcher, 2012), the more recent characterization of Aboriginal communities as vulnerable is a poignant example. Here researchers have suggested once again that some northern Aboriginal communities lack the capacity to adapt to the social, economic, and environment changes occurring in the North due to their over-reliance on narrow economic bases and a general diminution of human resources. Remote northern communities that continue to pursue subsistence-based ways of life are considered particularly vulnerable, as are high consumers of traditional foods. The policy recommendations stemming from this body of scholarship include: 1) diversifying local economies and creating more wage earning opportunities; 2) securing funding from government and other institutional support systems; 3) greater integration into the larger market economy; and 4) relocation of those communities deemed most vulnerable. To reverse the vulnerable conditions that northern Aboriginal communities are now experiencing, it is recommended that outside support and externally generated solutions are required. It should not escape the reader that these same recommendations were made nearly a century ago, and were informed by the theories of acculturation and modernization that we are so critical of today. For too long the conditions of Aboriginal subsistence economies has escaped critical critique, leaving theories of acculturation, modernization, and more recent variations to go unchallenged. Yet the characterization of Aboriginal people and their livelihoods as "real" or informal, vulnerable, or adaptive, are not unproblematic typologies, but rather have very real implications. Rather than trying to locate Aboriginal economies in one place or another, we must strive to account for the complexity and hybridity of Aboriginal economies in their contemporary form and, when appropriate, and in cooperation with Aboriginal peoples themselves, offer direction to governments on how best to be supportive.

Note

1   It must be noted that each of these scholars has made important contributions to scholarship and public policy. Tanner in particular is regarded as one of Canada's most eminent anthropologists and his commitment to advancing Aboriginal rights in Canada could never be called into question.

References

Aboriginal and Northern Affairs Canada, 2009. *Northern Strategy: Our North, Our Heritage, Our Future.* Ottawa: Aboriginal and Northern Affairs Canada.

Avrith-Wakeam, G., 1993. George Dawson, Franz Boas and The Origins of Professional Anthropology in Canada. *Canadian Journal of the History of Science, Technology and Medicine*, 17(1–2): 185–203.

Berger, T. R., 1977. *Northern Frontier, Northern Homeland: The Report of the Mackenzie Valley Pipeline Inquiry.* Ottawa: Minister of Supply and Services.

Bodley, J. H., 1975. *Victims of Progress.* Lanham, MD: Alta Mira Press.

Buchanan, C., 2006. Canadian Anthropology and Ideas of Aboriginal Emendation. In J. Harrison and R. Darnell (eds.), *Historicizing Canadian Anthropology*. Vancouver: University of British Columbia Press: 93–106.

Chabot, M., 2003. Economic Changes, Household Strategies, and Social Relations of Contemporary Nunavik Inuit. *Polar Record*, 39(208): 19–34.

Chance, N. (ed.), 1968. *Conflict in Culture: Problems of Developmental Change Among the Cree.* Ottawa: Canadian Research Centre for Anthropology, Saint Paul University.

Charest, J., 2011. Plan Nord: A Word from the Premier. Government of Quebec, available at: http://plannord.gouv.qc.ca/english/messages/jean-charest.asp (accessed 8 December 2015).

Collings, P., 2014. *Inummarik: Men's Lives in an Inuit Community.* Montreal: McGill-Queen's University Press.

Council of Canadian Academies, 2014. Aboriginal Food Security in Northern Canada: An Assessment of the State of Knowledge. The Expert Panel on the State of Knowledge of Food Security in Northern Canada, Council of Canadian Academies. Ottawa, ON.

Dombrowski, K., 2007. Lifestyle and Livelihood: Culture Politics and Alaska Native Subsistence. *Anthropologica*, 49(2): 211–230.

Duhaime, G. and N. Bernard, 2008. *Arctic Food Security.* Edmonton: Canadian Circumpolar Press.

Duhaime, G., M. Chabot, and M. Gaudreault, 2002. Food Consumption Patterns and Socio-economic Factors Among the Inuit of Nunavik. *Ecology of Food and Nutrition*, 41(2): 91–118.

Dyck, N., 2006. Canadian Anthropology and the Ethnography of "Indian Administration." In J. Harrison and R. Darnell (eds.), *Historicizing Canadian Anthropology*. Vancouver: University of British Columbia Press: 78–92.

Fall, J., 2014. *Subsistence in Alaska: A Year 2012 Update.* Anchorage: Department of Fish and Game, Division of Subsistence.

Fall, J., 1990. The Division of Subsistence of the Alaska Department of Fish and Game: An Overview of its Research Program and Findings. *Arctic Anthropology*, 27(2): 68–92.

Ferguson, M. and F. Messier, 1997. Collection and Analysis of Traditional Ecological Knowledge about a Population of Arctic Tundra Caribou. *Arctic*, 50(1): 17–28.

Ford, J. and L. Berrang-Ford, 2009. Food Security in Igloolik, Nunavut: An Exploratory Study. *Polar Record*, 45(234): 225–236.

232   *David Natcher*

Furgal, C., S. Hamilton, S. Meakin, and V. Rajdev, 2012. Policy Options and Recommendations for Addressing Food (In)Security in Nunavut: Synthesis Document. Iqaluit, (NU): Government of Nunavut Department of Health and Social Services.

Gilchrist, G., M. Mallory, and F. Merkel, 2005. Can Local Ecological Knowledge Contribute to Wildlife Management? Case Studies of Migratory Birds. *Ecology and Society*, 10(1): 20 [online], available at: www.ecologyandsociety.org/vol10/iss1/art20/.

George, J., 2011. Nunavik Wants Better Information on Plan Nord: KRG Nunatsiaq Online, November 30, available at: www.nunatsiaqonline. ca/stories/article/65674nunavik_wants_better_information_on_plan_nord_krg/ (accessed 8 December 2015).

Gombay, N., 2010. *Making a Living: Place, Food, and Economy in an Inuit Community*. Saskatoon: Purich Publishing.

Haalboom, B. and D. C. Natcher, 2012. The Power and Peril of "Vulnerability": Lending a Cautious Eye to Community Labels in Climate Change Research. *Arctic*, 65(3): 319–327.

Halperin, R. H., 1994. *Cultural Economies: Past and Present*. Austin: University of Texas Press.

Harder, M. and G. Wenzel, 2012. Inuit Subsistence, Social Economy and Food Security in Clyde River, Nunavut. *Arctic*, 65(3): 305–318.

Hedican, E., 1995. *Applied Anthropology in Canada: Understanding Aboriginal Issues*. Toronto: University of Toronto Press.

Hobart, C. W., 1982. Wage Employment and Cultural Retention: The Case of the Canadian Inuit. *International Journal of Comparative Sociology*, 23(1): 47–61.

James Bay and Northern Quebec Native Harvest Research Committee (JBNQNHRC, 1988. Final Report: Research to Establish Present Levels of Harvesting for the Inuit of Northern Quebec, 1976–1980. Quebec James Bay and Northern Quebec Native Harvest Research Committee, Quebec City.

Jenness, D., 1978. The Economic Situation of the Eskimo. In V. F. Valentine and F. G. Vallee (eds.), *Eskimo of the Canadian Arctic*. Toronto: Macmillan.

Kishigami, N., 2008. A Cultural Anthropological Study of Subsistence Activities with Special Focus on Indigenous Hunting, Fishing and Gathering in the Arctic Regions. Paper presented at the 2008 International Conference on Arctic Social Science, Nuuk, Greenland.

Kuhnlein, H. V., 2009. Back to the Future: Using Traditional Food and Knowledge to Promote a Healthy Future among Inuit. In H. V. Kuhnlein, B. Erasmus, and D. Spigelski (eds.), *Indigenous Peoples' Food Systems: the Many Dimensions of Culture, Diversity and Environment for Nutrition and Health*. Rome: United Nations Food and Agriculture Organization.

Kulchyski, P. and F. J. Tester, 2008. *Kiumajut: Game Management and Inuit rights, 1900–70*. Vancouver: University of British Columbia Press.

Labrador Inuit Land Claims Agreement (LILCA), 2003. Queen's Printer for Canada, Ottawa.

Lonner, T. D., 1980. Subsistence as an Economic System in Alaska: Theoretical and Policy Implications. Alaska Department of Fish and Game, Division of Subsistence, Technical Paper Number 67. Anchorage, Alaska.

Lotz, J., 1970. *Northern Realities: The Future of Northern Development in Canada*. Toronto: New Press.

Natcher, D. C., D. Castro, and L. Felt, 2015. Hunter Support Programs and the Northern Social Economy. In Chris Southcott (ed.), *Northern Communities Working Together: The Social Economy of Canada's North*. Toronto: University of Toronto Press: 189–203.

Nunavut Land Claims Agreement (NLCA), 1993. Queen's Printer for Canada, Ottawa.

Nunavut Wildlife Management Board (NWMB), 2004. The Nunavut Wildlife Harvest Study: Final Report. Iqaluit, Nunavut.

Polanyi, K., 1957. *The Great Transformation: The Political and Economic Origins of Our Time*. Boston: Beacon Press.

Sahlins, M., 1971. The Intensity of Domestic Production in Primitive Societies: Social Inflections of the Chayanov Slope. In G. Dalton (ed.), *Studies in Economic Anthropology*. Washington, DC: American Anthropological Association: 30–51.

Sandlos, J., 2007. *Hunters at the Margin: Native People and Wildlife Conservation in the Northwest Territories*. Vancouver: University of British Columbia Press.

Tanner, A., 1966. *Trappers, Hunters and Fishermen: Wildlife Utilization in the Yukon Territory*. Department of Northern Affairs and National Resources, Northern Coordination and Research Centre. Ottawa: Canada.

Thornton, T., 2001. Subsistence in Northern Communities: Lessons from Alaska. *The Northern Review*, 23: 82–102.

Usher, P., 1993. Northern Development, Impact Assessment, and Social Change. In N. Dyck and J. B. Waldram (eds.), *Anthropology, Public Policy, and Native Peoples in Canada*. Montreal and Kingston: McGill-Queen's University Press: 98–130

Wein, E. E. and M. M. R. Freeman, 1995. Frequency of Traditional Food Use by Three Yukon First Nations Living in Four Communities. *Arctic*, 48(2): 161–171.

Weiner, M., 1966. *Modernization: The Dynamics of Growth*. New York and London: Basic Books, Inc.

Wenzel, G. W., 1991. *Animal Rights, Human Rights: Ecology, Economy and Ideology in the Canadian Arctic*. Toronto, ON: University of Toronto Press.

Wenzel, G., 2005. Canadian Inuk Subsistence and Economy. In L. Muller-Wille, M. C. S. Kingsley and S. S. Nielson (eds.), *Socio-Economic Research on Management Systems of Living Resources: Strategies, Recommendations and Examples. Proceedings of the Workshop on Social and Ecological Research Related to the Management of Marine Resources in West Greenland*. Nuuk: Greenland Institute of Natural Resources: 146–151.

Wheeler, P. and T. Thornton, 2005. Subsistence Research in Alaska: A Thirty-Year Retrospective. *Alaska Journal of Anthropology*, 3: 69–103.

Williamson, T., 1997. *From Sina to Sikujâluk: Our Footprint. Mapping Inuit Environmental Knowledge in the Nain District of Northern Labrador*. Prepared for the Labrador Inuit Association. Nain, Labrador.

Wynn, G., 2007. Foreword: The Enigmatic North. In J. Sandlos (ed.), *Hunters at the Margin: Native People and Wildlife Conservation in the Northwest Territories*. Vancouver: University of British Columbia Press: xi–xx.

Young, R. A. and P. McDermott, 1988. Employment Training Programs and Acculturation of Native Peoples in Canada's Northwest Territories. *Arctic*, 41(3): 195–202.

13 Traditional knowledge and resource development

Henry P. Huntington

What is today called "traditional knowledge"[1] has been throughout most of human history the primary way that people have understood their surroundings and acted on that knowledge. In much of the world and in many respects, traditional knowledge remains the main source of understanding (e.g., Johannes 1981; Berkes 1999). Scientific knowledge, which we may broadly contrast with traditional knowledge by saying that it derives from formal and explicitly structured inquiry rather than from repeated experience, is simply not available to inform most ways that people interact with their environment (e.g., Johannes 1998). At the same time, where scientific knowledge is available, it is typically regarded as the optimal source of objective and reliable information, with the result that traditional knowledge is pushed aside or regarded as a potential source of data that can be incorporated within a scientific paradigm (e.g., Van de Velde, Stirling and Richardson 2003).

Thanks in large part to political and intellectual activism by indigenous peoples' organizations (e.g., Brooke 1993), together with leadership from scholars willing to listen to local expertise (e.g., Ferguson and Messier 1997; Carmack and Macdonald 2008), traditional knowledge has become a popular topic in the Arctic. Many studies have been carried out to document or otherwise examine traditional knowledge, major assessments such as the Arctic Climate Impact Assessment (ACIA 2005) have made space for reviewing or reporting at least some components traditional knowledge, and much has been written about the philosophy, practice, and politics of traditional knowledge (e.g., Agrawal 1995; Cruikshank 1998; Ellis 2005, Huntington 2005). In one sense, this level of activity has been beneficial, drawing serious attention to systems of knowledge that have served Arctic peoples well through millennia, and that offer alternative ways of looking at the world and the place of humans.

In another sense, the recent emphasis on traditional knowledge has obscured the deeper roots of collaboration between visitors and locals in much of the Arctic. Explorers such as Roald Amundsen were careful to pay close attention to local practices (Amundsen 1908). Laurence Irving recognized the expertise that the Iñupiaq Simon Paneak from the Brooks Range in northern Alaska offered in the field, and acknowledged it publicly by making Paneak a co-author on scientific publications (Irving and Paneak 1954). More troublesome than historical amnesia, the politicization of traditional knowledge as a concept and as a field of practice has made it difficult to offer sensible critiques of when and how traditional knowledge is a valuable resource to draw upon (e.g., the exchange in Gilchrist, Mallory and Merkel 2005; Brook and McLachlan 2005; Gilchrist and Mallory 2007), leading on occasion to papers that posit dire consequences for allowing traditional knowledge to achieve formal standing in legal and regulatory matters (e.g., Howard and Widdowson 1996). When incorporating traditional knowledge becomes a requirement of every study or activity, there is a risk that it becomes an item on a checklist to be noted only in passing. This kind of tokenism can be seen where the use of traditional knowledge is required, whether traditional knowledge is likely to be relevant or not.

This chapter examines the ways in which traditional knowledge has been applied to resource development, specifically to identify where further research would be useful. With this in mind, we can divide the practice of engaging traditional knowledge into four categories. First, traditional knowledge is used by its holders in everyday life. This is where it is generated, where it is refined, and where it is first applied. For the most part, and with some exceptions, such practices of traditional knowledge have been ignored by outside researchers (e.g., discussion in Ingold and Kurttila 2000; Nageak 1991). Second, traditional knowledge has been documented by many researchers, either for its own sake as a means of recording valuable information, or as a starting point for further analyses of the physical or biological environment (e.g., Lewis et al. 2009; Weatherhead, Gearheard and Barry 2010). There has been much activity in this area, including the development of suitable methods (e.g. Huntington 1998, 2000; Huntington et al. 2004, 2011), recognition of the depth and relevance of traditional knowledge (e.g., Berkes 1999; Mallory et al. 2003, 2006; Fraser et al. 2006), and, as seen in the mention of traditional knowledge in land claims settlements and elsewhere, increasing awareness among the holders of traditional knowledge that what they know has value to others as well.

Third, traditional knowledge has been applied to the management of hunting and fishing or to addressing challenges such as climate change (e.g., Huntington et al. 1999; Mymrin et al. 1999; Berkes and Jolly 2001; George et al. 2004; Gearheard et al. 2006; Dowsley 2007; Noongwook et al. 2007). This, too, is a robust area of activity, closely related to documenting traditional knowledge, as many studies have focused on establishing the sizes of populations or the details of local uses to help support sound management and especially the continuation of traditional practices (e.g., Ferguson and Messier 1997; Noongwook et al. 2007). Finally, traditional knowledge has been applied to various aspects of resource development. Here, we refer specifically to the commercial exploitation of mineral as well as living resources. As Keeling et al. point out in their chapter dealing with environmental impacts, traditional knowledge has primarily been applied to renewable resources, though it has occasionally been involved in non-renewable resource development. Traditional knowledge has helped identify development prospects and, to a greater degree, been drawn upon for assessing and mitigating environmental impacts associated with industrial activity (e.g., Nakashima 1990; Stevenson 1996; Usher 2000; Ellis 2005). This category is the focus of this chapter, but the relatively recent surge of activity in this area makes it helpful to refer also to other categories by way of analogy to the case of resource development.

The chapter begins with a review of research and related activities concerning traditional knowledge, particularly as it relates to resource development. Then it examines the ways in which traditional knowledge has been and is being applied in this context, noting where engaging with traditional knowledge is seen to be important for various reasons. Next, it considers how the recent emphasis on traditional knowledge has or has not benefited northern communities, before concluding with the identification of key areas for further research and the prospects for carrying out such studies. It is important to emphasize that this review and analysis is not about traditional knowledge itself, but instead about the policies and practices of studying and drawing upon traditional knowledge in the course of resource development.

Research and related activities concerning traditional knowledge

Scientists have relied on traditional knowledge holders in the Arctic for a long time, whether for support in the field or for insights into the workings of the natural world. Laurence Irving was unusual for his time in acknowledging the intellectual contributions made by his partner Simon Paneak. More typically, the role of local experts was relegated to the acknowledgments or omitted altogether. In the 1970s and 1980s, large-scale studies of land use and occupancy gave more prominence to how well indigenous peoples know their home landscapes, often on a vast geographical scale (e.g., Freeman 1976). From there, it was a relatively short step to documenting observations of animals and other phenomena, and then to discussing animal behavior and other aspects of biology and ecology (e.g., Ferguson and Messier 1997; Berkes 1999; Huntington et al. 1999).

In Alaska, the involvement of the International Whaling Commission in the management of the bowhead whale hunt led first to a political crisis (the hunt was initially curtailed, and then opened again but under a quota system that remains in place today), and then to a vigorous research program to better count the whale stock and learn more about its behavior (Huntington 1992; Albert 2001). The whale census benefited greatly from the advice of experienced whaling captains, who suggested counting when the sea was covered with ice and also looking farther offshore than could be seen from observation posts along the edge of the landfast sea ice. The result was a much improved—and higher—population estimate, and thus less pressure to reduce or end the hunt. At the time, the fact that this sequence of events depended heavily on "traditional knowledge" was recognized by the participants but not particularly noted in publications or otherwise in the public eye. Only later was the story re-cast in terms of the integration of traditional and scientific knowledge. Nonetheless, it remains a prominent example of the power of collaboration.

In Canada, the work of Nakashima and Murray (1988) in the 1980s, documenting Inuit knowledge of eiders in Hudson Bay, was an early explicit acknowledgment of the expertise held by hunters and the value of studies that tried to access that expertise and make it available for a wider audience. Their example helped spur interest in the field and support for further studies of this kind. It also helped make it acceptable for biologists and others to acknowledge more fully what they had learned from indigenous hunters and how they had used those insights to guide their scientific inquiries and thus learn even more. More studies were undertaken across Canada (e.g., Ferguson and Messier 1997; Lemelin et al. 2010), in Greenland (e.g., Thomsen 1993), in Alaska (e.g., Huntington et al. 1999; Noongwook et al. 2007), and in Russia (e.g., Mymrin et al. 1999), a trend that has continued in most regions.

Nakashima and Murray's work was in the context of the James Bay Hydro Project, assessing impacts of large-scale river diversions on the waters of Hudson Bay and, in this case, the eiders that inhabited those waters. The James Bay Hydro Project acknowledged the role of traditional knowledge in environmental assessment (e.g., Mailhot 1993), an application of traditional knowledge explored as well by Stevenson (1996) in the Northwest Territories. For the most part, however, the stated purpose of traditional knowledge studies related more to wildlife management, as in the case of the Alaska bowhead whales, than to resource development.

Not long after traditional knowledge became a topic of academic as well as practical interest, other scholars began providing commentaries on the nature of traditional knowledge and then on the political and sociological context in which traditional knowledge studies were carried out. Specifically, critics noted that the documentation of traditional knowledge and calls for its use in resource management typically entailed taking that knowledge out of its own context (e.g., Cruikshank 2001), comparing it with scientific sources (at times under the guise of "validating" traditional knowledge by making sure it agreed with scientific knowledge), and then using it to claim that management decisions were done in cooperation with local hunters and thus culturally legitimate (e.g., Nadasdy 1999; Spaeder 2005). A chief component of such critiques was the observation of unequal power relations between government agencies and trained scientists on one hand, and indigenous hunters with little expertise in bureaucratic processes on the other (e.g., Nadasdy 2003; Ellis 2005; Tyrrell 2008).

Further criticism centered around the legitimacy of comparing knowledge or attempting to confirm or refute one system of knowledge in terms of the other (e.g., Cruikshank 1998, 2001). Not surprisingly, the attempts to critique scientific knowledge for failing to match traditional knowledge were typically confined to anecdotes and informal interactions among traditional knowledge holders and those who sought to identify with them, whereas the evaluations of traditional knowledge in scientific terms were often published in scientific journals (e.g., Gilchrist, Mallory and Merkel 2005). This is not to say that no comparisons are valid, nor that traditional knowledge is always correct, but simply to underscore the point that power relations between practitioners of traditional knowledge and practitioners of scientific research or government resource management are un-equal, and that this inequality has a bearing on how traditional knowledge is used and discussed.

Despite these potential pitfalls, indigenous organizations have continued to advocate greater use of traditional knowledge in science, in resource management, and in resource development, especially for environmental impact assessment and mitigation. To date, however, little research has been done to examine how traditional knowledge is actually used in practice in resource development and environmental assessment. There are some examples, as noted in the next section, but documentation is typically sparse, and the degree to which information from traditional knowledge influenced the outcomes has not been evaluated.

The application of traditional knowledge to resource development

In Canada, the James Bay Hydro Project was an early example of attention to traditional knowledge in the planning and impact assessment of resource develop-ment. The project dammed several major rivers flowing into James and Hudson Bays to generate electricity, affecting the lands and waters of Cree and Inuit. Nakashima (1990) and Mailhot (1993) conducted studies or examined the role of traditional knowledge in environmental assessment for this project.

Building on that example, the Inuvialuit Final Agreement in the Northwest Territories and Yukon established five co-management boards to govern wildlife management, fisheries management, and environmental impacts (Smith 2001). The last of these is covered by the Environmental Impact Screening Committee and the Environmental Impact Review Board. As with the other co-management bodies, membership on these two is divided evenly between Inuvialuit (appointed by the Inuvialuit Game Council) and government (territorial and federal). The work of the boards is supported by a Joint Secretariat based in Inuvik. Subsequent land claims agreements in Northern Canada have incorporated similar provisions, giving at least some mention to traditional knowledge and establishing co-management bodies in one form or another. In this region, community conservation plans developed in each of the six Inuvialuit communities identify important areas where resource development and other intensive non-local human activity should be limited or prohibited, a useful contribution of traditional knowledge to the management of development activities that may affect local practices and the species and habitats that support those practices.

Other land claim agreements in Northern Canada have had similar provisions for the use of traditional knowledge and the establishment of co-management or local management bodies (e.g., the Gwichin Comprehensive Land Claim Agree-ment in 1992, the Sahtu Dene and Métis Comprehensive Land Claim Agreement of 1993, the Champagne and Aishihik First Nations Final Agreement in 1995, and others that followed).

A fundamental premise of the co-management system is that the indigenous or local members are holders of traditional knowledge and thus bring with them the values, perspectives, and insights inherent to that system of knowledge. Decisions are therefore based on both scientific knowledge (presumably represented by government agency personnel as well as available publications) and traditional knowledge. In practice, scientific knowledge is more readily available via publications, and it is unlikely that even the most experienced among the indigenous participants can in fact provide from his or her own experience all the relevant traditional knowledge needed to inform a given decision about resource development. And yet, each person can rarely bring more than his or her experience, because little is formally documented and available to a wider audience. Nonetheless, the system provides a formal mechanism for the direct involvement of traditional knowledge holders in the oversight of resource development, a practice that does not exist outside of Canada (e.g., Usher 2000).

In Nunavut, the territorial government has emphasized traditional knowledge (known in Nunavut as *Inuit Qaujimajatuqangit* or IQ) and its documentation and use in resource management and resource development. This work stems from the Nunavut Land Claims Agreement and is often carried out as part of the Inuit Impact and Benefit Agreements that must be completed for each development project. The Nunavut Wildlife Management Board, the Nunavut Planning Commission, and other institutions of public governance in the territory have invested in traditional knowledge studies, the creation of a traditional knowledge database, and other such steps to facilitate greater use of IQ. In practice, although much basic research has been carried out, the results are harder to see. Few studies have been made public, and the implementation of the policy of utilizing IQ has been criticized in various ways over the years. Consistent with the Inuvialuit experience, Bonesteel (2006) observed that real progress on the incorporation of IQ into governance will require more Inuit holding positions of senior management in government, where they can both put the policies into practice and also apply their own knowledge to the decisions they make.

In Alaska, traditional knowledge has gained visibility over the past two decades, and is now mentioned in many environmental impact assessments, science plans, and other such documents. The Environmental Impact Statement for the Northstar oil prospect in the Beaufort Sea (U.S. Army Corps of Engineers 1999) was one of the first to make a concerted effort to document and incorporate traditional knowledge. Subsequent attention to traditional knowledge has perhaps fallen short of expectations in indigenous communities and of what may be possible (e.g., Holland-Bartels and Pierce 2011), but much work continues to be done. Building on this experience, the Bureau of Ocean Energy Management recently highlighted its efforts to document traditional knowledge and to incorporate the results of such studies into its decision-making process (BOEM 2012). That document includes an interview with Fenton Rexford, Tribal Administrator for the Native Village of Kaktovik. Rexford emphasizes the importance of government-to-government[2] consultation as a means of ensuring that traditional knowledge is given its proper role in resource management decisions, again consistent with experiences in Canada.

The government-to-government consultation process in Alaska led to the documentation of traditional knowledge for the U.S. Environmental Protection Agency's (EPA) permitting process for discharges from oil and gas platforms in Cook Inlet, southern Alaska. The tribes of the region asked that traditional knowledge of the impacts of such discharges play a role in the decision that was to be made. The EPA hired a consultant to conduct interviews in the Cook Inlet tribal communities and report the findings. The results described the dependence of tribal members on clean waters in the Inlet, so that the fish, marine mammals, seabirds, and other food they obtained from the sea were healthy and nutritious, allowing tribal members to carry on their traditional cultural practices. While important to the overall context of resource development in the Inlet, this information was not directly relevant to the formulation of technical regulations concerning discharges. The tribes were successful in that they attained a role for traditional knowledge in the overall permitting process, but less successful in that the influence of traditional knowledge on the ultimate decision created an indirect role for tribal governments rather than a direct one, as is the case for Inuvialuit in Canada. The EPA has, however, continued to invest in traditional knowledge studies on the North Slope of Alaska, also in regard to offshore oil and gas activity (e.g., Stephen R. Braund & Associates 2011).

Not all resource management and resource development agencies are open to the application of traditional knowledge to their decisions. The Scientific and Statistical Committee of the North Pacific Fisheries Management Council, for example, has been skeptical about the reliability of traditional knowledge and its objectivity when it comes to setting and allocating fishing quotas (e.g., Scientific and Statistical Committee 2010). This reluctance is expressed in terms of a need to maintain the highest scientific standards, as is appropriate when considerable economic interests affecting tens of thousands of people are at stake. At the same time, the standards for peer review seem to make an *a priori* distinction between scientific knowledge (subject to careful scrutiny but assumed to be objective and reliable) and traditional knowledge (assumed to be unreliable until "verified and validated"). While there is no doubt that traditional knowledge can be flawed and that it can entail a conflict of interest among those who have a stake in the resulting resource management or resource development decisions, the suggestion that traditional knowledge must be verified and validated (presumably within a scientific paradigm, rather than, say, through review by appropriate peers) appears to be an unusual case in the Arctic today.

By contrast, proponents of the Pebble Mine, a major mining project in southwestern Alaska, have commissioned a number of studies to assess potential impacts to the region and especially its rich salmon runs (www.pebbleresearch. com). One of the studies focused on subsistence and traditional knowledge (Chapter 23 of the study available at the website above), and was carried out by the same company that conducted the assessment of cultural resources (Chapter 22) and assisted with the assessment of socioeconomics (Chapter 21). The report on subsistence and traditional knowledge emphasizes subsistence data, which is also the case for a report prepared by the same contractor for the EPA for northern

Alaska (Stephen R. Braund & Associates 2011). In fall 2012, the Keystone Center ran a series of independent science panels to review the studies commissioned by the Pebble Partnership. One of these looked at the traditional knowledge studies that have been undertaken, along with other social and cultural chapters of the overall study. Videos of the science panel discussions are available (https:// keystone.org/policy-initiatives-center-for-science-a-public-policy/environment/ 441-pebble-dialogue-home.html), but the reports have not yet been posted on the Keystone website. In short, "traditional knowledge" as a general concept is being taken seriously, but the definition of what constitutes "traditional knowledge" is still unclear. The impact that traditional knowledge studies have on final decisions is similarly unclear at this point.

In Greenland, the governance system is different. In 1979, Denmark's colonial administration ended with the granting of a home rule system to the island and its inhabitants. In 2009, the Greenland Government's powers were expanded under the Act on Greenland Self-Government. In principle, Greenlanders now manage their own affairs, and in that sense, the questions of power relations between managers and managed should disappear. In practice, however, there remains a gap between government agency personnel, typically university educated, and hunters and fishers, still practicing variations on traditional practices. Power relations are still uneven, and actual reliance on local observations and understanding remains an emerging practice in Greenland, even if it has long been supported in theory by the Greenland Government (1999). A significant feature of the Greenland experience, however, is that greater involvement of local hunters and fishers has led to more rapid management action (e.g., Danielsen et al. 2005), a notable point when it comes to the application of traditional knowledge alongside local participation.

In terms of resource development, however, decisions in Greenland remain largely top down (Huntington et al. 2012). Environmental impact assessment is a developing field in Greenland (Hansen and Kørnøv 2010), and so the role of traditional knowledge is still uncertain, as is the role of "civil society" in Greenland's internal affairs (Aqqaluk Lynge, pers. comm., 2010).

Benefits to northern communities

Resource development offers economic benefits to northern communities, and also entails risks of environmental, social, and cultural disruption (e.g., AMAP 2010). Anything that can help maximize benefits and minimize risks is a plus. The greater use of traditional knowledge in resource development planning and regulation is often promoted on the grounds that it will improve outcomes, while also offering economic benefits to communities while studies are carried out, e.g., via payments to participants. There is also a benefit associated with having one's knowledge respected, with being acknowledged as an important contributor to collective understanding and sound decision-making. For people who have often been ignored in the process of setting regulations or making major decisions, this latter benefit should not be underestimated.

In this way, traditional knowledge can, theoretically at least, increase local control of the project in the sense that issues the community feels are important, that arise out of traditional knowledge, can be more adequately dealt with. Traditional knowledge may inform communities and developers of potential problems that would lead to significant negative impacts if not dealt with. By including full attention to traditional knowledge as part of the planning process, there is less of a chance that these concerns would be ignored or overlooked. It is thus a means of increasing local voice and as such increasing local control of a project.

These benefits are real, in the sense that one can point to examples where local engagement has produced all of these outcomes. But they are also theoretical, in the sense that it remains unclear how widespread these benefits are and whether what has been accomplished is really the realization of the full potential of traditional knowledge to contribute to resource development and its regulation.

Through co-management boards, institutions of public governance, and other mechanisms, Arctic peoples have become increasingly engaged in the process of making decisions that concern the lands, waters, and species they use. This engagement ranges from providing advice to actually making the final decisions. What role traditional knowledge plays in the actions of local members of such bodies is not clear. How often the use of traditional knowledge has led to better decisions is similarly unknown. Whether co-management approaches lead to real engagement of indigenous peoples on their own terms, or to some degree of co-optation as decisions are given the semblance but not the substance of local input, remains to be evaluated.

The idea of paying local participants in traditional knowledge studies has becoming increasingly accepted, though is far from universal. This source of income is one clear benefit, at least to those who are selected to take part, but such payments typically account for the time taken during an interview or workshop, rather than the depth of expertise being offered. Researchers and research assistants conducting the project (who may also be local residents) are more often paid a regular wage or salary, and thus typically earn far more than those who are interviewed. To be fair, this discrepancy reflects the amount of work required of each side, as the researcher had to obtain funding, manage the project, analyze responses, and write up the results. Nonetheless, the person being interviewed has accumulated his or her expertise over a long period. So what is the appropriate rate of compensation for that expertise? And are community members given opportunities to play larger roles in the planning, conduct, and follow-up of such projects? In other words, how do actual payments reflect potential payments, and to what degree is this benefit being achieved from the point of view of local residents?

Finally, ample anecdotal evidence suggests intangible benefits such as increased pride, self-respect, and community standing as a result of being recognized as an expert and a valuable contributor to an important task. But this benefit has not been examined formally, nor have potential drawbacks been considered, such as the difficulty of standing out in societies that value humility and cooperation over spotlights and personal ambition. Co-management approaches have been criticized for creating two classes in communities that previously did not make such distinctions, at least not as formally (e.g., Caulfield 1997).

Research needs and prospects

The battle to establish traditional knowledge as a respectable field of study has been won. Many studies of traditional knowledge have been funded, many articles written, and much recognition provided in terms of stated expectations about the inclusion of traditional knowledge in Arctic research, management, and policy. This advance extends to the application of traditional knowledge to resource management especially, but also to some extent to the resource development. None of these steps is without some resistance and controversy, but the balance of both theory and practice is undoubtedly on the side of greater inclusion of traditional knowledge.

What remains to be seen is how secure these gains are. With the growing political clout of Arctic indigenous peoples, it seems unlikely that traditional knowledge will fade from its current prominence. At the same time, however, there remains a gap between rhetoric and outcomes, and the failure of traditional knowledge studies and applications to deliver on the promises of their advocates would undermine at least some of the gains made in the past two or three decades. Enough experience has been accumulated by now to allow an evaluation of what the study and engagement of traditional knowledge have actually provided to communities, managers, and others; what obstacles need to be addressed; and what approaches have been most effective in realizing the potential benefits of engaging a different system of understanding in making sound decisions. Here, I suggest six inter-related areas for further study, to better understand how far the field of traditional knowledge has come, and which avenues are most promising for further growth. While these questions focus on the role of traditional knowledge in resource development, experiences in resource management may shed additional light where resource development cases are few or otherwise difficult to study.

How has traditional knowledge influenced resource development in the Arctic?

A great deal of effort has gone into promoting the idea that traditional knowledge has a lot to offer and should, as a matter of justice to indigenous peoples or of intellectual openness, be fully engaged in making important decisions that affect Arctic peoples and their environment. The result has been a widespread effort to, at a minimum, conduct studies and include mention of traditional knowledge in environmental impact assessments and the like. Many of these efforts, however, have not been publicly reported, and for those that have, little has been done to evaluate whether in fact the attention to traditional knowledge has led to recognizable differences in decision outcomes. In other words, is the attention given to traditional knowledge having an impact? If so, what is the nature of that impact, and how has it been achieved? If not, what barriers appear to be blocking the path towards greater use of traditional knowledge? A review of major development projects in Arctic areas in the past decade would help shed light on this question, and ideally would help identify effective ways to realize the greatest benefits from applying traditional knowledge to meeting the challenges of Arctic resource development.

What is the role of power relations in determining if and how traditional knowledge influences decision outcomes?

Part of the rationale for greater inclusion of traditional knowledge in research and in management is that it can help address the power imbalance between representatives of the dominant society and members of marginalized communities. Through land claims settlements, co-management, and other political achievements, many Arctic peoples have made substantial strides in the past few decades towards greater influence over their own futures. But, the resources available to government and industry remain far greater than those commanded by most indigenous governments and organizations. Comparing the availability of scientific information and traditional knowledge offers a case in point: far more scientific studies than studies of traditional knowledge have been published or otherwise reported, and thus are available for reference when making decisions. The inclusion of traditional knowledge holders on co-management boards is a welcome development, but even when represented in equal numbers, they are likely to lack the institutional support available to other members of the co-management bodies. In this view, traditional knowledge risks being treated as a source of data to be used in a scientific paradigm, rather than as a coherent system of understanding in its own right. Sufficient case studies are available to allow a comparative analysis of power relations and outcomes, to better understand the degree to which power imbalance is an obstacle to greater acceptance of and reliance on traditional knowledge in the resource development arena.

How do holders of traditional knowledge view their experiences in influencing resource development?

Both the proponents of using traditional knowledge and those who are trying to meet those expectations have claimed success in much of the work done to date. When these claims have been made on behalf of the actual holders of traditional knowledge, however, their foundations are not clear, as few if any attempts have been made to assess how the participants in traditional knowledge studies or management activities feel about their experiences. Has reality matched their expectations when they first became involved? Are they satisfied with the outcomes they have seen? Do they have recommendations for improvements? Without direct input from this diverse group, it is impossible to develop a complete picture of the state of current use of traditional knowledge in resource development.

The results, however, must be treated with caution, as the holders of traditional knowledge may have few alternatives with which to compare their own experiences. As noted in the previous question, imbalance of power can lead to reduced expectations and, potentially, satisfaction with suboptimal results. The potential for co-optation must be kept in mind as participants' views are interpreted.

What infrastructure is needed, and what is available, to support the wider engagement of traditional knowledge?

The generation, dissemination, and archiving of scientific knowledge is supported by considerable infrastructure, in the form of research agencies, scientific journals, conferences, professional associations and networks, data systems, and more. Some of this is accessible to traditional knowledge, at least when documented as part of the wider scientific enterprise. Less is available for traditional knowledge on its own terms, a deficit that includes the use and perpetuation of traditional knowledge as a way of life. When scientists participate in, say, an environmental impact assessment, they can readily access large amounts of data and information to address key questions and allow careful analyses. When traditional knowledge holders participate in the same exercise, they are often on their own, or at best have far fewer resources to draw upon. Assessing the significance of this deficit and identifying what is needed to better support traditional knowledge holders can help traditional knowledge play a larger and more consistent role in resource development.

How do indigenous proponents of the use of traditional knowledge view progress to date?

Many indigenous leaders have called for greater use of traditional knowledge in research, resource management, and resource development. These calls have resulted in formal language in land claims agreements, research plans, and environmental impact assessments. It is less clear, however, that the gains in recent years have achieved the aspirations of those leaders, or that the use of traditional knowledge was, in itself, the ultimate goal. Documenting traditional knowledge, as with generating scientific findings, is typically a slow and laborious process. Resource development typically requires a considerable amount of information on many topics. There is unlikely to be time to conduct new studies on demand as decisions take shape. Co-management approaches offer an alternative, adding traditional knowledge holders to the decision-making group rather than simply trying to insert their knowledge into a decision-making process. If the end goal is greater influence over decisions that affect their future, then indigenous leaders are likely to want to shape the outcomes, not just the information that is put in. Use of traditional knowledge is an input, not necessarily an outcome, and it seems unlikely that indigenous leaders will long be satisfied with advocating for better inputs. Understanding where political leadership is likely to go next will be important in determining what kinds of studies are most needed. Otherwise, the research community is left trying to anticipate indigenous aspirations rather than knowing with confidence what those aspirations are.

How well-suited is traditional knowledge for resource development decision-making?

The basic premise of using traditional knowledge in resource development is that it has something unique to add. It is less clear what, exactly, that "something" is. Traditional knowledge holders are intimately familiar with their lands and waters, and with the species they use as well as aspects of the ecosystems that support those species. Resource development decisions, especially concerning environmental impact assessments, have become a highly technical, specialized field. The translation of traditional knowledge into such technical analyses is problematic at best. More often, traditional knowledge seems to call into question some of the assumptions upon which resource development is based: that mitigation is possible, that environmental impacts can be anticipated and contained, that it is possible to develop resources and protect intact ecosystems and cultures. Relegating traditional knowledge to a short commentary on technical points in an environmental impact assessment appears to miss the larger opportunity to re-consider the human-environment relationship. This is not to say that traditional knowledge and resource development are always in opposition, but merely that the nature of traditional knowledge needs to be better understood in order to apply it most effectively to making sound decisions concerning resource development. Much has been written comparing and contrasting traditional and scientific knowledge, but most of this has been done in the abstract, as lists of different principles. An analysis in concrete terms, examining the entire decision-making process and assessing where traditional knowledge most directly addresses both the premises and the technicalities of those decisions, would make a significant advance in our understanding of how one system of knowledge can be brought to bear on a process based on another system of knowledge.

Acknowledgments

I thank Chris Southcott and the ReSDA project for the opportunity to contribute this chapter, and for their support of the research that led to it. Katie Royer carried out the majority of the literature search and similar research, for which I am very grateful. Any mistakes in fact, presentation, or interpretation are, however, mine alone.

Notes

1   Known also by many related terms, such as indigenous knowledge, local and traditional knowledge, traditional ecological knowledge, *Inuit Qaujimajatuqangit*, etc. While there are some differences in how and where these terms are used, the basic idea is

similar: knowledge that has been gained by experience and shared among members of a group or community, often across generations. Agrawal (1995) pointed out that neither "traditional knowledge" nor "scientific knowledge" are monoliths, and that most attempts to distinguish one from the other rely on simplistic differentiations that do not hold up to scrutiny. I acknowledge Agrawal's argument, but also retain the distinction between traditional and scientific knowledge as a convenient way of separating two sources of information that are usually thought of differently and treated differently in policy and management.

2   Under U.S. law, tribal governments enjoy a government-to-government relationship with the federal government, and as such are entitled to formal consultation on matters that affect them, such as resource development in nearby lands and waters.

References

ACIA. 2005. *Arctic Climate Impact Assessment*. Cambridge, UK: Cambridge University Press.

Agrawal, A. 1995. Dismantling the Divide between Indigenous and Scientific Knowledge. *Development and Change* 26(3): 413–439.

Albert, T. F. 2001. The Influence of Harry Brower, Sr., an Inupiaq Eskimo Hunter, on the Bowhead Whale Research Program Conducted at the UIC-NARL Facility by the North Slope Borough. In: D. W. Norton, ed. *Fifty More Years Below Zero: Tributes and Meditations for the Naval Arctic Research Laboratory's First Half Century at Barrow, Alaska*. Calgary: Arctic Institute of North America. pp. 265–278.

AMAP. 2010. Assessment 2007: Oil and Gas Activities in the Arctic—Effects and Potential Effects. Oslo: Arctic Monitoring and Assessment Program.

Amundsen, R. E. G. 1908. *The Northwest Passage*. London: Archibald Constable.

Berkes, F. 1999. *Sacred Ecology: Traditional Ecological Knowledge and Resource Management*. Philadelphia, PA: Taylor & Francis.

Berkes, F., and D. Jolly. 2001. Adapting to Climate Change: Social-Ecological Resilience in a Canadian Western Arctic Community. *Conservation Ecology* 5(2): 18 [online] URL: www.consecol.org/vol5/iss2/art18.

BOEM. 2012. Special Issue on Traditional Knowledge. *BOEM Ocean Science* 9(2): 1–16.

Bonesteel, S. 2006. Use of Traditional Inuit Culture in the Policies and Organization of the Government of Nunavut. In: B. Collignon and M. Therrien, eds. *Orality in the twenty-first Century: Inuit Discourse and Practices. Proceedings of the 15th Inuit Studies Conference*. Paris: CERLOM. www.inuitoralityconference.com/art/Bonesteel.pdf [accessed 6 October 2012].

Brook, R. K., and S. M. McLachlan. 2005. On Using Expert-Based Science to "Test" Local Ecological Knowledge. *Ecology and Society* 10(2): r3 [online] URL: www.ecologyandsociety.org/vol10/iss2/resp3/.

Brooke, L. F. 1993. *The Participation of Indigenous Peoples and the Application of their Environmental and Ecological Knowledge in the Arctic Environmental Protection Strategy*. Vol. 1. Ottawa: Inuit Circumpolar Conference.

Carmack, E., and R. Macdonald. 2008. Water and Ice-Related Phenomena in the Coastal Region of the Beaufort Sea: Some Parallels Between Native Experience and Western Science. *Arctic* 61(3): 265–280.

Caulfield, R. A. 1997. *Greenlanders, Whales, and Whaling: Sustainability and Self-Determination in the Arctic*. Hanover, NH: University Press of New England.

Cruikshank, J. 1998. *The Social Life of Stories: Narrative and Knowledge in the Yukon Territory*. Lincoln: University of Nebraska Press.

Cruikshank, J. 2001. Glaciers and Climate Change: Perspectives from Oral Tradition. *Arctic* 54(4): 377–393.

Danielsen, F., A. E. Jensen, P. A. Alviola, D. S. Balete, M. M. Mendoza, A. Tagtag, . . . and M. Enghoff. 2005. Does Monitoring Matter? A Quantitative Assessment of Management Decisions from Locally-Based Monitoring of Protected Areas. *Biodiversity and Conservation* 14: 2633–2652.

Dowsley, M. 2007. Inuit Perspectives on Polar Bears (*Ursus maritimus*) and Climate Change in Baffin Bay, Nunavut, Canada. *Research and Practice in Social Sciences* 2(2): 53–74.

Ellis, S. C. 2005. Meaningful Consideration? A Review of Traditional Knowledge in Environmental Decision Making. *Arctic* 58(1): 66–77.

Ferguson, M. A. D. and F. Messier. 1997. Collection and Analysis of Traditional Ecological Knowledge About a Population of Arctic Tundra Caribou. *Arctic* 50(1): 17–28.

Fraser, D. J., T. Coon, M. R. Prince, R. Dion, and L. Bernatchez. 2006. Integrating Traditional and Evolutionary Knowledge in Biodiversity Conservation: A Population Level Case Study. *Ecology and Society* 11(2): 4 [online] URL: www.ecologyandsociety. org/vol11/iss2/art4/.

Freeman, M. M. R., ed. 1976. *Inuit Land Use and Occupancy Project*. 3 vols. Ottawa: Department of Supply and Services Canada.

Freeman, M. M. R. 1989. Graphs and Gaffs: A Cautionary Tale in the Common Property Resource Debate. In: F. Berkes, ed. *Common Property Resources: Ecology and Community-Based Sustainable Development*. London: Belhaven Press. pp. 92–109.

Gearheard, S., W. Matumeak, I. Angutikjuaq, J. Maslanik, H. P. Huntington, J. Leavitt, . . . and R. G. Barry. 2006. "It's Not that Simple": A Collaborative Comparison of Sea Ice Environments, Their Uses, Observed Changes, and Adaptations in Barrow, Alaska, USA, and Clyde River, Nunavut, Canada. *Ambio* 35(4): 203–211.

George, J. C., H. P. Huntington, K. Brewster, H. Eicken, D. W. Norton, and R. Glenn. 2004. Observations on Shorefast Ice Failures in Arctic Alaska and the Responses of the Inupiat Hunting Community. *Arctic* 57(4): 363–374.

Gilchrist, G. and M. L. Mallory. 2007. Comparing Expert-Based Science with Local Ecological Knowledge: What Are We Afraid Of? *Ecology and Society* 12(1): r1 [online] URL: www.ecologyandsociety.org/vol12/iss1/resp1/.

Gilchrist, G., M. L. Mallory, and F. Merkel. 2005. Can Local Ecological Knowledge Contribute to Wildlife Management? Case Studies of Migratory Birds. *Ecology and Society* 10(1): 20 [online] URL: www.ecologyandsociety.org/vol10/iss1/art20/.

Greenland Government. 1999. Landstingslov nr. 12 af 29. oktober 1999 om fangst og jagt, ß2 stk. 3. Available at: http://dk.nanoq.gl (accessed 20 February 2012).

Hansen, A. M. and L. Kørnøv. 2010. A Value-Rational View of Impact Assessment of Mega Industry in a Greenland Planning and Policy Context. *Impact Assessment and Project Appraisal* 28(2): 135–145.

Holland-Bartels, L. and B. Pierce, eds. 2011. An Evaluation of the Science Needs to Inform Decisions on Outer Continental Shelf Energy Development in the Chukchi and Beaufort Seas, Alaska. U.S. Geological Survey Circular 1370.

Howard, A. and F. Widdowson. 1996. Traditional Knowledge Threatens Environmental Assessment. *Policy Options* 17(9): 34–36.

Huntington, H. P. 1992. *Wildlife Management and Subsistence Hunting in Alaska*. London: Belhaven Press.

Huntington, H. P. 1998. Observations on the Utility of the Semi-Directive Interview for Documenting Traditional Ecological Knowledge. *Arctic* 51(3): 237–242.

Huntington, H. P. 2000. Using Traditional Ecological Knowledge in Science: Methods and Applications. *Ecological Applications* 10(5): 1270–1274.

Huntington, H. P. 2005. "We Dance Around in a Ring and Suppose": Academic Engagement with Traditional Knowledge. *Arctic Anthropology* 42(1): 29–32.

Huntington, H.P., and the Communities of Buckland, Elim, Koyuk, Point Lay, and Shaktoolik. 1999. Traditional Knowledge of the Ecology of Beluga Whales (*Delphinapterus leucas*) in the Eastern Chukchi and Northern Bering Seas, Alaska. *Arctic* 52(1): 49–61.

Huntington, H. P., P. K. Brown-Schwalenberg, K. J. Frost, M. E. Fernandez-Gimenez, D. W. Norton, and D. H. Rosenberg. 2002. Observations on the Workshop as a Means of Improving Communication between Holders of Traditional and Scientific Knowledge. *Environmental Management* 30(6): 778–792.

Huntington, H. P., S. Gearheard, A. Mahoney, and A. K. Salomon. 2011. Integrating Traditional and Scientific Knowledge through Collaborative Natural Science Field Research: Identifying Elements for Success. *Arctic* 64(4): 437–445.

Huntington, H. P., A. Lynge, J. Stotts, A. Hartsig, L. Porta, and C. Debicki. 2012. Less Ice, More Talk: The Benefits and Burdens for Arctic Communities of Consultations Concerning Development Activities. *Carbon and Climate Law Review* 1(2012): 33–46.

Ingold, T. and T. Kurttila. 2000. Perceiving the Environment in Finnish Lapland. *Body and Society* 6(3–4): 183–196.

Irving, L. and S. Paneak. 1954. Biological Reconnaissance along the Ahlasuruk River East of Howard Pass, Brooks Range, Alaska. *Journal of the Washington Academy of Sciences* 44(7): 201–211.

Johannes, R. E. 1981. *Words from the Lagoon: Fishing and Marine Lore in the Palau District of Micronesia*. Berkeley, CA: University of California Press.

Johannes, R. E. 1998. The Case for Data-Less Marine Resource Management: Examples from Tropical Nearshore Finfisheries. *TREE* 13(6): 243–246.

Lemelin, R. H., M. Dowsley, B. Walmark, F. Siebel, L. Bird, G. Hunter, . . . and the Weenusk First Nation at Peawanuck. 2010. Wabusk of the Omushkegouk: Cree-Polar Bear (*Ursus maritimus*) Interactions in Northern Ontario. *Human Ecology* 38(6): 803–815.

Lewis, E., M. O. Hammill, M. Power, D. W. Doidge, and V. Lesage. 2009. Movement and Aggregation of Eastern Hudson Bay Beluga Whales (*Delphinapterus leucas*): A Comparison of Patterns Found through Satellite Telemetry and Nunavik Traditional Ecological Knowledge. *Arctic* 62(1): 13–24.

Mailhot, J. 1993. Traditional Ecological Knowledge: The Diversity of Knowledge Systems and Their Study. Great Whale Environmental Assessment: Background Paper No. 4, Great Whale Public Review Support Office.

Mallory, M. L., H. G. Gilchrist, B. M. Braune, and A. J. Gaston. 2006. Marine Birds as Indicators of Arctic Marine Ecosystem Health: Linking the Northern Ecosystem Initiative to Long-Term Studies. *Environmental Monitoring and Assessment* 113(1–3): 31–48.

Mallory, M. L., H. G. Gilchrist, A. J. Fontaine, and J. A. Akearok. 2003. Local Ecological Knowledge of Ivory Gull Declines in Arctic Canada. *Arctic* 56(3): 293–298.

Mymrin, N. I., The Communities of Novoe Chaplino, Sireniki, Uelen, and Yanrakinnot, and H.P. Huntington. 1999. Traditional Knowledge of the Ecology of Beluga Whales (*Delphinapterus leucas*) in the Northern Bering Sea, Chukotka, Russia. *Arctic* 52(1): 62–70.

Nadasdy, P. 1999. The Politics of TEK: Power and the "Integration" of Knowledge. *Arctic Anthropology* 36(1): 1–18.

250   *Henry P. Huntington*

Nadasdy, P. 2003. *Hunters and Bureaucrats: Power, Knowledge, and Aboriginal-State Relations in the Southwest Yukon*. Vancouver: UBC Press.

Nageak, J. M. 1991. *An Unusual Whaling Season: An Interview with Waldo Bodfish, Sr.* In: *Bodfish, Waldo Sr. Kusiq*. Fairbanks: University of Alaska. pp. 237–246.

Nakashima, D. J. 1990. *Application of Native Knowledge in EIA: Inuit, Eiders, and Hudson Bay Oil*. Ottawa: Canadian Environmental Assessment Research Council.

Nakashima, D. J. and D. J. Murray. 1988. The Common Eider (*Somateria millissima sedentaris*) of Eastern Hudson Bay: A Survey of Nest Colonies and Inuit Ecological Knowledge. Ottawa: Environmental Studies Revolving Fund Report No. 102.

Noongwook, G., the Native Village of Savoonga, the Native Village of Gambell, H.P. Huntington, and J.C. George. 2007. Traditional Knowledge of the Bowhead Whale (*Balaena mysticetus*) around St. Lawrence Island, Alaska. *Arctic* 60(1): 47–54.

Scientific and Statistical Committee. 2010. Draft report to the North Pacific Fisheries Management Council, February 8–10, 2010. Available at: www.fakr.noaa.gov/npfmc/PDFdocuments/minutes/SSC210.pdf.

Smith, D. 2001. Co-management in the Inuvialuit Settlement Region. In: CAFF. *Arctic Flora and Fauna*. Helsinki: Edita. pp. 64–65.

Spaeder, J. J. 2005. Co-management in a Landscape of Resistance: The Political Ecology of Wildlife Management in Western Alaska. *Anthropologica* 47(2): 165–178.

Stephen R. Braund & Associates. 2011. Chukchi and Beaufort Seas National Pollutant Discharge Elimination System Exploration General Permits Reissuance: Report of Traditional Knowledge Workshops—Point Lay, Barrow, Nuiqsut, and Kaktovik. Prepared for Tetra Tech. and the U.S. Environmental Protection Agency. March 11, 2011.

Stevenson, M. G. 1996. Indigenous Knowledge in Environmental Assessment. *Arctic* 49(3): 278–291.

Thomsen, M. L. 1993. Local Knowledge of the Distribution, Biology, and Hunting of Beluga and Narwhal: A Survey among Inuit Hunters in West and North Greenland. Nuuk: Greenland Hunters' and Fishermen's Association, Greenland Home Rule Authorities, and Inuit Circumpolar Conference.

Tyrrell, M. 2008. Nunavik Inuit Perspectives on Beluga Whale Management in the Canadian Arctic. *Human Organization* 67(3): 322–334.

U.S. Army Corps of Engineers. 1999. Beaufort Sea Oil and Gas Development/Northstar Project: Final Environmental Impact Statement. 4 volumes + appendices. Prepared by the U. S. Army Corps of Engineers, Alaska District.

Usher, P. J. 2000. Traditional Ecological Knowledge in Environmental Assessment and Management. *Arctic* 53(2): 183–193.

Van de Velde, F., I. Stirling, and E. Richardson. 2003. Polar Bear (*Ursus maritimus*) Denning in the Area of the Simpson Peninsula, Nunavut. *Arctic* 56(2): 191–197.

Weatherhead, E., S. Gearheard, and R. G. Barry. 2010. Changes in Weather Persistence: Insight from Inuit Knowledge. *Global Environmental Change* 20(3): 523–528.

14 Gender in research on northern resource development

Suzanne Mills, Martha Dowsley, and David Cox

As other authors in this series of gap analyses have demonstrated, the effects of resource development in the north are profound and complex. Perspectives about how to minimize negative impacts and increase benefits to communities are therefore diverse. In this chapter, we review the literature that explores how resource development is gendered. Indigenous women across the circumpolar north have often voiced their concerns over, and ideas about, whether or how development should proceed. While some women have sought to slow or stop development, others have sought to influence the trajectory of development to both minimize negative socio-economic consequences and ensure that resource wealth is equally distributed throughout the community.

Often women have voiced their concerns about development in submissions throughout Environmental Assessment processes. For example, in their response to the 1995 Environmental Impact Statement for a proposed diamond mine in the NWT, women from remote communities highlighted their concerns that the mine would have negative impacts for family and community relations while not providing substantial employment benefits to women (Brockman and Argue, 1995). More often, however, Indigenous women have participated in EAs, not to oppose development outright, but ask that proponents account for unequal gender outcomes. For example, in a response to the EA statement for the Lower Churchill Hydro Development in Labrador, the Mokami Satus of Women Council outlined how the proponent did not set detailed targets for the training and hiring of women or provide adequate information about child care, housing, and response to alcohol and drug abuse (Hallett and Baikie, 2011). The lack of attention by resource companies to gender has therefore lead many Indigenous women's organizations to suggest that institutions undertake gender-based analyses when considering resource development (Pauktuutit, 2012; TIA, 1997). Three themes emerge in Indigenous women's submissions to EA processes and in reports from women's organizations about development: that women are concerned about whether women will have access to benefits from resource development; that women are concerned about the impacts of development for individuals, families, and communities; and that they want to ensure that gender is considered in resource governance.

In this chapter, we draw on community reports and academic articles to suggest that adopting a gendered lens is critical to understanding the flows of wealth to, and the social and cultural changes within, northern communities and therefore to the larger project of increasing benefits and reducing the negative impacts of resource development. The literature that exists suggests that resource development is reshaping gender relations in northern communities, altering the flow of wealth through families and kin networks, the status and power relations between women and men, and social and cultural practices and beliefs. Our overview encompasses research published on northern resource development and gender relations in the Arctic from the 1980s through to 2015, the date the chapter was first submitted. Although this chapter draws on literature from the global circumpolar region, our focus is on Canada's north since that is where we are situated. We adopt a constructivist understanding of gender, seeing gender as a range of attributes associated with the categories "women" and "men" that are culturally specific and that change over time. We are cognizant, however, that much of the literature pertaining to gender focuses on Indigenous women and their relations with men. The review therefore focuses on research about Indigenous women and resource development despite our earlier intentions to adopt a broad perspective on gender. A more critical gap, perhaps, is that during the time period examined, there were few Indigenous scholars conducting empirical research about gender and northern resource development. Though the research summarized adopts a variety research approaches, the paucity of Indigenous scholars has meant that few of the studies below adopt Indigenous perspectives, perspectives that are critical to decolonizing northern research (Smith 1999).

With these limitations in mind, this review charts three approaches to research about gender and resource development. We begin by describing research that focuses on how the distribution of economic benefits of resource development are gendered. We then describe research that documents the gendered impacts of resource development on women, community, and family. Last we examine women's participation in resource development, decision-making, and governance. We suggest that aside from a handful of careful analyses of gender and resource development, most more narrowly adopt an "impacts of resource development on women" approach. Future research should adopt a broader lens to examine the full diversity of changing gender roles and relations in the north, including men and masculinity and how these are connected to changing participation in traditional, social, and capitalist economies.

Gendered distribution of the economic benefits from resource development

Indigenous women living in the north have often voiced concerns about the distribution of resource wealth within their communities. In EA hearings and discussions prior to development, women's groups have been wary of resource development, suggesting that rising individual incomes can heighten economic disparities within and among communities and families. Northern women have also been concerned that material wealth resulting from resource development often flows disproportionately to male individuals, rather than to women, families, and community members in need (Status of Women Council of the NWT, 1999; National Aboriginal Health Organization, 2008). Some of the academic literature suggests that these concerns are warranted. Studies from disparate areas of the world have found that men benefit disproportionately from mining while women shoulder a greater amount of the costs (Noronha and Nairy, 2005; Mishra, 2009). Though proponents of development often distinguish three types of benefits – wages, cash transfers and business development opportunities – most of the literature about gender has focused on employment, neglecting other flows of resource wealth. The literature examining how cash transfers resulting from profit-sharing arrangements or paid as compensation for environmental impacts are gendered is limited while research about who does and does not benefit from procurement policies that support Indigenous businesses is almost non-existent.

Employment

Wages from jobs in high-paying natural resource sectors are a critical way that northern communities have sought to harness resource rents for local economies. Yet several studies have suggested that Indigenous women are less likely to be employed in resource work than Indigenous men. Although this gender disparity also applies to non-Indigenous women and poor women internationally (Lahiri-Dutt, 2001; 2006), exclusion from resource employment is particularly critical for Indigenous women living in northern communities with often few other employment opportunities (Reed, 1999; Mills, 2006; Rude and Deiter, 2004). For example, a study by Southcott (2003) found that women in Northern Ontario had lower participation rates than men in Northern Ontario and than women in Ontario as a whole. When Indigenous women are working in resource employment, they are often over-represented in jobs that are low-waged and precarious. For example, in a study of occupational and industry segregation in forestry in Canada, Mills (2006) found that Indigenous women were concentrated in lower-paid and less stable occupations such as cooking and cleaning or tree planting. Indigenous women were underrepresented in male-dominated sawmill and pulp mill work, and almost completely absent from female-typed clerical and secretarial work where white women predominated. Research on other industries has found similar patterns. For example, a report by Women in Mining Canada (2010) also found that women in Canada were underrepresented in all resource sectors in 2006. Women's underrepresentation ranged from a high of 20.2 % in forestry and logging to a low of 14% in mining. Reported wage gaps between men and women in the mining sector are substantial and are likely the result of occupational segregation within the sector. Reported statistics from Northern Canada are similar. Gibson (2008) found that women only accounted for 16% of employment in the diamond industry where they had a 46% employment rate in the NWT as a whole

(Gibson, 2008). Similarly, at the Voisey's Bay nickel mine in Labrador, women represented 17.5% of the workforce in 2009, and were concentrated in catering and housekeeping positions (Cox and Mills, 2015).

Qualitative research studies have identified several barriers to the participation of women in resource industries and linked these to the construction of resource work as masculine. Most of this research, however, has focused on the experiences of non-Indigenous women and as such has not addressed the gendered effects of colonial policies and practices on women's work. Research on women working in forestry and mining has found that women in non-traditional occupations often feel heightened pressure to prove themselves on the job and are tokenized by male co-workers and employers (Mills, 2011, Tallichet, 1995, 2000). Tallichet (2000) found that informal practices such as harassment and sexualization and gender stereotyping worked in conjunction with formal training and employment practices to limit women's access to many mining jobs. Both Women in Mining Canada (2010) and Sharma (2010) suggest that supply side factors, such as a potential reluctance of women to enter the field of mining because of gendered representations or inflexible work arrangements, also act as barriers to the entry of women. This position, which positions women's choices as the principle reason for their underrepresentation has been challenged by others authors who have adopted more structured explanations for women's low participation. These scholars have often pointed to the embeddedness of hegemonic forms of masculinity in resource work (Reed, 1999; Brandth and Haugen, 2000; Tallichet, 2000). Several authors have charted practices that reproduce and legitimize white working-class masculinity serves to exclude women and racialized men from many resource jobs (Dunk, 1994). Others, such as Lahiri-Dutt (2012) deconstruct the association of mining and working-class masculinity by highlighting the historical roles that women played in mining.

Research about the effectiveness of Impact and Benefit Agreements (IBA) has also touched on the question of whether these agreements are able to facilitate the employment of Indigenous women. As contractual relationships between Indigenous governments and companies, IBAs can help to challenge colonial patterns of employment that exclude Indigenous workers from high-paying resource employment. Though most of this literature focuses on Indigenous employment as a whole, some authors have also address gender discrepancies in employment (O'Faircheallaigh, 1999; Gibson, 2008). Reflecting larger debates in the IBA literature, there is disagreement over whether negotiated agreements are able to address gender inequality in employment. O'Faircheallaigh (1999) suggested that IBAs have the potential to promote the training and hiring of women and points to the IBAs negotiated by the Labrador Inuit Association at Voisey's Bay and The Cape Flattery mine in Australia that included specific employment provisions for Indigenous (O'Faircheallaigh, 1999). Others, however, have suggested that programmes to promote the participation of Indigenous peoples have been less successful for women than for men (Gibston, 2008, Cox and Mills 2015). Research on the Diavik Mine found that despite negotiated agreements, a combination of factors continued to impede the entry of Indigenous women into non-traditional jobs. These included: discriminatory hiring, inadequate training, and cultural barriers (Status of Women Council of NWT, 1999). Women in the NWT also perceived Diavik Mine to have discriminatory hiring practices, that women were only hired in low-paying traditional occupations such as housekeeping, cleaning, and cooking (National Aboriginal Health Organization, 2008). The work of Gibson (2008) also suggests that while IBAs increase access to employment, the benefits largely accrue to men. In northern communities affected by mining, unemployment dropped from above 50% to below 30% after the beginning of mining operations. And, although all incomes rose as a result of mining, men's incomes rose more than women's as they were more likely to obtain employment. Furthermore, research in forestry and mining by Egan and Klausen (1998) and Butler and Menzies (2000) found that employment and training programmes for Indigenous workers did not address the unique challenges faced by Indigenous women. Little (2005) studied a training programme for low-income Indigenous and non-Indigenous women to become apprentice carpenters in Regina and concluded that tailored training programmes were critical so that women could develop strategies to overcome barriers to employment such as sexism, racism, and harassment. Little (2005) argued that the success of the program was not just in the number of women who become carpenters but in the programme's ability to improve self-esteem and mental health. While Little's book focuses on low-income, predominantly Indigenous women in Regina, there are lessons to be learned that can be applied to training programs for women in the North.

Though the often policy-oriented research about Indigenous women and work has undoubtedly provided some useful insights for practitioners, in its pragmatism, it has often overlooked the effects of colonialism on gendered participation in resource employment. One exception is Butler and Menzies' (2000) work on the participation of Tsimshian women in forestry in British Columbia after the Hudson's Bay company established a fort in 1834. The authors recount how a succession of colonial policies slowly pushed women to the sidelines of resource production. Indigenous women were initially included in harvesting and selling wood but they were later excluded as the industry became more capitalized and as colonial policies limited First Nations ownership and access to waged work (Butler and Menzies, 2000). Understanding the social and political context, and in particular the impact of colonial policies and practices in reproducing hegemonic forms of masculinity is critical to increasing women's influence and participation in resource industries.

Cash transfers

IBAs and Socio-economic Agreements also provide for income transfers directly to Indigenous governments. Despite the importance of cash transfers, there is almost no literature examining how the income generated by resource development through IBAs might benefit women, for example by increasing the funding for services used differentially by women. One such study by O'Fairchealleigh (2007) assesses three models of income distribution in Australian IBAs for their gendered effects. He concludes that only one model respects the varying degree of risk carried by different subsets of the population (such as women) in the allocation formula. With the first model the entire community would meet and make decisions each year on how to allocate funds in three categories: a holding mechanism, commercial investments, and immediate regular cash payments made to members. The holding mechanism distributed money to a tax-exempt trust fund that pro-vided payments to individuals at age 18 and to those who needed medical and social services. Under this model, women would receive direct payments from the revenues gained from mining and could use this money at their own discretion. However, this model has not been used in recent years since it provides benefits in the short term but not the long term. The second model, and the most widely used in recent years, is the establishment of a trust fund that is embedded in the IBA to ensure long-term benefits. This model can benefit women positively, especially in the long term, but it depends on how the money is allocated, what is prioritized, and whether women are involved in the allocation process when the trust fund is established. The third model has a one-time lump sum payment that was divided 50/50 between a women's fund and a men's fund. Communities were also given fixed annual payments and profit related annual payments that were split between several funds (sustainability fund, special purpose fund and partnership fund). This model had obvious benefits to women as a large one-time lump sum was given to a fund for the use of Indigenous women. However, this model is unique in Australia and no Canadian IBAs contain such a model.

The above research, while suggesting that women benefit equally from resource development, fails to provide an overall picture of how the income flows resulting from resource development are gendered. Particular attention should be paid to two research areas that are largely void, the gender distribution of cash transfers resulting from negotiated agreements and of the profits from Indigenous-owned firms. While there is little research on women's participation in joint-ventures resulting from resource development, since women in northern communities are often more educated than men, they are often highly involved in administrative work. One business leader in a Taicho community commented:

> We are truly blessed in the Dogrib region, for we have very strong Dogrib women. The head of the education is a Dogrib woman holding a degree. The head of Social Service is a Dogrib woman. The head of our Lands Adminis-tration is sitting right here, our Office Manager. In almost every department or every social or government group the head is a woman. The majority of young people getting an education are Dogrib women. Right now as we speak 56% of the total staff of our schools is Dogrib women. . . . Right now as I speak we have 92 young aboriginal people attending post secondary education. Out of that 92 there are 64% of them are aboriginal women of the Dogrib region.

(quoted in Gibson, 2008: 226)

Therefore, despite the exclusion of women from many forms of resource employment, women in northern communities may be able to garner economic benefits from development through employment in social services or administrative and management capabilities or through participation in business ventures. Once again, the individualized approach to gender adopted in the research about income flows does not reflect the collective and community-centred approaches often articulated by Indigenous women.

Gendered social and cultural impacts

Research about socio-cultural impacts has tended to focus on how the gendered nature of work in resource industries contributes to a shift in the economic activities of northern communities towards market activities. This shift in economic activities in turn influences gender relations and cultural norms. Since this research documents potential gendered impacts at multiple interconnected scales, there is a danger of contributing to victimizing portrayals of Indigenous women and to negative portrayals of Indigenous communities more broadly. The tendency for research to portray Indigenous women as victims can contribute to a colonial narrative that contrasts the experiences of Indigenous women with those of Western women who become the model for emancipated and educated women hood (Parpart, 1993). In contrast, Indigenous women have often emphasized how the roles of women in the north are valued equally to those of men, unlike the case in the south.

> Socio-culturally speaking, indigenous Arctic women seem not to have experienced loss of importance in the family setting to the same degree as their southern sisters, or indeed by comparison with their men-folk. In fact, in writings about equity issues in the Arctic, the notion that each human being is valued in much the same way is rather common.
>
> (Williamson et al., 2004)

Notwithstanding these interventions, documenting the gendered impacts of resource development has served an important political function for some Indigenous women's groups, who have used reports to challenge or gain influence over proposed development initiatives. The need to provide communities and policy makers with information needs to be balanced against a desire to not impose a liberal/modernist vantage that depicts Indigenous populations as deficient (Salee, 2006). Although the research examined below often uses pre-development conditions or communities that have not experienced resource development as comparators, some of the research examining health outcomes adopts Western methodologies rather than Indigenous approaches to health and wellbeing. Overall, the literature suggests that the reshaping of local economies toward resource work and away from subsistence production and the gender division of labour that accompanies resource development influences gender roles and relations in ways that may be detrimental to the wellbeing of Indigenous women and men.

Effects of masculinized resource employment

A sizable body of research documents the effects of resource extraction on both resource workers and those who live in communities that host a resource extraction industry. Studies range from community reports to more comprehensive studies that combine interviews with quantitative data and the perspectives of elders. Taken together, these sources suggest that resource development can increase the prevalence of alcohol consumption, gambling, sexual exploitation, teen pregnancy, single parenthood, and family violence (Brubacher and Associates, 2002; Kuyek, 2003; Gibson and Klinck, 2005; National Aboriginal Health Organization, 2008; Gibson, 2008; Government of NWT, 2009; Davidson and Hawe, 2012). Though some studies cite the positive effects that resource wages have on the wellbeing of families, most present a more negative picture, for example describing the social effects of rising income inequality and the separation of predominantly male workers from their communities while on work rotations at remote sites (Brubacher and Associates, 2002; Gibson, 2008). These conclusions are similar to an early comparative study conducted by Schaefer (1983) comparing general health parameters of two Inuit communities, one in the Western Arctic greatly affected by the D. E. W. line construction and oil exploration, and the other in the Eastern Arctic, which was largely traditional at the time of the study. Schaefer (1983) found that the Inuit community affected by Western development had poorer health on a number of measures as well as a greater incidence of family break downs, substance abuse, and deaths due to violence compared to the community in the Eastern Arctic. He then concluded that there were physical health consequences related to increased psychological stress of participating in resource development.

Many of the gendered effects of resource development emanate from the masculinized nature of resource industries and, consequently, the overrepresentation of men in resource employment (Gibson and Klinck, 2005; Weitzner, 2006; Gibson, 2008). For example, Sharma and Rees (2007) found that women who are at home while their male spouses are employed by the resource industry experience psychological stress as a result of increased pressure to perform domestic labour. Researchers have also linked the increasing income inequality within households arising from resource development to increased family strain and greater incidence of domestic violence and family breakdown (Weitzner, 2006; Gibson, 2008; Czyzewski et al., 2014). Similar issues have been reported elsewhere in Canada and abroad for remote site labour (Storey et al., 1988; Pollard, 1990; Houghton, 1993; Noronha and Nairy, 2005; Noble and Bronson, 2006; Davidson and Hawe, 2012). Additionally, Sharma (2010) suggests that the gendered employment structure in resource industries creates a patriarchal culture in the community and family life that often varies substantially from previous social organization. In Gibson's (2008) research in the Northwest Territories, Dene women often described their husbands as a "once a month Santa Claus" (Gibson 2008: 181), who they had to "train" to uphold their portion of the workload each time upon return from the mine. This has adverse effects on children as they often have a parent (usually the father) who misses out on important events such as graduations, birthdays, and family events, as well as day-to-day contact. The impacts of resource development often continue after the project is over. Women have expressed concerns of the "feast and famine" system that is associated with resource development projects; debts are acquired when jobs are available and after development ends, payments on loans create stress that results in psychological and physical impacts (Yukon Conservation Society, 2000).

Research exploring the effects of long-distance commuting, however, has been mixed. For example, Gibson (2008) found that some Dene women preferred the rotational schedule, as they enjoyed being at the mine, away from conflict at home. O'Faircheallaigh (1995) has also suggested that cultures of reciprocity that include support from extended families help the partners of men working in fly-in/ fly-out operations cope with prolonged separation. He qualifies this, however, by saying that these supports are in decline as a result of economic development.

Resource development and subsistence production

Research charting the gendered effects of a shifting relationship between wage employment and subsistence production has tended to cast a critical light on resource development. Scholars such as Kuokkanen (2011) and Rude and Deiter (2004) have argued that resource development can undermine the important roles played by women in subsistence harvesting by shifting the balance of economic activities towards male-dominated resource sectors. Additionally, since the subsistence economy is an important way that many households in Canada's North feed themselves, Indigenous women have also often expressed concern about the impacts that resource development will have on fish and wildlife in their submissions to environmental assessments (Tongamiut Inuit Annait, 1997)

The relationship between subsistence harvesting is complex, however. On one hand, studies have shown that income from wage sources allows individual households to purchase equipment needed to participate in subsistence production and therefore households with moderate wage income levels have greater production than those with low incomes (Koke, 2008). On the other hand, as wages increase, the perceived value of subsistence activities declines as does an individual's probability of selling or exchanging subsistence produce. Additionally, many researchers have found that the negative environmental consequences of resource development spur a shift towards wage employment that affects families, extended families, and communities (Stern, 2005, Bjerregaard, et al. 2004). Declines in harvesting influence the livelihoods and gender roles and relations of all community members. For example, Stern's (2005) work in community of Holman, located in the Inuvialuit Settlement Region, found that waged work weakened Inuit social values and created tensions within the community. This shift undermined the social support systems that women had previously found helpful.

Several studies have sought to document both the importance and the fluidity of women's roles in subsistence production. Kuokkanen (2011) highlights that women contribute to family survival through subsistence activities in a number of ways that go beyond the man-as-hunter, woman-as-gatherer dichotomy, including processing and preparing meat, fur, and hides. Similarly, Bielawski's (2003) research found that while gender roles are rather stringent during hunting trips with both men and women; women often hunt and butcher caribou themselves, showing that this knowledge is not held by men alone. Participation in the subsistence economy involves both men and women, and the loss of one partner due to employment schedules can result in a severe decline of a family's harvesting activities. The significance of the subsistence economic activities for women go beyond economic benefit and is intertwined with health, spirituality, and culture. By investigating berry harvesting practices of Teetl'it Gwich'in women, Parlee et al. (2005) identified nine different values, of which the commercial value of berries was not one. The interwoven and complex manner in which women partake in berry harvesting includes a network of sharing, sustainable harvesting, spirituality, relationship to the land, and health and wellbeing of the family. This underscores the importance of mitigating negative environmental effects of resource development in order to maintain the cultural economies of northern communities.

Further, the introduction of Western cultural views and behaviours through contact with resource extraction industries and non-local workers can have negative impacts on local cultures. Women in British Columbia raised concerns that with development comes Western cultural beliefs about nature that clash with the traditional roles of women being stewards of the land (Rude and Deiter, 2004). These same women also spoke of being distressed and grieved while seeing the destruction caused by development which deterred them from berry picking. This study suggests that declining participation in subsistence harvesting and the prominence of capitalist economic ideas relating to resource development negatively affects the power and roles of women in their communities.

Resource development and the role of women in resource governance

There is a small but growing body of literature that looks at the participation of Indigenous women in EA and IBA processes (Cox and Mills, 2015; O'Faircheallaigh, 1995). The literature falls mainly into three areas of governance: EA and IBA processes; wildlife management; and community governance. There is a growing literature on women's participation in resource decision-making including Environmental Assessments and Impact Benefit Agreements (Cox and Mills, 2015, 1999; Lawrence, 1997; O'Faircheallaigh, 2011) as well as women's involvement in wildlife management at the community, regional, and territorial level, particularly in relation to traditional knowledge (Berkes and Jolly, 2001; Parlee et al., 2005). There is comparatively less literature examining how resource development influences the involvement of women in formal governance, including Indigenous governments, band councils, and national and local boards and organizations. Our review supports the findings of Kuokkanen (2011), however, in suggesting that since resource development can hinder traditional economies, it also helps to undermine the important roles that women have in traditional economies and knowledge systems (Kuokkanen, 2011). Consequently, the role of women in governance may be affected by both the declining importance of traditional economies and by their underrepresentation in formal resource governance and management.

Women's participation in EAs and IBAs

As mentioned in the introduction, women's groups have often voiced their concerns about resource development through their participation in EA processes. Only a few authors writing about the EA process in Canada have considered the participation of women or the gendered outcomes of EAs (Cox and Mills, 2015; Bielawski, 2003; George, 1999; Lawrence, 1997; O'Faircheallaigh, 2011). These studies have suggested that the participation of women in EAs has had little success in altering development trajectories.

Studies and reports documenting the perspectives of Indigenous women towards EA processes have suggested that EAs often do not adequately consider the gendered impacts of proposed projects. For example, in their submission to the review of the NWT diamond project, Brockman and Argue (1995) presented concerns raised by Inuit women during the process. Inuit women felt that the Environmental Impact Statement did not contain the information needed to mitigate negative impacts on women from the proposed diamond mine. Brockman and Argue (1995) also note that Indigenous women were interested and concerned about more than what was relegated to "women's issues" during the socio-economic impact part of the EA process. Inuit women involved in the EA for the Voisey's Bay Mine in Labrador similarly critiqued the EIS statement for neglecting to conduct a gendered analysis and were adamant in their demands for programmes to increase employment in non-traditional jobs and support for women in communities (O'Reilly and Eacott, 1998: 26). This participation only had some effect on the final recommendations, however, and little effect on final development trajectory through IBAs and land claims negotiations (Archibald and Crnkovich, 1999; Cox and Mills, 2015).

A small number of studies have examined the involvement of women in IBA negotiations (Weitzner, 2006; O'Faircheallaigh, 1995; Kuyek, 2003; Cox and Mills, 2015). These authors highlight how considering gender in IBAs is particularly important given the increasing important role that IBAs play in helping communities secure benefits and mitigate negative socio-cultural impacts from development. Community consultation, however, is more difficult during IBA negotiations because the negotiation process requires a degree of secrecy. For this reason O'Faircheallaigh (1995) suggests that including women on the negotiating team is essential. Additionally, these authors have noted that although women have sometimes been excluded from IBA negotiations, there is an increasing trend to include women in a formal capacity and gender as an important issue for discussion. Both Weitzner (2006) and Archibald and Crnkovich (1999) noted that when IBA negotiations were rushed they provided less opportunity for participation from a broad section of the community, including women, youth, and elders. O'Faircheallaigh (1995) also suggested that attention be paid to the physical orientation of the room, how information is communicated and should take childcare into consideration by holding several shorter meetings rather than one or two long meetings so that women were not excluded from the conversation. Kuyek (2003) notes that women in the Yukon are often not equipped with the information necessary to feel confident enough to speak up about issues effecting women's health.

Alternatively, some research has suggested that women often have substantial influence over IBA negotiations. O'Faircheallaigh (2011) argues that Indigenous women may play larger roles in pre-project agreements and assessments than acknowledged in the literature. He suggests that women are more likely to use informal mechanisms of negotiation. Weitzner (2006) reported that throughout the more recent Lutsel K'e negotiations, women in the community felt represented and claimed to have a strong voice. Although there was no formal organized women's group, there have been women Chiefs, and Band Councilors, there are many women in the workforce, and negotiations on mining issues were headed by women (Weitzner, 2006).

Few studies have examined the contents of IBAs through a gendered lens. A recent report for Pauktuutit examining the gendered effects of mining in Qamani'tuaq found that IIBA for the Meadowbank Gold Mine signed in 2011 did "not contain any sections that deal specifically with the needs of women in relation to the expected socio-economic and cultural impacts of the mine" (Czyzewski et al., 2014: 33). The only consideration for social impacts was that the company was committed to create a Community Wellness Report and Implementation Plan by March 31 of 2012. This report would "address matters related to mental and physical health, alcohol and drug abuse, family relations, migration in or out of the community, the prevalence and use of Inuktitut, culture, job satisfaction and the management of personal finances by residents of Qamani'tuaq" (Czyzewski et al., 2014: ii). Additionally, at the time of study, no funds from a Community Development Fund, created as an outcome of the IIBA, had been spent in Qamani'tuaq. Concerns that IBAs do not sufficiently address social and cultural impacts of resource development had been expressed several decades earlier by Martha Flaherty, the past president of the Pauktuutit. In an address at the 1993 Annual General Meeting of Tungavik Federation of Nunavut she stated:

> We have a concern that Inuit Impact and Benefit Agreements may be negotiated too narrowly. We would like to see the contents of these agreements broadened to include more requirements on the developer to support community development initiatives in communities affected by the specific development project . . . In our meetings, we have heard what women have said about development and its effects on the environment, their lives and their families, the delivery of goods and services, transportation, and housing.

This information can help ensure Impact and Benefit Agreements address the needs of all Inuit in the community, not just those who will be working for the developer.

(quoted in Archibald and Crnkovich, 1999)

This demand to include all members of the community, not just those who will be gaining employment, is an important one, but one that often gets ignored due to industry concerns of being responsible for what governments are traditionally responsible for. For example, at the time of Czyzewski et al.'s (2014) study, no women were on the board of KIA, the organization responsible for negotiating IIBAs in the Kivalliq region of Nunavut. Shanks (2006) points out that industry is reluctant to include funding for community needs such as child care, housing, or social services since they fear government wants to download their responsibilities onto industry. This limitation further disadvantages women, who are often the greatest users and providers of social services in communities.

Indigenous women, Indigenous knowledge, and subsistence production

Several authors have highlighted Indigenous women's contributions to the production of Indigenous knowledge and to traditional economic activities (Berkes and Jolly, 2001; Parlee et al., 2006). Some of this work has aimed to deconstruct the notion that Indigenous economies have a rigid gendered division of labour that excludes women from wildlife harvesting. For example, Berkes and Jolly (2001) describe the fluidity of gender roles in Inuit society, finding that women can hunt and men can sew skins. The authors provide an example of three sisters who were raised by their mother who took over the duties of their father after his death (Berkes and Jolly, 2001). Kafarowski (2004) similarly notes that although subsistence fishing in Nunavut follows a gendered division of labour, there is also a degree of interchangeability in tasks. This is not the case in commercial fisheries in Nunavut where women are underemployed in offshore fishing and hold low-paying jobs in the processing industry (Kafarowski, 2004).

Other research has also sought to challenge the privileging of men's harvesting activities by highlighting the roles that Inuit women play in procuring and processing food and in environmental management. Shannon (2006) found that most studies focus on how Inuit women support men in harvesting, rather than on activities that do not have a gendered division of labour such as fishing. She argues that such activities, where women are full participants rather than complementary or contributory participants, should be garner more attention from policy makers and academics. The research of Kafarowski (2005) emphasizes how Inuit women in Nunavik and Nunavut possess knowledge that is different from that of men but equally important, involving food preparation and the short- and long-term health of the family. The traditional knowledge of women is an important factor in monitoring the effects of resource development. Women in Yukon communities affected by mining draw on local knowledge to assess changes in the environment and have drawn water and soil samples themselves to test for contaminants (Yukon Conservation Society, 2000). Parlee et al. (2006) note that the traditional knowledge of women is also used in localized resource management decisions regarding fish and berries. The traditional knowledge of women in the Northwest Territories is used in year to year decision-making, influencing rules for resource access, sharing information and harvest sharing.

Despite its importance, however, the traditional knowledge of women is often ignored in formal decision-making processes involving resource development in the north. As a result, women's interventions in environmental decision-making and monitoring activities have often sought to emphasize both the importance of traditional economic activities and the need to adopt holistic monitoring mechanisms. Women from the Denesoline community of Lutsel K'e Dene First Nation, for example, thought the socio-economic and human health indicators developed by the government during the environmental assessment period for the Ekati Diamond Mine were too narrow. They collaborated with academics and released a study to develop their own framework and indicators for monitoring the impacts of mining on the health of the community (Parlee, O'Neil and Lutsel K'e Dene First Nation, 2007). Women were also involved in critiquing the superficial incorporation of traditional knowledge in the EA leading up to the Ekati Diamond Mine (Bielawski, 2003).

Further, both Natcher (2013) and Kafarowski (2005) found that despite being involved in fishing and hunting, women are largely absent from the boards of hunters and trapper organizations in the Arctic. Kafarowski (2005) argued that men and women have different traditional knowledge(s) in regards to hunting and fishing and that the traditional knowledge of women is left out of decision-making processes. Because hunting and fishing is seen as the man's domain, women are largely underrepresented in this decision-making process. Kafarowski (2006) notes that women are also underrepresented on hunter and trapper organizations in Nunavut, with representation as low as 5% in the Kitikmeot region. This under-representation of women on decision-making boards was also evident in Nunavut Wildlife Management Board and the Baffin Fisheries Coalition, where, in 2003, women held 11% (1) and 7.6% (23) respectively of the seats on those boards (Kafarowski, 2004). Because of article 23 of the Nunavut Land Settlement Act, beneficiaries of the settlement receive priority hiring over non-beneficiaries.

Women's participation in governance beyond resource development institutions

Although there is some research on the participation of Indigenous women in formal governance, there are few studies that link women's participation to resource development. The research that does exist suggests that resource development may amplify a move towards Western systems of governance that hinder the political influence of northern women.

The most publicized case of Indigenous women and community governance came in 1997 when the Nunavut Implementation Committee suggested a two-member constituency system, whereby one man and one woman would be elected from each constituency (Williamson, 2006). The proposal was sent to a plebiscite where 57% of Nunavut voters voted against the proposal (Dahl, 1997). Dahl (1997) and Williamson (2006) understood the vote against gender parity to be a result of regional differences intertwined with the declining role of Inuit men as economic provider through hunting and trapping. The parity proposal and the Nunavut Land Claim Agreement are offered by Bennett (2011) as two missed opportunities to address women's inequality in Nunavut, where Inuit men dominate hamlet councils, senior levels of government, and territorial politics, including land management. Based on interviews with members of Nunavut's legislative assembly, White (2006) argues that the Nunavut government system, which combines a Westminster cabinet-parliamentary system with traditional Inuit culture, is not a fundamental alteration to the system but a rather substantial adaptation that left many Western values intact. Without the voices of Inuit women on these formal decision-making bodies, Inuit women will not be able to influence decision-making regarding resource development.

A paper by the Women's Native Association of Canada also argued that self-government of Indigenous people is fundamental to addressing concerns held by Indigenous women (Native Women's Association of Canada, 2007). However, even with Indigenous self-government, there is no guarantee that the voices of women will be heard. As Hipwell et al. (2002) note, the imposition of the Indian Act resulted in male-dominated band councils that usurped the traditional power of women. As a result, women's voices have been marginalized in many land claim negotiations (Hipwell et al., 2002). In 2010, a book published by the Language and Culture Program of Nunavut Arctic College documented the stories of 12 Inuit women involved in leadership and governance in Inuit communities around Canada. Madeleine Redfern, who resides in Nunavut, wrote about applying for positions on the Nunavut Wildlife Management Board, the Nunavut Impact Review Board, and the Kakivak Association, all of which she was denied which troubled her because she knew she was qualified (Redfern, 2010: 106).

Complaints that women in the north have expressed towards the land claim resolution process reflect institutional biases in the process. There is little doubt that institutions governing the north have changed in a direction towards more Indigenous control with the resolution of land claims and establishment of Indigenous governments. However, Natcher and Davis (2007) argue that the devolution of control over natural resource management in the Yukon has resulted in new institutions that, while having Indigenous involvement, do not reflect Indigenous forms of management. The authors argue that rather than a true devolution of control there has been a "deconcentration of preexisting forms of state management and the perpetuation of values that support them" (Natcher and Davis, 2007: 277). This has serious consequences for Indigenous women in the north. Even if more Indigenous women participate in the negotiation process of land claims and management processes for natural resource extraction, they have to contend with preexisting patriarchal values. While Canada's comprehensive land claims process has resulted in co-management boards of natural resources that allow for more Indigenous control, there are conflicts that can occur when different cultural groups with different values enter into a coordinated resource management process (Natcher, Davis, and Hickey, 2005). Patriarchal values of resource management are likely to continue even after the resolution of land claims agreements.

Conclusions and areas for future research

The research on gender and resource development is varied yet shows some similar trends. Most studies suggest that a shift away from traditional economies towards wage-based economies has negative outcomes for women, families, and communities and that northern women and families have benefited less from resource development than men. Furthermore, there are some indications that the shift towards industrial resource extraction has hampered the political participation of Indigenous women in institutions governing resource development and in their communities more broadly.

The research agenda of gender and resource development, however, is far from complete. Most studies have adopted a case study methodology and understand gender narrowly as impacts on Indigenous women. Few studies explore how the colonial disruption of gender relations interacts with resource development. Additionally, little of the research is led by Indigenous authors, and that which exists is often located in community reports rather than academic papers. One consequence of the absence of Indigenous perspectives is that much of the research is atomistic and fails to draw connections between economic, social, cultural, and governance spheres.

Our hope is that future research is sensitive not just to women, but to men and diverse forms of Indigenous masculinity, and that it accounts for the complex, intersecting importance of gender, sexuality, Indigeneity, and colonization. It would also involve a critical examination of taken-for-granted understandings of the gendered dimensions of Indigenous community life, many of which are based on outdated (and thoroughly gendered) anthropological scholarship. Most importantly, however, we feel that the paucity of research adopting an Indigenous lens to gender and resource development in the north constitutes a significant research gap. An Indigenous lens would centre colonialism in examinations of the complexity of community change resulting from resource development. Supporting scholarship by Indigenous scholars in Canada's north is therefore of paramount importance.

References

Archibald, L. and Crnkovich, M. (1999). *If Gender Mattered: A Case Study of Inuit Women, Land Claims and the Voisey's Bay Nickel Project*. Ottowa: Status of Women Canada.

Bennett, M. (2011). Lost Opportunities and Future Pomise: Gender Eequality in Nunavut. M.A. thesis. Athabasca: Athabasca University.

Berkes, F. and Jolly, D. (2001). Adapting to Climate Change: Social-Ecological Resilience in a Canadian Western Arctic Community. *Conservation Ecology, 5*(2), 18.

Bielawski, E. (2003). *Rogue Diamonds: The Rush for Northern Riches on Dene Land.* Vancouver: Douglas and McIntyre.

Bjerregaard, P., Young, T. K., Dewailly, É., and Ebbensson, S. O. (2004). Indigenous Health in the Arctic: An Overview of the Circumpolar Inuit Population. *Scandinavian Journal of Public Health, 32*(5), 390–395.

Brandth, B. and Haugen, M. (2000). From Lumberjack to Business Manager: Masculinity in the Norwegian Forestry Press. *Journal of Rural Studies, 16*: 343–355.

Brockman, A. and Argue, M., BHP Diamond Mine Environmental Assessment Panel and BHP Diamonds Inc. (1995). Review of NWT Diamonds Project Environmental Impact Statement: Socio-Economic Impacts on Women. 1–41. Yellowknife, NWT: Status of Women Canada. http://publications.gc.ca/site/eng/293449/publication.html.

Brower, C. (2015). Testimony of Charlotte Brower, Mayor of North Slope Borough US Senate. Washington, DC: Committee on Energy and Natural Resources, United States Senate.

Butler, C. F. and Menzies, C. R. (2000). Out of the Woods: Tsimshian Women and Forestry Work. *Anthropology of Work Review, 21*(2): 12–17.

Cox, D. and Mills, S. (Forthcoming). Women's Participation in the Voisey's Bay EA: Impacts on EA Recommendations, Impact and Benefit Agreements and Women's Employment Outcomes. *Arctic.*

Czyzewski, K., Tester, F., Aaruaq Qamani'tuaq, N., and Blangy, S. 2014. The Impact of Resource Extraction on Inuit Women and Families in Qamani'tuaq, Nunavut Territory: A Qualitative Assessment Prepared for the Canadian Women's Foundation, Ottawa. Pauktuutit, Inuit Women of Canada and Vancouver, School of Social Work, University of British Columbia. January, p. 188. www.pauktuutit.ca/wp-content/uploads/Quantitative-Report-Final.pdf.

Dahl, J. (1997). Gender Parity in Nunavut? *Indigenous Affairs, 3/4*(July/December), 8.

Davidson, C. M. and Hawe, P. 2012. All that Glitters: Diamond Mining and Tłįchǫ Youth in Behchokǫ̀ Northwest Territories. *Arctic,* 65(2): 214–228.

Dunk T. W. (1994) *It's a Working Man's Town: Male Working-Class Culture.* Montreal and Kingston: McGill-Queens University Press.

Egan, B. and Klausen, S. (1998). Female in a Forest Town: The Marginalization of Women in Port Alberni's Economy. *BC Studies: The British Columbian Quarterly, 118*: 5–40.

George, C. (1999). Testing for Sustainable Development Through Environmental Assessment. *Environmental Impact Assessment Review, 1999*(19), 175–200.

Gibson, V. V. (2008). *Negotiated Spaces: Work, Home and Relationships in the Dene Diamond Economy* (Doctoral dissertation, University of British Columbia).

Gibson, G. and Klinck, J. (2005). Canada's Resilient North: The Impact of Mining on Aboriginal Communities. *Pimatisiwin: A Journal of Aboriginal and Aboriginal Community Health, 3*(1).

Government of the Northwest Territories. (2009). Communities and Diamonds, 2009 Annual Report, retrieved January 31, 2013. www.iti.gov.nt.ca/publications/2010/diamonds/2009CommunitiesDiamondsReport_2010–10–22.pdf.

Hallett, V. and Baikie, G. (2011). Out of the Rhetoric and into the Reality of Local Women's Lives. *Submission to the Environmental Assessment Panel on the Lower Churchill Hydro Development.* 1–10: FemNorthNet. http://criaw-icref.ca/sites/criaw/files/Out%20of%20twentiethe%20Rhetoric-FNN-March%2028%202011.pdf.

Hipwell, W., Mamen, K., Weitzner, V., and Whiteman, G. (2002). Aboriginal Peoples and Mining in Canada: Consultation, Participation and Prospects for Change. 1–57. Ottawa: The North-South Institute.

Houghton, D. S. (1993). Long-Distance Commuting: A New Approach to Mining in Australia. *The Geographical Journal, 159*(3), 281–290.

Kafarowski, J. (2004). Canada. In L. Sloan (Ed.), *Women's Participation in Decision-Making Processes in Arctic Fisheries Resource Management.* Kvinneuniversitetet Nord: Forlaget Nora.

Kafarowski, J. (2005). "Everyone Should Have a Voice, Everyone's Equal": Gender, Decision Making and Environmental Policy in the Canadian Arctic. *Canadian Woman Studies, 24*(4): 12–17.

Kafarowski, J. (2006). Gender, Decision-Making and Co-Management in Arctic Fisheries and Wildlife. In L. Sloan (Ed.), *Women and Natural Resource Management in the Rural North: Artic Council Sustainable Development Working Group 2004–2006.* Kvinneuniversitetet Nord: Forlaget Nora.

Koke, P. E. (2008). The Impact of Mining Development on Subsistence Practices of Indigenous Peoples: Lessons Learned from Northern Quebec And Alaska. M.A., Prince George: The University of Northern British-Columbia.

Kuokkanen, R. (2011). Indigenous Economies, Theories of Subsistence, and Women: Exploring the Social Economy Model for Indigenous Governance *The American Indian Quarterly, 35*(2), 215–240.

Kuyek, J. N. (2003). Overburdened: Understanding the Impacts of Mineral Extraction on Women's Health. *Canadian Woman Studies, 23*(1): 121–123.

Lahiri-Dutt, K. (2001). From Gin Girls to Scavengers: Women in Raniganj Collieries. *Economic and Political Weekly, 36*(44), 4213–4221.

Lahiri-Dutt, K. (2006). Mainstreaming Gender in the Mines: Results from an Indonesian Colliery. *Development in Practice, 16*(2), 215–221.

Lahiri-Dutt, K. (2012). Digging Women: Towards a New Agenda for Feminist Critiques of Mining. *Gender, Place and Culture, 19*(2), 193–212.

Lawrence, D. 1997. Integrating Sustainability and Environmental Impact Assessment. *Environmental Management, 21*(1), 23–42.

Little, M. H. (2005). *If I had a Hammer: Retraining that Really Works.* Vancouver: UBC Press.

Mills, S. E. (2006). Segregation of Women and Aboriginal People within Canada's Forest Sector by Industry and Occupation. *Canadian Journal of Native Studies, 26*(1), 147–171.

Mills, S. E. (2011). White and Aboriginal Women Workers' Perceptions of Diversity Management Practices in a Multinational Forest Company. *Labour/Le Travail, 67,* 45–76.

Mishra, P. P. (2009). Coal Mining and Rural Livelihoods: Case of the Ib Valley Coalfield, Orissa. *Economic and Political Weekly, 44,* 117–123.

Natcher, D. (2013). Gender and Resource Co-Management in Northern Canada. *Arctic, 66*(2), 218–221.

Natcher, D. C. and Davis, S. (2007). Rethinking Devolution: Challenges for Aboriginal Resource Management in the Yukon Territory. *Society and Natural Resources, 20*(3), 271–279.

Natcher, D. C., Davis, S., and Hickey, C. G. (2005). Co-Management: Managing Relationships, Not Resources. *Human Organization, 64*(3), 240–250.

National Aboriginal Health Organization. (2008). *Resource Extraction and Aboriginal Communities in Northern Canada: Gender Considerations.* Ottawa: NAHO.

Native Women's Association of Canada. (2007). Aboriginal Women and Self-Determination: An Issue Paper. 6. Corner Brook: National Aboriginal Women's Summit.

Noble, B. F. and Bronson, J. E. (2006). Integrating Human Health into Environmental Impact Assessment: Case Studies of Canada's Northern Mining Resource Sector. *Arctic*, *58*(4), 395–406.

Noronha, L. and Nairy, S. (2005) Assessing Quality of Life in a Mining Region. *Economic and Political Weekly*, *41*(1), 72–78.

O'Faircheallaigh, C. (1995). Long Distance Commuting in Resource Industries: Implications for Native Peoples in Australia and Canada. *Human Organization*, *54*(2), 205–213.

O'Faircheallaigh, C. (1999). Making Social Impact Assessment Count: A Negotiation-Based Approach for Aboriginal Peoples. *Society and Natural Resources*, *12*(1), 63–80.

O'Faircheallaigh, C. (2007). Reflections on the Sharing of Benefits from Australian Impact Benefit Agreements (IBAs). Paper presented at a forum on devolution, funded by the Gordon Foundation, Fort Good Hope, May 2–4, 2007.

O'Faircheallaigh, C. (2011). Indigenous Women and Mining Agreement Negotiations: Australia and Canada. In K. Lahiri-Dutt (Ed.), *Gendering the Field: Towards Sustainable Livelihoods for Mining Communities*. Vol. 6. 87–110. Canberra: Australian National University.

O'Reilly, K. and Eacott, E. (1998). Aboriginal Peoples and Impact and Benefit Agreement: Report of a National Workshop. Northern Minerals Program Working Paper No. 7.

Parlee, B., Berkes, F., and Teetl'it Gwich'in Renewable Resources Council (2005). Health of the Land, Health of the People: A Case Study on Gwich'in Berry Harvesting in Northern Canada. *EcoHealth*, *2*(2), 127–137.

Parlee, B., Berkes, F., and Teetl'it Gwich'in Renewable Resources Council (2006). Aboriginal Knowledge of Ecological Variability and Commons Management: A Case Study on Berry Harvesting from Northern Canada. *Human Ecology*, *34*(4), 515–528.

Parlee, B., O'Neil, J., and Nation, L. K. D. F. (2007). "The Dene Way of Life": Perspectives on Health from Canada's North. *Journal of Canadian Studies*, *41*(3), 112–133.

Parpart, J. L. (1993). Who is the "Other"?: A Postmodern Feminist Critique of Women and Development Theory and Practice. *Development and Change*, *24*(3), 439–464.

Pauktuutit (2012). Resource Extraction Workshop Report. Pauktuutit Annual General Meeting Resource Extraction, March 28, 2012. Ottawa. http://pauktuutit.ca/wp-content/blogs.dir/1/assets/Resource-Extraction-Workshop-Report.pdf.

Pollard. L. (1990). Fly-in/Fly-out: Social Implications for Remote Resource Development in Western Australia. Perth, Western Australia: Social Impact Unit, Department of State Development.

Reed, M. (1999). "Jobs Talk": Retreating from the Social Sustainability of Forestry Communities. *The Forestry Chronicle*, *75*(5), 755–763.

Redfern, M. (2010). Supporting Civil Development. In L. McComber and S. Partridge (Eds.), *Arnait Nipingit: Inuit Women in Leadership and Governance*. Iqaluit: Language and Culture Program of Nunavut Arctic College.

Rude, D. and Deiter, C. (2004). *From the Fur Trade to Free Trade: Forestry and First Nations Women in Canada*. Ottawa: Status of Women Canada.

Salee, D. (2006) Quality of Life of Aboriginal People in Canada: An Analysis of Current Research. *IRPP Choices*, *12*(6), ISSN 0711–0677 Montreal.

Schaefer, O. (1983). Psycho-social Problems Experienced by Native Population Groups in the Canadian Arctic Involved in Resource Development. *Canadian Journal of Community Mental Health (Revue canadienne de santé mentale communautaire)*, *2*, 53–55.

Shanks, G. (2006). *Sharing in the Benefits of Resource Development. A Study of First Nations and Industry Impact Benefit Agreements*. Ottawa: Public Policy Forum.

Shannon, K. A. (2006). Everyone Goes Fishing: Understanding Procurement for Men, Women and Children in an Arctic Community. *Études/Inuit/Studies*, *30*(1), 9–29.

Sharma, S. (2010). The Impact of Mining on Women: Lessons from the Coal Mining Bowen Basin of Queensland, Australia. *Impact Assessment and Project Appraisal*, *28*(3), 201–215.

Sharma, S. and Rees, S. (2007). Consideration of the Determinants of Women's Mental Health in Remote Australian Mining Towns. *Australian Journal of Rural Health*, *15*(1), 1–7.

Smith, L. T. (1999). *Decolonizing Methodologies: Research and Indigenous Peoples*. Dunedin: University of Otago Press.

Southcott, C. (2003). Women in the Workforce in Northern Ontario 2001 Census Research Paper Series: Report# 8. *Northern Ontario Training Boards*.

Status of Women Council of the Northwest Territories. (1999). Review of Diavik Diamonds Project Socio-Economic Environmental Effects Report: Impacts on Women and Families. Retrieved from www.statusofwomen.nt.ca/download/review_diavik.pdf.

Stern, P. (2005). Wage Labor, Housing Policy, and the Nucleation of Inuit Households. *Arctic Anthropology*, *42*(2), 66–81.

Storey, K. Lewis, J., Shrimpton, M., and Clark, D. (1988). Working Offshore: Family Adaptations to Rotational Work in Newfoundland's Oil and Gas Community. In Brealey, T. B., Neil, C. C., and Newton, P. W. (Eds.). *Resource Communities: Settlement and Workforce Issues*. Australia: CSIRO, 353–363.

Tallichet, S. (1995). Gendered Relations in the Mines and the Division of Labour Underground. *Gender and Society*, *9*(6), 697–711.

Tallichet, S. (2000). Barriers to Women's Advancement in Underground Coal Mining. *Rural Sociology*, *65*(2), 234–252.

Tongamiut Inuit Annait (1997). 52% of the Population Deserves a Closer Look: A Proposal for Guidelines Regarding the Environmental and Socioeconomic Impacts on Women from the Mining Development at Voisey's Bay. Environmental assessment of the Voisey's Bay mine and mill undertaking. Unpubl. ms. Available from the Canadian Environmental Assessment Agency, 22nd Floor, Place Bell, 160 Elgin Street.

Weitzner, V. (2006). "Dealing Full Force": Lutsel K'e Dene First Nation's Experience Negotiating with Mining Companies: The North-South Institute and Lutsel K'eDene First Nation. Retrieved on July 26, 2012 from www.nsi-ins.ca/wp-content/uploads/2012/10/2006-Dealing-full-force-Lutsel-ke-Dene-first-nations-experience-negotiating-with-mining-companies.pdf.

White, G. (2006). Traditional Aboriginal Values in a Westminster Parliament: The Legislative Assembly of Nunavut. *The Journal of Legislative Studies*, *12*(1), 8–31.

Williamson, K., J., Hoogensen, G., Lotherington, A. T., Hamilton, H., Savage, S., Koukarenko, N., . . . and Poppel, M. (2004). Gender Issues. In Arctic Council (Ed.) *Arctic Human Development Report*. Akureyri: Stefansson Arctic Institute.

Williamson, L. (2006). Inuit Gender Parity and Why It Was Not Accepted in the Nunavut Legislature. *Etudes/Inuit/Studies*, *30*(1), 51–68.

Women in Mining Canada. (2010). Ramp Up: A Study on the Status of Women in Canada's Mining and Exploration Sector. Retrieved on July 26, 2012 from http://0101.nccdn.net/1_5/1f2/13b/0cb/RAMP-UP-Report.pdf.

Yukon Conservation Society. (2000). Gaining Ground: Women, Mining and the Environment. Retrieved from www.yukonconservation.org/library/pdf/gain.pdf.

15 Resource development and climate change

A gap analysis

Chris Southcott

Over the past decade, Arctic resource development is often discussed as being directly related to climate change. In the media and in much of the grey literature, increased resource development in the Arctic is often portrayed as the result of climate change. There seems to be an intuitive recognition that climate change is melting the ice covering the Arctic and, as a result, resources that would have previously been inaccessible are now available for exploitation. An associated belief is that climate change, by melting ice, is making transportation easier and, as such, making it easier to ship technology in and resources out of the region. Yet, despite this seemingly intuitive idea, the limited research that has been done on the subject seems to indicate otherwise. Indeed, climate change may be making resource development more difficult – at least in the immediate term.

It is not immediately evident that looking at the relationship between resource development and climate change is likely to help us find out if and how resource development can help bring more benefits to Arctic communities. As such, some could question why ReSDA would devote any attention to the relationship between resource development and climate change. Research on climate change in the Arctic has increased substantially over the past 15 years and it is unlikely that ReSDA could add much to what current researchers are doing. Still, the fact that at least some people believe that an important causal relationship exists between the warming of the Arctic and increased interest in the development of Arctic resources means that ReSDA should at least try and discover what relationship, if any, exists. A gap analysis looking at this relationship should allow us to better understand if further research is necessary in order to assist in helping communities find ways to get more out of resource developments. At the very least, we need to help communities understand if climate change will increase resource development pressures or prove to be an obstacle to attempts by communities to use resource development as a tool to bring about a more sustainable future.

This gap analysis will consist of a literature review of the relationship between climate change and Arctic resource development. Climate change is now the most researched challenge facing the Arctic and as such there have already been a number of literature reviews of both climate change research and the social impacts of climate change. We borrow heavily from these previous studies.

We summarize the existing research in an attempt to allow us to better understand what, if any, relationship exists between climate change and resource development in an attempt to understand its implications for Arctic communities.

Climate change and resource development

An article in Foreign Affairs in 2008 by Scott G. Borgerson can be used as an example of the current beliefs concerning the relationship between climate change and resource development. Borgerson starts out "Global warming has given birth to a new scramble for territory and resources among the five Arctic powers" (Borgerson, 2008: 63). The image portrayed is that Arctic ice has, in the past, been an obstacle to resource development in the region. One part of the perspective is that ice is seen as a cover to hide the resources and a barrier to access them. "the great melt is likely to yield more of the very commodities that precipitated it" (Borgerson, 2008: 64). Another part of the perspective is that warmer temperatures will lead to easier access. "Less ice also means increased access to Arctic fish, timber, and minerals, such as lead, magnesium, nickel, and zinc" (Borgerson, 2008: 64). This access will be primarily through increased maritime transportation possibilities caused by melting sea ice.

This view is often seen in the global media. *The New York Times* published a front page story in its September 19, 2012 issue with the title "Race is On as Ice Melt Reveals Arctic Treasures". The article stated,

> With Arctic ice melting at record pace, the world's superpowers are increasingly jockeying for . . . economic position in outposts like this one, previously regarded as barren wastelands. At stake are the Arctic's abundant supplies of oil, gas and minerals that are, thanks to climate change, becoming newly accessible.
>
> (Rosenthal, 2012)

The same year *The Economist*, in a special report called "The Melting North", noted that the melting of Arctic ice, caused by climate change, was "going to make a lot of people rich" (*The Economist*, 2012). They state that as the "the frozen tundra retreats northwards . . . much more valuable materials will become increasingly accessible". Oil and gas also becomes more accessible. "Oil companies do not like to talk about it, but this points to another positive feedback from the melt. Climate change caused by burning fossil fuels will allow more Arctic hydrocarbons to be extracted and burned." Canada's *Globe and Mail* noted in 2014 that the melting caused by climate change is "opening a new energy frontier" (Ritter, 2014).

Variations of this general perspective are seen in many of the articles, books, television shows, and films dealing with Arctic climate change. What is interesting is that there are very few studies that have even looked at the issue in any sort of depth. Those that do exist point to a more nuanced perspective on the relationship between climate change and resource development. Some even suggest that climate change should actually be seen as a barrier to resource development.

The "human dimensions" of climate change

A history of the research dealing with the relationship between climate change and resource development is generally seen through the work on the "human dimensions" of climate change. While initially the work on climate change in the Arctic was dominated by the physical and biological sciences, over the past 15 years there has been a wealth of literature produced on the topic. One recent study noted that at least 117 refereed articles dealing with the human dimensions of climate change in the Eastern Canadian Arctic alone were published between 2000 and 2010 (Ford et al., 2012).

Despite this wealth of research, the same study also notes that there is actually very little solid information on specific impacts. Most of the literature is

> dominated by studies examining the safety of using semi-permanent trails for hunting and recreational travel, particularly how changing snow and ice regimes, less-predictable weather, and changing wind patterns are an increasing danger, making routes less dependable and compromising the ability to engage in harvesting activities.
>
> (Ford et al., 2012: 292).

These same studies, generally based on indigenous accounts and traditional knowledge relating to climate change, also note that these impacts are currently being managed by these communities through adaptations but that barriers could arise in the future to the ability of communities to adapt.

Few studies exist that have looked in depth at other current human dimensions of climate change, including resource development. The only research noted by the above study relating to climate change and resource development indicated that climate change is causing problems for mining in that hydro-electric power is becoming less dependable due to climate change-related stream flow changes (Ford et al., 2012: 293).

The early IPCC reports

While little concrete research exists, there is a long history of commentary on the potential impacts of global warming on Arctic resource development. When the Intergovernmental Panel on Climate Change (IPCC) was created by the World Meteorological Association and the United Nations Environmental Programme in 1988, earlier reports had already indicated that climate change would impact higher latitudes more severely than lower latitudes (Bolin, 2007). As a result the first assessment published in 1990 contained an extensive discussion of potential impacts on the Arctic, including social impacts. At that time all impacts were however hypothetical as there had been no concrete evidence of social impacts of climate change on Arctic communities.

The report from Working Group II of the IPCC, which dealt with the assessment of impacts, listed a number of potential impacts of climate change in Arctic communities (IPCC FAR WG II, 1990). It highlighted the possible negative impacts on the traditional harvesting activities of Indigenous peoples – especially in Canada (IPCC FAR WG II, 1990: 5–6). This first report also noted that melting sea ice may increase transportation possibilities in the region which would have certain security implications. It also inferred that less sea ice may have positive implications for resource development when its states that the "possibility of climate-induced changes in access and prospects for resource development may result in altered national positions and competing claims" (IPCC FAR WG I, 1990: 6–20).

Despite this comment, much of the analysis contained in these particular reports seems to stress the problems for resource development in the Arctic that would be caused by climate change. It noted the problems for infrastructure caused by melting permafrost (IPCC FAR WG II, 1990: 5–8), problems caused by loss of winter roads (IPCC FAR WG II, 1990: 5–20), and problems for hydro-electric projects, necessary for resource development, that would be caused by changes in run-off regimes (IPCC FAR WG II, 1990: 7–16). There is little evidence in the first impact report of the IPCC that resource development would be significantly enhanced by climate change. In 1992 the IPCC published a supplementary report that did not directly mention resource development in the Arctic but did note, once again, that melting sea ice in the Arctic could lengthen the shipping season in the Arctic (Houghton, Callander, and Varney, 1992: 35).

The Second IPCC Assessment of 1995 discussed in greater detail the potential impacts of climate change on humans in the Arctic. Once again the potential impacts on traditional harvesting activities were listed as the most likely negative consequence of climate change (IPCC SAR WG I, 1996: 257). The greatest economic impact of climate change in the Arctic "is likely to stem from decreases in ice thickness and bearing capacity, which could severely restrict the size and load limit of vehicular traffic" (IPCC SAR WG I, 1996: 257). While this comment seemed to refer to vehicles involved in traditional harvesting, one would presume it could also refer to vehicles involved in resource development.

More detail was included concerning potential impacts on resource development and once again, while certain potentially positive impacts were mentioned, the report seems to indicate just as many, if not more, negative impacts. The report stresses the potentially negative impacts of thawing permafrost on mining and other resource development infrastructure (IPCC SAR WG I, 1996: 257). Mining roads would be negatively affected by erosion (IPCC SAR WG I, 1996: 205).

The oil and gas industry could benefit from reduced sea ice. This would allow greater transportation possibilities (IPCC SAR WG I, 1996: 259) and "shorter winters disrupting construction, exploration, and drilling programs" (IPCC SAR WG I, 1996: 259). This may serve to reduce operating costs (IPCC SAR WG I,

1996: 383). At the same time the report mentions several potentially negative impacts of climate change on oil and gas development. In particular they cite a report on the Beaufort Sea where a longer open water season would see the frequency of extreme wave heights increase. "Present design requirements for long-lived coastal and offshore structures, such as oil installations, will be inadequate under these conditions" (IPCC SAR WG I, 1996: 259). The report also noted that oil and gas pipelines would be negatively affected by melting permafrost (IPCC SAR WG I, 1996: 382).

While these reports stress that climate change could have both positive and negative impacts on resource development, a regional report published by the IPCC indicated that the impacts are likely to be more positive. In a brief statement the report noted that while climate change related changes would be "particularly important for indigenous peoples leading traditional lifestyles" it would also mean "new opportunities for shipping, the oil industry, fishing, mining, tourism, and migration of people. Sea ice changes projected for the Arctic have major strategic implications for trade, especially between Asia and Europe" (Watson, Zinyowera, and Moss, 1997: 8).

The 2001 Report of the IPCC repeated once again the same observations of climate change having both positive and negative potential impacts on human societies and resource development in the Arctic. Interestingly, while echoing earlier comments on the likely negative impacts of climate change on Indigenous communities, it also stressed, for the first time, that Indigenous communities "may be sufficiently resilient to cope with these changes" (IPCC TAR WG II, 2001: 805). As had the previous reports, the 2001 noted the potentially positive aspects of less sea ice on maritime transportation and oil and gas industry operating costs IPCC TAR WG II, 2001: 805).

It is interesting that much more time was spent in this particular report discussing potential problems for extractive resource development production that could be caused by climate change as opposed to potential benefits. The issue of infrastructure problems caused by melting permafrost was repeated ((IPCC TAR WG II, 2001: 946) as was the issue of increased wave action ((IPCC TAR WG II, 2001: 828). A new detail that was added was that coastal erosion was increasingly a problem for communities and for oil and gas production facilities ((IPCC TAR WG II, 2001: 827). This issue of storm surges and the negative impacts that this could have on resource industry production as also introduced ((IPCC TAR WG II, 2001: 828).

The report went into greater detail when discussing how climate change could represent problems for resource development. It noted that although less sea ice would mean that the oil and gas industry could utilize more convention drilling techniques, these "likely changes are not without concerns" ((IPCC TAR WG II, 2001: 828). While drill ships could reduce operating costs "increased wave action, storm surges, and coastal erosion may necessitate design changes to conventional offshore and coastal facilities" ((IPCC TAR WG II, 2001: 828). The costs of pipeline construction would increase because extensive trenching may be needed to combat the effects of coastal instability and erosion. As concerns the oil and gas industry, the report concludes that the "impact of climate change is likely to lead to increased costs in the industry associated with design and operational changes" ((IPCC TAR WG II, 2001: 828).

The Arctic Climate Impact Assessment

Following the 2001 IPCC Assessment, work started on a separate climate impact assessment for the Arctic. The Arctic Climate Impact Assessment, organized under the mandate of the Arctic Council, is, to this day, the most important attempt to assess the impact of climate change on the Arctic. It entailed a much more detailed report of potential impacts of climate change on Arctic communities and on resource development. It is also unique in that while the previous discussions of climate change impacts were hypothetical, the ACIA included studies that started to see real impacts occurring on Arctic societies.

The scientific report of the ACIA is organized by chapters based on the different sectors of the Arctic environment (ACIA, 2005). Several of these chapters mention resource development relating to climate change in brief hypothetical statements. Such is the case in the chapter on Cryosphere and Hydrology, which notes that freshwater ice uncertainty could cause problems for extractive resource development (ACIA, 2005: 229), or the chapter on Marine Systems which noted that less sea ice will result in more possibilities in terms of maritime transportation and that this may lead to "increased exploration for reserves of oil and gas, and minerals" (ACIA, 2005: 519). The chapter dealing with wildlife management and conservation discusses the increased importance of resource development in the Arctic during the last half of the twentieth century but does not link it to climate change. Indeed the chapter seems to indicate that the negative impacts from resource development may be more important to the Arctic than those linked to climate change. They claim that the negative impacts of this resource development on the environment makes it more difficult to isolate the specific impacts caused by climate change (ACIA, 2005: 621). The chapter on health impacts does not refer to extractive resource development directly but does state that increased shipping due to climate change may cause problems for Arctic communities because of the threat of "the sudden catastrophic release of hazardous materials into the local, regional, and eventually circumpolar environment" (ACIA, 2005: 892). According to the chapter, this is increasingly likely to happen due to the increase in extreme weather events caused by climate change.

The two chapters of the ACIA where the relationship between resource development and climate change are discussed most extensively are in the chapter dealing with renewable resources and the chapter on infrastructure. It is interesting that the renewable resources chapter does not specifically state that climate change is having an impact on Arctic communities – only that as "this assessment shows, these activities and relationships appear to be threatened by severe climate change" (ACIA, 2005: 651). Despite the hypothetical nature of the above statement, the chapter refers to two studies that claim to show actual impacts. They note a study by Fugal et al. (2002) which shows that local anxieties over environmental changes are currently being experienced by communities in northern Quebec and Labrador, and that climate change is having an impact on human health and the availability of important traditional/country foods (ACIA, 2005: 656). In addition they refer to an Alaskan study by Callaway et al. (1999) which shows that climate change is increasing the monetary costs of hunting for North Slope Indigenous communities (ACIA, 2005: 656). Elsewhere, in addition to the above mentioned studies, they also point out that climate change related erosion has caused enormous problems for certain costal Indigenous communities such as Shishmaref (ACIA, 2005: 660).

In the rest of the chapter their discussion of resource development is built largely around the observation, already noted above in relation to the chapter dealing with wildlife management and conservation, that resource development may be more important than climate change in terms of challenges currently facing wild-life management and conservation. They continue the chapter by underlining the importance of global economic impacts on Arctic societies and state: "(T)hus, for some indigenous communities climate change may not be the most immediate issue of local concern" (ACIA, 2005: 657). They go further in stating that there "is scientific difficulty in stating how far climate change alone has affected arctic marine ecosystems in the past fifty years, for instance, as the impacts of overfishing and over-hunting may be far greater" (ACIA, 2005: 658). In their discussion of the potential impacts of climate change on the Inuvialuit of Canada they state that when looking at the impacts of other historical trends "the impact of climate change is relatively minor, at least so far, and it is not beyond the ability of the community to adapt" (ACIA, 2005: 670).

The authors of the chapter note the scientific studies indicating the Arctic environment will be impacted by climate change under the existing models and they state that this will likely affect subsistence activities. But at the same time, they stress that it is important to contextualize climate change impacts with reference to other changes experienced by Arctic residents. Being able to access traditional food resources and ensuring food security will be a major challenge in an Arctic affected increasingly by both climate change and global processes (ACIA, 2005: 658).

The chapter authors note how the Indigenous peoples have successfully adapted to climate change in the past but state that other factors may impact their ability to do so in the future and perhaps the most important factor is resource develop-ment. Resource development can be a barrier or it can enhance their ability to adapt. The chapter seems to stress the positive impacts of resource development.

In addition, significant steps have been taken with innovative co-management regimes that allow for the sharing of responsibility for resource manage-ment between indigenous and other uses and the state (Huntington, 1992; Osherenko, 1988; Roberts, 1996). Examples include the Alaska Eskimo

Whaling Commission, the Kola Saami Reindeer Breeding Project, the Inuvialuit Game Council, and the North Atlantic Marine Mammal Commission. Self-government is about being able to practice autonomy. The devolution of authority and the introduction of co-management allow indigenous peoples opportunities to improve the degree to which management and the regulation of resource use considers and incorporates indigenous views and traditional resource use systems.

(ACIA, 2005: 665, 666)

While the previous chapter suggests that at least some aspects of resource development may be a source of increased capacity to deal with potential negative impacts caused by climate change, the chapter dealing with infrastructure impacts seems to suggest that these possibilities may be limited by negative impacts of climate change on resource development. Resource development depends upon stable infrastructures. Climate change represents a threat to these stable infrastructures. The chapter cites an earlier study by Esch and Osterkamp which summarizes most of the engineering concerns related to permafrost warming. These include an increase in creep rate of existing piles and footings, increased creep of embankment foundations, eventual loss of bond support for pilings, thaw settlement during seasonal thawing, increased frost-heave forces on pilings, increased total and differential frost heave during winter, decrease in effective length of piling located in permafrost, progressive landslide movements, and progressive surface settlements (ACIA, 2005: 914).

In terms of coastal impacts, the chapter confirms that less ice may lead to more shipping and ease of access to resources but cautions that it will also cause increased wave generation which will lead to more erosion and other threats to coastal infrastructure. More extreme weather caused by climate change will further exacerbate the problems (ACIA, 2005: 922). Increased water flow in river systems may increase the transportation of materials on these rivers but increased erosion will threaten the infrastructures that permit river transportation. River ice breakup and ice jams are going to become harder to predict (ACIA, 2005: 923).

The chapter discusses a series of infrastructure problems for resource industries due to climate change. Northern pipelines are likely to be negatively impacted by frost heave and slope instability, shallow pile foundations settlement in permafrost could possibly be accelerated, and large tailings disposal facilities might be affected by permafrost thaw caused by climate change. Transportation routes such as roads, airports, and railways are going to be negatively impacted by destructive frost action and "will experience increasing failure rates both in the continuous and discontinuous permafrost zones" (ACIA, 2005: 931).

In summarizing their discussion of the impacts of climate change on resource development infrastructure they note that impacts so far have been small. As far as future impacts, the two most important potentially positive impacts are that less sea ice would mean cost reductions in the construction of oil and gas platforms that must withstand ice force and longer maritime transportation seasons.

Transportation costs would be lessened by "increased drafts in harbors and channels as sea level rises, a reduced need for ice strengthening of ship hulls and off shore oil and gas platforms, and a reduced need for icebreaker support" (ACIA, 2005: 937). These potential positive impacts could be countered by less ability to use ice roads, the deterioration of existing infrastructure, higher costs for future infrastructure, increased wave activity, and erosion.

Subsequent studies have failed to alter the inconclusive findings of the Arctic Climate Impact Assessment as regards the potential impacts of climate change on resource development. The Forth IPCC Assessment Report repeats many of the same inconclusive statements that were contained in the earlier reports (although with less detail) (IPCC AR4 WG II, 2007). It notes the ability of the Indigenous peoples of the Arctic to be resilient in the face of potential challenges linked to climate change but adds that "stresses in addition to climate change . . . will challenge adaptive capacity and increase vulnerability" (IPCC AR4 WG II, 2007: 56). In other words, normally Indigenous peoples would be able to adapt to climate change but other, one would assume more important, pressures are reducing their ability to do so.

The Forth Report states that there is a very high likelihood of climate change producing what they call "positive economic benefits" for some Arctic communities and that these may include "reduced heating costs, increased agricultural and forestry opportunities, more navigable northern sea routes and marine access to resources" (IPCC AR4 WG II, 2007: 56). Elsewhere the report is more specific in terms of positive economic impacts as far as resource development is concerned in that reduced sea ice will probably result in increased navigation and "possibly also a rise in offshore oil operations . . . but there are no quantitative data to support this" (IPCC AR4 WG II, 2007: 88).

While the Forth Report seems uncertain about positive impacts of climate change on resource development, it is fairly confident about the likelihood of negative impacts. It notes that the melting of permafrost

> is a risk to gas and oil pipelines, electrical transmission towers, nuclear-power plants and natural gas processing plants in the Arctic region . . . Structural failures in transportation and industrial infrastructure are becoming more common as a result of permafrost melting in northern Russia, the effects being more serious in the discontinuous permafrost zone.
>
> (IPCC AR4 WG II, 2007: 367)

Recent research on climate change and resource development

More recent discussions of the relationship between climate change and resource development tends to confirm the discussions found in the IPCC reports and the ACIA. While the grey literature tends to highlight the positive impacts of climate change on the potential for resource development in the Arctic, the scientific literature tends to highlight the negative consequences. While, as noted earlier, there is not a great deal of research on the topic, there are several recent findings that are relevant.

In 2007 the Government of Canada undertook a survey of the existing literature in an attempt to determine the likely impacts of climate change on the various regions of the country. This review included a chapter on Northern Canada (Furgal and Prowse, 2008). The section dealing with economic impacts starts out discussing hydro-electric development which, the chapter states, are increasingly important for mining operations. Hydro-electric developments will likely be negatively affected by climate change in that flows necessary for the production of power will be reduced (Furgal and Prowse, 2008: 79). In addition, the increasing in-stability of water flows caused by climate change would require increased monitoring of hydro-electric structures and could cause safety concerns.

Oil and gas exploration could be positively impacted by climate change in that some areas would become easier to access due to reduction in sea ice. At the same time, design changes to drilling platforms will be necessary to deal with increased wave action and storm surges. Land-based infrastructure will be negatively impacted by thawing permafrost and the unpredictability of the winter season. According to the report one of the biggest problems is that the current practise of dealing with drilling wastes by storing them in perma-frost sumps will have to be stopped in order to ensure there is no contamination. More expensive ways to deal with the waste will have to be introduced (Furgal and Prowse, 2008: 79). Coastal structures will be threatened by erosion and this would require mitigation technologies. Finally, pipelines necessary to transport the oil and gas would have their integrity threatened by thawing permafrost and this could represent an important danger to the environment (Furgal and Prowse, 2008: 80).

Mining would likely experience several negative impacts from climate change. Winter ice roads are necessary for the cost-effective transportation of structures and supplies to isolated mine sites. These roads would become more unpredict-able with climate change forcing companies to consider more expensive options. This could negatively impact the viability of mines in Canada's Arctic (Furgal and Prowse, 2008: 80). The other concern relates to mine waste. According to the report, the "stability of waste-rock piles, tailings piles and tailings-containment impoundments often depends on maintenance of frozen conditions to ensure that contaminants and acid-metal leachate (or acid-rock drainage) are not discharged to the environment" (Furgal and Prowse, 2008: 80).

It is interesting that while most of the research-based considerations of impacts stress the negative implications of climate change, a text box in the report stresses "Opportunities for Growth in the Mining and Transportation Sector" (Furgal and Prowse, 2008: 81). According to the argument in the text box, a longer shipping season would benefit the mining sector. With a longer season new port infra-structure would become more economically feasible and this would allow new ways of accessing resource deposits.

The Suzuki Foundation study

In 2008 a Canadian consulting contract from the David Suzuki Foundation undertook to survey the mining industry on their perceptions of climate change and its impact on mining (Ford, Pearce, Duerden, Prno, and Marshall, 2008). The results of this study were published as a report (Pearce, Ford, Prno, and Duerden, 2009) and were reproduced in a number of articles in academic journals (Ford et al., 2010; Ford et al., 2011; Pearce et al., 2011). The report describes their literature review which notes a limited scientific discourse on climate change impacts relating to mining (Pearce et al., 2011). Their review of the scientific discourse echoes many of the observations mentioned above in that researchers suggest that climate change would have a negative impact on infrastructure, transportation, processing operations, and other features while there is the possibility that "(m)elting sea ice also presents a number of potential opportunities for the mining sector including longer shipping seasons, shorter transportation routes, and increased exploration opportunities" (Pearce et al., 2011: 26).

Despite the increased attention given to climate change lately, the mining industry seemed to be little concerned about climate change impacts. In their review of trade journals in Canada between 2002 and 2009 only four articles can be seen to discuss impacts: two that deal with the early closure of ice roads in the Northwest Territories in 2006, one that suggests that infrastructure may be impacted in the future by climate change, and one that warned the mining industry that the public was becoming concerned about climate change and that the industry may need to develop "a green advantage and appeal to shareholders" ((Pearce et al., 2011: 18). According to the authors of the report:

> Absent in the trade journal literature, however, is a thorough discussion and rigorous evaluation of what climate change will mean for the physical aspects of mine operations both today and in the future, and capacity of the sector to adapt to change.
>
> (Pearce et al., 2011: 18)

In addition to the literature review, the report details two surveys of mining industry representatives and a series of case studies examining specific mine sites. It should be pointed out that the surveys and the case studies concerned the whole of Canada and were not specific to the Arctic. At the same, as many companies have multiple mine sites which often include the north, the results of these studies are interesting as a means of better understanding the relationship between climate change and resource development in the Arctic.

The first survey that was done by the consultants was questionnaire-based interview survey that was carried out at the 2008 Prospectors and Developers Association of Canada International Convention (Pearce et al., 2009). This association is one of the main mining industry organizations in Canada. While the survey was limited in terms of the number of respondents (Pearce et al., 2009: 42) the results do help us better understand the feeling of the mining industry in Canada towards climate change. Generally speaking, the results point to a mild degree of concern regarding climate change. When asked if climate change was currently affecting their operations, 20 (45%) of 42 said yes (Pearce et al., 2009: 39). When asked if these impacts were good or bad, 9 (45%) of these 20 said the impacts were bad while only 3 (15%) said they were good.

The respondents were then asked whether concerns exist about perceived future impacts of climate change. At this point 21 (50%) of the 42 respondents answered yes while 19 (43%) answered no. Of these 21 respondents 12 (58%) thought the future impacts would be bad for their industry while only 2 (10%) thought future impacts would be good (Pearce et al., 2009: 40). When asked if their company was taking action to deal with future climate change impacts only 12 respondents stated that they were.

The respondents at the Prospectors and Developers Association of Canada convention represented all facets of the mining industry including consultants and suppliers. Indeed, only 38% of respondents to this first survey actually worked for a mining company. As a result the consultants decided to undertake a follow-up questionnaire that would be directed exclusively at "mining practitioners working 'on the ground' at mine sites across the country" (Pearce et al., 2009: 49). This second survey used a variation of the initial questionnaire and was based on 62 telephone interviews of randomly selected mining practitioners conducted during the summer of 2008.

The second survey generally portrayed a lower level of concern about climate change than in the first survey.[1] According to the consultants, while 55% indicated that their company or organization was concerned about climate change, only 34% indicated that climate change was currently having an impact on their operations (Pearce et al., 2009: 53). This was significantly lower than in the first survey. Of those respondents who believed climate change was currently having an impact, the largest percentage (76%) thought these impacts concerned the transportation sector of their company. When asked if their company was currently taking steps to deal with the negative impacts of climate change only 21% responded yes (Pearce et al., 2009: 54). Only 8% responded that their company was currently taking action to take advantage of the positive impacts of climate change.

When asked about the future, 58%, or 36 respondents, expected climate change to impact their company sometime in the future (Pearce et al., 2009: 54). Of these 36 respondents, 44% expected future impacts to be bad for their company while 35% thought these future impacts would be good. Most thought future impacts would concern transportation. In regards to company plans regarding climate change, 40% thought their companies were currently taking actions to plan for future climate change (Pearce et al., 2009: 56).

These two surveys were supplemented by six case studies of specific mines in operation in Canada. Of these six, one was the diamond mines in the Northwest Territories, one was the mine at Voisey's Bay, Labrador, and another looked at mining in the Yukon. These case studies were based primarily on literature reviews but also included 46 semi-structured interviews (Pearce et al., 2009: 66).

The case study concerning the diamond mines in the Northwest Territories was the only one of the three that detailed a climate change-related impact which has already or is currently occurring. The study noted that in 2006 warm weather forced the early closure of the ice roads essential for transporting goods into the mine sites adding to operational costs (Pearce et al., 2009: 70).

The report noted several adaptations that companies were undertaking that would help them deal with climate change in the Northwest Territories but at the same time the consultants noted that many of these adaptations may not be due to any concerns about climate change.

> It is difficult to determine exactly how many of these initiatives are linked to climate change directly. In many cases, "energy management" initiatives are undertaken primarily with cost savings in mind. Energy for the diamond mines is produced 100% from diesel fuel and this fuel must be trucked in over the ice road or flown in by plane. Both these transportation options have significant costs attached and it is thus in the interest of mine operators to reduce fuel use wherever they can.
>
> (Pearce et al., 2009: 72)

In terms of the mining representatives in the Northwest Territories that were interviewed, climate change was a concern but a minor one (Pearce et al., 2009: 73)

This was similar to the findings in Labrador. The report noted that the Voisey's Bay mine has no formal plan for dealing with or adapting to potential future climatic changes (Pearce et al., 2009: 98). When the initial environmental assessment was done for the mine it was expected that impacts from climate change would be minimal over the life of the mine. While more recent models of climate change suggest that impacts may be greater than previously thought, the operations of the mine have seemingly not been discernibly impacted by climate change. When mine representatives were interviewed the belief was expressed that "planning will only occur when significant climatic changes are documented and start to have a discernible impact on company operations" (Pearce et al., 2009: 98).

The Yukon case study did not detect any current climate change-related impacts on mining operations in this territory but it did suggest that impacts could occur in the future. Opinions expressed by researchers point to potential future negative impacts from melting permafrost, such as threats to the transportation system and other similar types of infrastructure, as well as extreme weather events. Others note potential positive influences stemming from climate change (Pearce et al., 2009: 101). Under certain conditions, permafrost makes mining more difficult and, as such, the melting of permafrost may reduce production costs. The particular historical conditions of the Yukon and the presence of a large number of abandoned mines meant that quite a bit of concern was expressed about the impact of climate change on these sites (Pearce et al., 2009: 103).

When the results of this particular case study analysis was re-examined in a journal article, one of the main conclusions was that most in the mining industry view climate change as only a minor concern (Pearce et al., 2011: 363). This is due to the fact that there are much more important influences affecting their operations and their industry than climate change. According to the authors:

> most respondents involved in mine administration and managerial roles (24 of 31 respondents, 77%) viewed climate change as only a minor concern, particularly when compared to more pressing issues such as meeting regulatory and human resource requirements, and managing fluctuating market conditions.
>
> (Pearce et al., 2011: 363)

That Canadian mining companies are only marginally interested in climate change and that this interest is more often than not one of concern that it will negatively impact production, is similar to the findings of a recent study looking at the potential future of the oil and gas industry in the Arctic (Harsem, Eide, and Heen, 2011). The study attempted to understand which current changes are most likely to have an impact on future Arctic oil and gas development. These changes included climate change and as such the study tried to determine the impact of climate change on whether or not oil and gas development in the region will increase in the future.

The researchers isolate three major changes related to climate change that are likely to have an impact on oil and gas projects: ice structure and extent; extreme weather; and longer summer seasons and the Arctic tundra. They note that global warming has the potential to make offshore Arctic oil and gas development easier. They quote an earlier study by Dell and Pasteris that less ice and snow may reduce production costs "as oil companies may be able to replace ice based construction with lower cost conventional construction equipment" (Harsem, Eide, and Heen, 2011: 8040). Summer drilling seasons could be lengthened to allow more exploration and production. Reduced ice will also create the possibilities of new shipping routes.

At the same time, the potential positive impacts of this aspect of Arctic climate change could be countered by other possible negative impacts. Rather than dealing with constant and strong ice cover, the weakening of sea ice could mean that the ice becomes more moveable and unpredictable – especially in strong winds. This weaker ice "can therefore be expected to move at a greater speed compared to the older, more stable ice. This represents a potential risk for the oil and gas industry, as increased ice movement could interrupt drilling" (Harsem, Eide, and Heen, 2011: 8039). Increased iceberg activity could also damage off-shore production facilities. While technologies are currently being developed to deal with these eventualities, these innovations do tend to add to production costs.

While melting sea ice is a commonly understood result of global warming, another impact which is less understood is that of extreme weather. In the case of the Arctic, researchers have suggested that climate change will result in an increase in extreme weather. According to some researchers, the number of storms and hurricanes are likely to increase in the Arctic. While there is not unanimous agreement on the extent of an increase in extreme weather due to climate change, any increase in these events would be bad for resource development. According to the authors of the study, "(a)n increase in storm frequency could prove a serious risk to the oil industry, as it has the potential to disrupt drilling, production, and transportation" (Harsem, Eide, and Heen, 2011: 8039). Extreme weather would also mean an increase in risks relating to oil spills and the potential to contain them. One of the past advantages of the Arctic region in terms of resource production was relative stability of weather systems. Increased instability makes planning more difficult and increases both actual and potential costs related to developments.

Other negative impacts on oil and gas development are related to longer summers and the impacts on Arctic tundra. Generally speaking, the frozen tundra serves as an efficient production base for oil and gas development. A melting tundra causes a wide range of problems for land-based oil and gas development. According to the authors, "a lengthy summer season will shorten exploration activities, since drilling in the marshy Arctic tundra is not desirable, nor politically acceptable ... a milder climate will affect transportation equipment and infrastructure such as roads and pipelines" (Harsem, Eide, and Heen, 2011: 8039). They note that a "marshy Arctic tundra makes it almost impossible to conduct large scale transportation during the summer months, as the tundra is not able to support motorised vehicles" (Harsem, Eide, and Heen, 2011: 8040).

In their overall analysis the authors come to the conclusion that climate change changes are difficult to assess in terms of their impacts on future oil and gas developments but they are likely to have more negative impacts than positive ones. In terms of whether oil and gas development in the Arctic occur, at best climate change plays a relatively minor role. The most important factor influencing future oil and gas development in the future is global demand.

> the state of the global economy is the single most important driving factor for increased attention to the Arctic. Considering the counterfactual, even though ice melting continues and governments are handing out new drilling licenses, a downturn in the global economy will make increased production in the Arctic less likely to occur. Furthermore, as production cost is higher in the Arctic compared to other oil and gas producing regions, the region is even more vulnerable to changes in the global economy.
> (Harsem, Eide, and Heen, 2011: 8042)

Discussion

While there is an intuitive belief that climate change is currently impacting resource development in the Arctic, the actual research on the subject, although limited, tends to be inconclusive. Aside from a study looking at shifts in water flows in a hydro-electric project and an early closure to winter mining roads in

Canada in 2006, there is limited evidence-based research indicating that climate change is currently having any sort of an impact on resource development. Most of the commentary in the "scientific" discourse is hypothetical in nature. Based on their assessments of current climate change models, scientists believe climate change will impact resource development in the future. They are not sure however whether these impacts will be positive or negative for resource development.

Some project that climate change will have a positive impact in two instances: increased potential for off-shore oil and gas exploration and production, and improved ease of access to the Arctic through increased shipping. According to climate change researchers, loss of sea ice will make off-shore oil and gas operations in the Arctic more likely. Less sea ice and longer ice-free seasons will mean that cheaper conventional off-shore platforms can be used in the Arctic and used for longer periods of time. This makes both exploration and production cheaper and, as such, make Arctic oil and gas development more viable. Likewise, melting sea ice will make shipping easier in the Arctic. This will decrease the costs of those in both mining and oil and gas who use shipping to get equipment in and production out of the Arctic. Regions will become increasing accessible from a profitability perspective and this will lead to increased resource development in the Arctic.

At the same time, many of these same researchers also point to potentially negative impacts on resource development coming from climate change. Indeed, as far as the main scientific discourse on climate change in the Arctic is concerned, it does appear that more space is devoted to discussing potentially negative impacts than potentially positive ones. In terms of land-based transportation related to resource development, most, if not all, statements made stress the problems that would be caused by climate change. Ice roads essential for both oil and gas and mining would become less useful. Permafrost melt would make building and maintaining pipelines more difficult. As well, all land-based infrastructure would be negatively impacted due to melting permafrost and increased erosion. Variation in water flows would be problematic for hydro-electric installations. In addition, the potentially positive impacts of climate change on off-shore oil and gas installations could be countered by increased wave action and increases in extreme weather. An increase in extreme weather occurrences would also have a negative impact on future Arctic shipping.

Whatever scientists may think of the potential future influence of climate change on resource development, it is the representatives of industry that will be making the most important initial decisions on future resource development in the Arctic. They will decide whether to invest in the region or not and as such their opinions on the impacts of climate change are the most important to understanding the relationship between climate change and resource development. Although research is limited, existing information indicates that industry is not yet certain what the impacts will be. While some seem to be aware of potential positive impacts, potentially negative impacts seem to be highlighted as concerns. What is most interesting is that industry does not seem to be that interested in the relationship between climate change and resource development. This may due to the fact, hypothesized by some researchers in the Arctic Climate Impact Assessment, that there are other more important factors, such as global demand for resources, that are more of a concern to them in so far as decisions regarding future resource development in the Arctic.

Conclusions

The analysis above is not an exhaustive review of all material discussing the relationship between Arctic resource development and climate change. The publications reviewed here do, however, represent the main arguments presented in the literature. These arguments point to an uncertain relationship between the two phenomena. There may be a relationship between climate change and a future increase in Arctic resource development but it is likely that this relationship is currently of less importance than other factors. In addition, the nature of the relationship at this point is uncertain. We do not know if climate change will have a positive or a negative impact. While potential positive impacts are noted, negative impacts tend to be highlighted in the academic and scientific material.

In the long term, climate change may lead to more Arctic resource development but in the short term it is more likely to be a barrier. While some may believe that climate change is the main reason for increased interest in resource development, there is evidence that the main determining factor in Arctic resource development is the needs of the global economy. So far, there is little proof that climate change has created a situation where increased ease of access will lead to more resource development in the Arctic, however this may change in the future.

Key questions

In terms of the objectives of the Resources and Sustainable Development in the Arctic project, the gap analysis here offers us little indication that there is any direct usefulness in studying the relationship between climate change and resources development. Climate change will likely only affect the benefits communities derive from resource development through its influence on decisions regarding resource development projects. For now this influence appears limited and it is uncertain whether climate change will have a positive or negative impact on these decisions. In addition, while the impacts of climate change on Arctic communities have been the most studied relationships in Arctic social science over the past decade, relatively little has been uncovered in terms of proven impacts. It is unlikely that ReSDA could add much to what others have already attempted to do.

At the same time, given the importance of climate change in the Arctic, ReSDA does need to follow developments in research involving social impacts. It may be that more direct links between the two phenomena will be discovered later. In the meantime however, rather than try and determine whether climate change is influencing resource development in the Arctic, some of the existing literature notes that it may be more useful to look at the impacts of resource development related to the ability of communities to deal with climate change. While we do not know the extent that climate change is influencing resource development and whether climate change is a positive or negative force for resource development in the Arctic, there does appear to be a consensus that resource development will have an impact on Arctic communities' abilities to deal with climate change impacts (or if climate change ends up not having a major impact – other forces). It can help or hinder. Most of the resource development in the past has reduced communities' abilities to deal with such issues. Can we change this so that it can have a positive impact? Can Arctic communities create a situation where resource development will provide them with the necessary tools to ensure that they are not "vulnerable" to climate change? Can resource development be used to help them to "adapt" to climate change? How can resource development be done in a manner that it helps Arctic communities become "resilient" to potential challenges such as climate change? These are key questions that should guide ReSDA research when it considers climate change impacts.

Note

1   It should be noted that, unlike the first survey, the consultants did not publish all the response totals for the questionnaire. We are forced therefore to rely on unclear totals. These are primarily percentiles with inadequate explanations of what total numbers the percentiles represent.

References

ACIA. (2005). *Climate Impact Assessment Scientific Report*. Cambridge, UK: Cambridge University Press.

Bolin, B. (2007). *A History of the Science and Politics of Climate Change: The Role of the Intergovernmental Panel on Climate Change*. Cambridge: Cambridge University Press.

Borgerson, S. G. (2008). Arctic Meltdown – The Economic and Security Implications of Global Warming. *Foreign Affairs*, 87(2), 63–77.

Callaway, D., Eamer, J., Edwardsen, E., Jack, C., Marcy, S., Olrun, A. . . . and Whiting, A. (1999). Effects of Climate Change on Subsistence Communities in Alaska. In: G. Weller and P. Anderson (eds.). *Assessing the Consequences of Climate Change for Alaska and the Bering Sea Region*, pp. 59–73. Center for Global Change and Arctic System Research, University of Alaska Fairbanks.

Economist, The (June 16, 2012). The Melting North. Retrieved from www.economist.com/node/21556798.

Ford, J. D., Bolton, K. C., Shirley, J., Pearce, T., Tremblay, M., and Westlake, M. (2012). Research on the Human Dimensions of Climate Change in Nunavut, Nunavik, and Nunatsiavut: A Literature Review and Gap Analysis. *Arctic*, 65(3), 289–304.

Ford, J., Pearce, T., Duerden, F., Prno, J., and Marshall, D. (April 6, 2008). Global Warming – Climate Change Impacts on the Canadian Mining Sector. *Canadian Mining Journal*. Retrieved from www.canadianminingjournal.com/news/global-warming-climate-change-impacts-on-the-canadian-mining-sector/.

Ford, J. D., Pearce, T., Prno, J., Duerden, F., Ford, L. B., Beaumier, M., and Smith, T. (2010). Perceptions of Climate Change Risks in Primary Resource use Industries:

A Survey of the Canadian Mining Sector. *Regional Environmental Change*, 10(1), 65–81.

Ford, J. D., Pearce, T., Prno, J., Duerden, F., Ford, L. B., Smith, T. R., and Beaumier, M. (2011). Canary in a Coal Mine: Perceptions of Climate Change Risks and Response Options among Canadian Mine Operations. *Climatic Change*, 109(3–4), 399–415.

Furgal, C. and Prowse, T. D. (2008). Northern Canada. In D. S. Lemmen, F. J. Warren, F. Lacroix, and E. Bush (eds.). *From Impacts to Adaption: Canada in a Changing Climate 2007*, pp. 57–118. Ottawa: Government of Canada.

Harsem, O., Eide, A., and Heen, K. (2011). Factors Influencing Future Oil and Gas Prospects in the Arctic. *Energy Policy*, 39(12), 8037–8045.

Houghton, J. T., Callander, B. A., and Varney, S. K. (eds.) (1992). *Climate Change 1992: The Supplementary Report to the IPCC Scientific Assessment*. Cambridge, UK: Cambridge University Press.

IPCC AR4 WG II. (2007). *IPCC Fourth Assessment Report: Working Group II Report "Impacts, Adaptation and Vulnerability"*. Cambridge, UK: Cambridge University Press.

IPCC FAR WG I. (1990). *Climate Change: The IPCC Scientific Assessment – Report Prepared for IPCC by Working Group I*. Cambridge, UK: Cambridge University Press.

IPCC FAR WG II. (1990). *Climate Change: The IPCC Impacts Assessment – Report Prepared for IPCC by Working Group II*. Canberra: Australian Govt. Pub. Service for the Department of the Arts, Sport, the Environment, Tourism and Territories.

IPCC SAR WG I. (1996). *Climate Change 1995: The Science of Climate Change*. Cambridge, UK: Cambridge University Press.

IPCC TAR WG II. (2001). *Climate Change 2001: Impacts, Adaptation and Vulnerability*. Cambridge, UK: Cambridge University Press.

Pearce, T., Ford, J., Prno, J., and Duerden, F. (2009). Climate Change and Canadian Mining: Opportunities for Adaptation. Vancouver: David Suzuki Foundation.

Pearce, T. D., Ford, J. D., Prno, J., Duerden, F., Pittman, J., Beaumier, M., . . . and Smit, B. (2011). Climate Change and Mining in Canada. *Mitigation and Adaptation Strategies for Global Change*, 16(3), 347–368.

Ritter, K. (2014, June 11, 2014). Cold War: How Climate Change is Heating up Arctic Nations' Spying Game. *The Globe and Mail*. Retrieved from www.theglobeandmail. com/news/world/cold-war-how-climate-change-is-heating-up-arctic-nations-spying-game/article19113804/.

Rosenthal, E. (September 19, 2012). Race Is On as Ice Melt Reveals Arctic Treasures. *New York Times*. Retrieved from www.nytimes.com/2012/09/19/science/earth/arctic-resources-exposed-by-warming-set-off-competition.html?pagewanted=all&_r=0.

Watson, R. T., Zinyowera, M. C., and Moss, R. H. (eds.) (1997). *The Regional Impacts of Climate Change: An Assessment of Vulnerability*. Cambridge, UK: Cambridge University Press.

16 How can extractive industry help rather than hurt Arctic communities?

Chris Southcott, Frances Abele,
David Natcher, and Brenda Parlee

Across Northern Canada, as well as places such as Alaska and Greenland, the political empowerment of Indigenous governments has provided greater control over the conditions of resource development and, in so doing, has enhanced the potential benefits flowing to northern communities. As noted in the introduction, this book is based on a central premise – that the conditions related to extractive resource development in Northern Canada have changed to the point where communities can benefit from resource development without bearing unacceptable social, economic and environmental costs. This is of course not a necessary outcome of any particular resource development, but rather a possibility that can be realized if appropriate regulatory, community and business actions are taken.

In Northern Canada the signing of comprehensive land claim agreements, beginning in the 1970s, ushered in an era of political change that enabled Indigenous governments to regain self-determination over the development process (White, 2002). Legal recognition of their inherent rights to benefit from the development of lands and resources in their homelands coupled with new kinds of arrangements and requirements for participation in natural resource planning, management and monitoring created opportunities for greater benefit and improved mitigation of disbenefits. Such arrangements created the foundation and mechanisms for long-term social, economic and environmental sustainability. The chapters in this book reflect those discussions and draw attention to the lessons learned as well as the questions that still remain as communities strive to find this critical balance.

Land claim agreements created very specific kinds of power-sharing arrangements such as land use planning, co-management of wildlife as well as requirements for participation in environmental assessment and water/land regulation. Such arrangements created the foundation for other kinds of power sharing around resource development revenues including joint-venture business activities, bilateral agreements between industry and Indigenous governments (i.e., impact and benefit agreements) and increased opportunities for northern communities to capture resource rents. In many parts of the north, a variety of nuances exist around the negotiation and implementation of impact and benefit agreements and the use of associated revenues.

A key question is, how can be revenues be equitably distributed and/or used in ways that ensure long-term sustainability? In terms of resource rents, new research has been tracking revenue flow with the aim of improving the local capture of resource revenues and limiting the amount of economic leakage outside communities and northern regions (Huskey, 2018; Huskey and Southcott, 2014; Huskey and Southcott, 2016; O'Faircheallaigh and Gibson, 2012; O'Faircheallaigh, 2018; Thistle, 2016). Other research examines the best practices for communities and regions to manage and distribute these new revenues (Briones et al., 2014; Coates, 2015; Coates and Crowley, 2013; NAEDB, 2015; O'Faircheallaigh, 2013).

One of the most effective tools for capturing resource rents have been Impact Benefit Agreements (IBAs), which have the potential to build financial capital, while strengthing human and social capital. Recent research has looked at improving IBAs to ensure that they more adequately address community needs (Bradshaw, Fiddler and Wright, 2014; Gibson and O'Faircheallaigh, 2015; Jones and Bradshaw, 2015). As Bradshaw, Fiddler and Wright noted, we have little knowledge on how effective these agreements are at meeting the long-term needs of communities. More work needs to be done in ensuring IBAs are effective at real power sharing and ensuring revenues meet the evolving needs of communities. Coates argues that this requires a much better understanding of the impacts of past developments, both direct and indirect, in order to inform future development paths.

Petrov et al. echos this need but also calls for more adequate indicators that can be more effectively measured, monitored and controlled by communities themselves. Through a more effective use of indicators the social and cultural impacts of extractive development can be monitored and managed for the benefit of communities (Cater and Keeling, 2013; Czyzewski et al. 2014; Davison and Hawe, 2012; Green, 2013; Ritsema et al. 2015; Rodon and Levesque, 2015; Schweitzer et al. 2015; Tester, Lambert and Lim, 2013). New research is being conducted on finding ways of tracking and measuring impacts through developing new indicators and monitoring systems (Edouard and Duhaime, 2013; NAEDB, 2013; Petrov, 2014). For example, recognition that the impacts of development are gendered is key to understanding inequities in northern communities and how they can be improved or exacerbated as large projects emerge (Cox and Mills, 2015; Mills, Dowsley, and Cameron, 2014; Natcher, 2013; Staples and Natcher, 2015; Stienstra, 2015).

Minimizing the negative environmental impacts of extractive development is a central concern for northern communities. Noble et al. note that while the establishment of Environmental and Social Impact Assessment regimes has improved the ability to ensure fewer negative impacts, changes must be made to these regimes to ensure the more meaningful participation of northern communities in the processes. Community capacity to say no to those proposed projects that present too much risk is an important part of ensuring benefits outweigh potential negative impacts (Bernauer, 2010; Procter, 2016). For projects that are approved, it is important to find and continually improve means to anticipate negative impacts and minimize these. The Environmental Impact Assessment (EIA) process had been recognized as an important means by which negative impacts can be considered and mitigated and by which communities can be better engaged. An important area of research in this regard is the barriers to the EIA system fulfilling these objectives and what can be done about it. In particular, beyond decision-making about the approval or rejection of projects, more research is needed on the most effective means to ensure effective monitoring and corporate account-ability after projects are initiated and concluded, in all Canadian jurisdictions (Cox and Mills, 2015; Noble, Hanna and Gunn, 2014; Noble and Udofia, 2016; Udofia, Noble and Poelzer, 2015).

The need for a proactive approach is acute. Past extractive development, such as gold and silver mining, have resulted in dramatic environmental legacies (e.g., the Faro mine). With regard to these legacies, Keeling et al. have argued that overcoming and managing these historic legacies requires that communities need to be more closely involved in the management of these legacies. For this reason, research is still needed on how communities can best deal with the environmental legacies of past extractive developments and how they can use site remediation for local benefit and improve remediation plans (Dance, 2015; LeClerc and Keeling, 2015; Sandlos and Keeling, 2016). It will be important to find ways to enable communities that have had some success in these areas to document their experience and to have practical ways to share their findings with other com-munities that are facing similar challenges.

In many of the chapters of this book, the authors have pointed to the cultural diversity of Indigenous communities; as development proceeds, Indigenous leaders and governments are challenged to find ways of ensuring cultural continuity – although seeking to benefit from an emerging global economy, northerners also seek to protect many aspects of their language, traditions and relationships to each other and the land for future generations. Schweitzer et al,, along with Rodon and Levesque, note the special need for northern communities to better understand and assess the cultural impacts of extractive development. While the Berger Inquiry highlighted the fact that subsistence activities can be nega-tively impacted by industrial development, in an era where communities exercise more control, and when the costs of subsistence activities are often prohibitive to many, several studies have pointed out that extractive industry development now has the potential to be a crucial support for the subsistence economy (Koke, 2008; Kruse, 1986, 1991). Researchers are now trying to understand the conditions under which extractive industry activity could be used to ensure the long-term survival of subsistence activities (Natcher et al. 2016; Southcott and Natcher, 2018).

Parlee argues that our understanding of the interconnected social, economic, cultural and ecological impacts of large-scale development projects, such as diamond mines, can be greatly improved by using the concept and local indicators (measures) of well-being. This idea of well-being which parallels many Indigenous articulations of "a good life" or "our way of life" offer different insights that can be gleaned by using generic government indicators and conventional siloed approaches to assessment and monitoring. As Abele notes, Indigenous culture practices both require, and enable, the development of more innovative bottom-up policy tools to ensure that extractive developments are positive for these communities. These policy tools must ensure that traditional subsistence harvesting is considered a "normal" part of the northern mixed economy and, as Natcher notes, support communities' use of extractive industry benefits to enhance these activities. Huntington points out that an increased ability to include traditional knowledge in the governance processes surrounding resource development allows not only a more adequate understanding of impacts but will assist in ensuring the continued existence of this knowledge and greater local control. Finally, Mills et al. remind us that any discussion of how to ensure extractive industry developments help rather than harm northern communities must include a consideration of potentially differential impacts on women and men. In particular, as research by Saxinger and Gartler, as well as Hall, shows, the fly-in, fly-out organization of work in resource-based industries changes family dynamics and places unusual pressures on all members (Hall, 2017; Saxinger and Gartler, 2017).

In short, it is difficult to determine the best way to maximize benefits of resource development to communities in the absence of an adequate and community-centric understanding of social well-being to both interpret results and guide future negotiations. Recent research has therefore tried to better understand what communities perceive as well-being in a manner that can contribute to better managing of extractive industry impacts and benefits (Edouard and Duhaime, 2013; Jones and Bradshaw, 2015; Parlee, 2016; Parlee and Furgal, 2012).

Path (in)dependence

The research program described in this book is the result of a collaboration between academic researchers and northern community members and their interlocutors. Our goal has been the co-production of knowledge about how best to benefit from resource development. There are many challenges facing any community with this goal. Given the importance of resource development to the overall Canadian economy, and long-standing relationships between the resource sector and all Canadian governments, path dependency –the tendency of decision direction and momentum, once established, to be maintained – is a factor that must be taken into account in any assessment of new development projects in Northern Canada. Over-dependence upon the (relatively) easy money available from resource development – sometimes referred to as the resource curse – is a real possibility (Bennett, 2016; Petrov, 2016). The short-term economic gains achieved through development projects can lead to unanticipated outcomes – greater dependency, uncertainty and conflict between social and ecological dynamics – that will diminish the likelihood of achieving community sustainability (Huskey and Southcott, 2016; Parlee, 2015).

As northern communities recover from the current bust cycle and prepare for the next resource boom, they can rely upon previous research to find the best ways to make new benefits possible. That said, there is a need for further research that builds upon this foundation while taking new conditions (like the accelerating impact of climate changes) into account to assist communities in both increasing benefits from extractive resource development and ensuring that these benefits lead to long-term sustainability. Collaborative research is needed to ensure the short-term economic needs of communities do not endanger the culture, livelihoods and general well-being of northern communities over the long term. In most regions of the North, development remains at a stage where communities are still able to avoid an over-dependence on extractive industries (Wilson and Stammler, 2016). They can find ways to use the short-term economic benefits derived from non-renewable industrial development to put themselves in a better position to diversify local economies. In collaborative research and community-based planning they have the opportunity to use resource development opportunities to strengthen local institutions and capacities rather than placing them at greater risk in the future.

References

Bernauer, W. (2010). Mining, Harvesting and Decision Making in Baker Lake, Nunavut: A Case Study of Uranium Mining in Baker Lake. *Journal of Aboriginal Economic Development*, 7(1), 19–33.

Bradshaw, B., Fiddler, C. and Wright, A. (2014). *Impact and Benefit Agreements & Northern Resource Governance: What We Know and What We Still Need to Figure Out*. Whitehorse: ReSDA.

Briones, J., Daitch, S., Dias, A., Lajoie, M., Fan Li, J. and Schwann, A. (2014). *A Question of Future Prosperity: Developing a Heritage Fund in the Northwest Territories*. Ottawa: Action Canada.

Cater, T. and Keeling, A. (2013). "That's Where Our Future Came From": Mining, Landscape, and Memory in Rankin Inlet, Nunavut. *Études/Inuit/Studies*, 37(2), 59–82.

Coates, K. (2015). *Sharing the Wealth: How Resource Revenue Agreements Can Honour Treaties, Improve Communities, and Facilitate Canadian Development*. Ottawa: Macdonald-Laurier Institute.

Coates, K. and Crowley, B. (2013). *New Beginnings: How Canada's Natural Resource Wealth Could Re-shape Relations with Aboriginal People*. Ottawa: MacDonald-Laurier Institute.

Cox, D. and Mills, S. (2015). Gendering Environmental Assessment: Women's Participation and Employment Outcomes at Voisey's Bay. *Arctic*, 68(2), 246–260.

Czyzewski, K., Tester, F. J., Aaruaq, N. and Blangy, S. (2014). *The Impact of Resource Extraction on Inuit Women and Families in Qamani'tuaq, Nunavut Territory*. Vancouver: School of Social Work, University of British Columbia.

Dance, A. (2015). Northern Reclamation in Canada: Contemporary Policy and Practice for New and Legacy Mines. *Northern Review*, 41, 41–80.

Davison, C. M. and Hawe, P. (2012). All That Glitters: Diamond Mining and Taicho Youth in Behchoko, Northwest Territories. *Arctic, 65*(2), 214–228.

Edouard, R. and Duhaime, G. (2013). The Well-Being of the Canadian Arctic Inuit: The Relevant Weight of Economy in the Happiness Equations. *Social Indicators Research, 113*(1), 373–392.

Gibson, G. and O'Faircheallaigh, C. (2015). *IBA Community Toolkit: Negotiation and Implementation of Impact and Benefit Agreements* (Summer 2015 ed.). Toronto: Walter & Duncan Gordon Foundation.

Green, H. (2013). State, Company, and Community Relations at the Polaris Mine (Nunavut). *Études/Inuit/Studies, 37*(2), 37–57.

Hall, R. (2017). *Diamonds are Forever: A Decolonizing, Feminist Approach to Diamond Mining in Yellowknife*. Northwest Territories. (Ph.D.), York, Toronto.

Huskey, L. (2018). An Arctic Development Strategy? The North Slope Inupiat and the Resource Curse. *Canadian Journal of Development Studies, 39*(1), 89–100.

Huskey, L. and Southcott, C. (2014). *Resource Revenue Regimes around the Circumpolar North*. Whitehorse: ReSDA.

Huskey, L. and Southcott, C. (2016). "That's Where My Money Goes": Resource Production and Financial Flows in the Yukon Economy. *The Polar Journal, 6*(1), 11–29.

Jones, J. and Bradshaw, B. (2015). Addressing Historical Impacts Through Impact and Benefit Agreements and Health Impact Assessment. *Northern Review, 41*, 81–109.

Koke, P. E. (2008). *The Impact of Mining Development on Subsistence Practices Of Indigenous Peoples: Lessons Learned From Northern Quebec and Alaska*. (Master of Arts), The University of Northern British Columbia, Prince George.

Kruse, J. (1986). Subsistence and the North Slope Inupiat: The effects of energy development. In S. Langdon (Ed.), *Contemporary Alaskan Native Economies* (pp. 121–152). Langham, MD: University Press of America.

Kruse, J. (1991). Alaska Inupiat Subsistence and Wage Employment Patterns: Understanding Individual Choice. *Human Organization, 50*(4), 317–326.

LeClerc, E. and Keeling, A. (2015). From Cutlines to Traplines: Post-industrial Land Use at the Pine Point Mine. *Extractive Industries and Society – An International Journal, 2*(1), 7–18.

Mills, S., Dowsley, M. and Cameron, E. (2014). *Gender in Research on Northern Resource Development*. Whitehorse: ReSDA.

NAEDB. (2013). *The Aboriginal Economic Benchmarking Report Core Indicator 3: Wealth and Well-Being*. Ottawa: National Aboriginal Economic Development Board.

NAEDB. (2015). *Enhancing Aboriginal Financial Readiness for Major Resource Development Opportunities*. Ottawa: National Aboriginal Economic Development Panel.

Natcher, D. (2013). Gender and Resource Co-Management in Northern Canada. *Arctic, 66*(2), 218–221.

Natcher, D., Shirley, S., Rodon, T. and Southcott, C. (2016). Constraints to Wildlife Harvesting among Aboriginal Communities in Alaska and Canada. *Food Security, 8*(6), 1153–1167.

Noble, B., Hanna, K. and Gunn, J. (2014). *Northern Environmental Assessment*. Whitehorse: ReSDA.

Noble, B. and Udofia, A. (2016). *Protectors of the Land: Toward an EA Process that Works for Aboriginal Communities and Developers*. Ottawa: Macdonald-Laurier Institute.

O'Faircheallaigh, C. (2013). Extractive Industries and Indigenous Peoples: A Changing Dynamic? *Journal of Rural Studies, 30*, 20–30.

O'Faircheallaigh, C. and Gibson, G. (2012). Economic Risk and Mineral Taxation on Indigenous Lands. *Resources Policy, 37*(1), 10–18.

O'Faircheallaigh, C. (2018). Using Revenues from Indigenous Impact and Benefit Agreements: Building Theoretical Insights. *Canadian Journal of Development Studies, 39*(1), 101–118.

Parlee, B. (2016). *Resource Development and Well-being in Northern Canada.* Whitehorse: ReSDA.

Parlee, B. and Furgal, C. (2012). Well-being and Environmental Change in the Arctic: A Synthesis of Selected Research from Canada's International Polar Year Program. *Climatic Change, 115*(1), 13–34.

Petrov, A. (2014). *Measuring Impacts: A Review of Frameworks, Methodologies and Indicators for Assessing Socio-Economic Impacts of Resource Activity in the Arctic.* Whitehorse: ReSDA.

Procter, A. (2016). Uranium and the Boundaries of Indigeneity in Nunatsiavut, Labrador. *Extractive Industries and Society-an International Journal, 3*(2), 288–296.

Ritsema, R., Dawson, J., Jorgensen, M. and Macdougall, B. (2015). "Steering Our Own Ship?" An Assessment of Self-Determination and Self-Governance for Community Development in Nunavut. *Northern Review, 41,* 205–228.

Rodon, T. and Levesque, F. (2015). Understanding the Social and Economic Impacts of Mining Development in Inuit Communities: Experiences with Past and Present Mines in Inuit Nunangat. *Northern Review, 41,* 13–39.

Sandlos, J. and Keeling, A. (2016). Aboriginal Communities, Traditional Knowledge, and the Environmental Legacies of Extractive Development in Canada. *Extractive Industries and Society-an International Journal, 3*(2), 278–287.

Saxinger, G. and Gartler, S. (2017). *The Mobile Workers Guide.* Whitehorse: Na-cho Nyak Dun First Nation and ReSDA.

Schweitzer, P., Stammler, F., Ebsen, C., Ivanova, A. and Litvina, I. (2015). *Social Impacts of Non-Renewable Resource Development on Indigenous Communities in Alaska, Greenland and Russia.* Whitehorse: ReSDA.

Southcott, C. and Natcher, D. (2018). Extractive Industries and Indigenous Subsistence Economies: A Complex and Unresolved Relationship. *Canadian Journal of Development Studies, Online,* 1–18.

Staples, K. and Natcher, D. C. (2015). Gender, Decision Making, and Natural Resource Co-management in Yukon. *Arctic, 68*(3), 356–366.

Stienstra, D. (2015). Northern Crises Women's Relationships and Resistances to Resource Extractions. *International Feminist Journal of Politics, 17*(4), 630–651.

Tester, F., Lambert, D. and Lim, T. (2013). Wistful Thinking: Making Inuit Labour and the Nanisivik Mine near Ikpiarjuk (Arctic Bay), Northern Baffin Island. *Études/ Inuit/Studies, 37*(2), 15–36.

Thistle, J. (2016). Forging Full Value? Iron Ore Mining in Newfoundland and Labrador, 1954–2014. *Extractive Industries and Society-an International Journal, 3*(1), 103–116.

Udofia, A., Noble, B. and Poelzer, G. (2015). Community Engagement in Environmental Assessment for Resource Development: Benefits, Enduring Concerns, Opportunities for Improvement. *Northern Review, 39,* 98–110.

White, G. (2002). Treaty Federalism in Northern Canada: Aboriginal-Government Land Claims Boards. *Publius-the Journal of Federalism, 32*(3), 89–114.

Wilson, E. and Stammler, F. (2016). Beyond Extractivism and Alternative Cosmologies: Arctic Communities and Extractive Industries in Uncertain Times. *Extractive Industries and Society – An International Journal, 3*(1), 1–8.

Index

Note: Page numbers in *italic* refer to Figures; those in **bold** refer to Tables